83 Springer Series in Solid-State Sciences
Edited by K. von Klitzing and H.-J. Queisser

Springer Series in Solid-State Sciences

Editors: M. Cardona P. Fulde K. von Klitzing H.-J. Queisser

Managing Editor: H. K. V. Lotsch

Volumes 1–49 are listed on the back inside cover

50 **Multiple Diffraction of X-Rays in Crystals**
By Shih-Lin Chang

51 **Phonon Scattering in Condensed Matter**
Editors: W. Eisenmenger, K. Laßmann, and S. Döttinger

52 **Superconductivity in Magnetic and Exotic Materials**
Editors: T. Matsubara and A. Kotani

53 **Two-Dimensional Systems, Heterostructures, and Superlattices**
Editors: G. Bauer, F. Kuchar, and H. Heinrich

54 **Magnetic Excitations and Fluctuations**
Editors: S. Lovesey, U. Balucani, F. Borsa, and V. Tognetti

55 **The Theory of Magnetism II**
Thermodynamics and Statistical Mechanics
By D. C. Mattis

56 **Spin Fluctuations in Itinerant Electron Magnetism** By T. Moriya

57 **Polycrystalline Semiconductors,**
Physical Properties and Applications
Editor: G. Harbeke

58 **The Recursion Method and Its Applications**
Editors: D. Pettifor and D. Weaire

59 **Dynamical Processes and Ordering on Solid Surfaces**
Editors: A. Yoshimori and M. Tsukada

60 **Excitonic Processes in Solids**
By M. Ueta, H. Kanzaki, K. Kobayashi, Y. Toyozawa, and E. Hanamura

61 **Localization, Interaction, and Transport Phenomena**
Editors: B. Kramer, G. Bergmann, and Y. Bruynseraede

62 **Theory of Heavy Fermions and Valence Fluctuations**
Editors: T. Kasuya and T. Saso

63 **Electronic Properties of Polymers and Related Compounds**
Editors: H. Kuzmany, M. Mehring, and S. Roth

64 **Symmetries in Physics** Group Theory Applied to Physical Problems
By W. Ludwig and C. Falter

65 **Phonons: Theory and Experiments II**
Experiments and Interpretation of Experimental Results By P. Brüesch

66 **Phonons: Theory and Experiments III**
Phenomena Related to Phonons
By P. Brüesch

67 **Two-Dimensional Systems: Physics and New Devices**
Editors: G. Bauer, F. Kuchar, and H. Heinrich

68 **Phonon Scattering in Condensed Matter V**
Editors: A. C. Anderson and J. P. Wolfe

69 **Nonlinearity in Condensed Matter**
Editors: A. R. Bishop, D. K. Campbell, P. Kumar, and S. E. Trullinger

70 **From Hamiltonians to Phase Diagrams**
The Electronic and Statistical-Mechanical Theory of sp-Bonded Metals and Alloys
By J. Hafner

71 **High Magnetic Fields in Semiconductor Physics**
Editor: G. Landwehr

72 **One-Dimensional Conductors**
By S. Kagoshima, H. Nagasawa, and T. Sambongi

73 **Quantum Solid-State Physics**
Editors: S. V. Vonsovsky and M. I. Katsnelson

74 **Quantum Monte Carlo Methods** in Equilibrium and Nonequilibrium Systems
Editor: M. Suzuki

75 **Electronic Structure and Optical Properties of Semiconductors**
By M. L. Cohen and J. R. Chelikowsky

76 **Electronic Properties of Conjugated Polymers**
Editors: H. Kuzmany, M. Mehring, and S. Roth

77 **Fermi Surface Effects**
Editors: J. Kondo and A. Yoshimori

78 **Group Theory and Its Applications in Physics**
By T. Inui, Y. Tanabe, and Y. Onodera

79 **Elementary Excitations in Quantum Fluids**
Editors: K. Ohbayashi and M. Watabe

80 **Monte Carlo Simulation in Statistical Physics**
An Introduction
By K. Binder and D. W. Heermann

81 **Core-Level Spectroscopy in Condensed Systems**
Editors: J. Kanamori and A. Kotani

82 **Introduction to Photoemission Spectroscopy**
By S. Hüfner

83 **Physics and Technology of Submicron Structures**
Editors: H. Heinrich, G. Bauer, and F. Kuchar

84 **Beyond the Crystalline State**
An Emerging Perspective
By G. Venkataraman, D. Sahoo, and V. Balakrishnan

85 **The Fractional Quantum Hall Effect**
Properties of an Incompressible Quantum Fluid
By T. Chakraborty and P. Pietiläinen

H. Heinrich
G. Bauer F. Kuchar (Eds.)

Physics and Technology of Submicron Structures

Proceedings of the
Fifth International Winter School,
Mauterndorf, Austria, February 22–26, 1988

With 190 Figures

Springer-Verlag Berlin Heidelberg New York
London Paris Tokyo

Professor Dr. Helmut Heinrich

Institut für Experimentalphysik, Universität Linz,
A-4040 Linz, Austria

Professor Dr. Günther Bauer

Institut für Physik, Montanuniversität Leoben,
A-8700 Leoben, Austria

Professor Dr. Friedemar Kuchar

Institut für Festkörperphysik, Universität Wien,
A-1090 Wien, Austria

Series Editors:

Professor Dr., Dres. h. c. Manuel Cardona
Professor Dr., Dr. h. c. Peter Fulde
Professor Dr. Klaus von Klitzing
Professor Dr. Hans-Joachim Queisser

Max-Planck-Institut für Festkörperforschung, Heisenbergstrasse 1
D-7000 Stuttgart 80, Fed. Rep. of Germany

Managing Editor: Dr. Helmut K.V. Lotsch

Springer-Verlag, Tiergartenstrasse 17,
D-6900 Heidelberg, Fed. Rep. of Germany

ISBN-13:978-3-642-83433-2 e-ISBN-13:978-3-642-83431-8
DOI: 10.1007/978-3-642-83431-8

Softcover reprint of the hardcover 1st edition 1988

2154/3150-543210 – Printed on acid-free paper

Preface

The winter school on the Physics and Technology of Submicron Structures was the fifth in the series of international winter schools on new developments in solid-state physics organized by the Austrian Physical Society. The school was held in the castle of Mauterndorf, near Salzburg in Austria, February 22–26, 1988, and was attended by more than 200 registered participants from Europe, the United States, Japan, and the People's Republic of China.

This is the third winter school in this series to have the proceedings published in the Springer Series in Solid-State Sciences. (The earlier proceedings appeared as Vols. 53 and 67.) Only the contributions of the invited speakers are included in this volume. They are arranged, according to subject, in five parts: the fabrication of microstructures; vertical transport and tunneling phenomena; quantum interference effects in submicron structures, and universal conductance fluctuations in mesoscopic systems; phenomena related to the transition from two- to one-dimensional conduction in semiconductors; and finally the physical background of future submicron devices.

Major breakthroughs in low-dimensional physics were reported at the school and we hope that this volume will preserve some of the spirit of excitement of the participants and lecturers.

The organizers are grateful to all the speakers for the trouble they took in preparing their manuscripts. It is a pleasure to acknowledge the generous financial support received from:

Bundesministerium für Wissenschaft und Forschung, Austria
Österreichische Physikalische Gesellschaft
Salzburger Landesregierung
Österreichische Forschungsgemeinschaft
US Air Force, EOARD, London
Office of Naval Research, Washington
Fonds zur Förderung der wissenschaftlichen Forschung, Wien
Gesellschaft für Mikroelektronik, Wien
BOMEM, Canada,
Bruker, FRG,

Instruments SA-Riber, France and FRG,
Oxford Instruments, Great Britain and FRG,
Karl Süss KG, FRG,
VTS-Joachim Schwarz, FRG

Mauterndorf,
March 1988

H. Heinrich G. Bauer F. Kuchar

Contents

Part III Quantum Interference Effects, Mesoscopic Systems

Part IV From Two Dimensions to One: Conductance Phenomena and Optical Properties

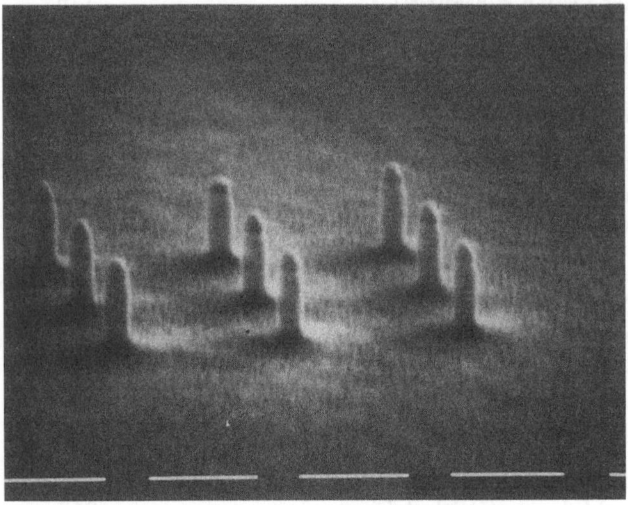

Scanning electron micrograph of anisotropically etched columns containing quantum dots.

(Courtesy of M.A. Reed, Central Research Laboratories, Texas Instruments Inc., Dallas, TX 75265, USA)

Part I

Fabrication of Microstructures

Fabrication of Ultra-Small Structures: Quantum Wires

H.G. Craighead, A. Scherer, and M.L. Roukes

Bell Communications Research, 331 Newman Springs Road,
Red Bank, NJ 07701, USA

Abstract. The ability to fabricate structures and devices with dimensions smaller than relevant physical length scales is leading to the discovery and exploitation of new physical phenomena. Electron beam lithography has been used to define most of the ultra-small objects studied to date. The limits of this method can be understood on the basis of electron scattering and electron beam resist characteristics. Except for rare cases, the electron beam defined pattern must be transferred into another material of interest, and this can impose significant limits on the minimum attainable feature size. In this paper we will describe physical and practical limits of ultra-small structure creation with an emphasis on III-V compound semiconductors. Examples of small object fabrication are given with particular consideration of the fabrication of quasi-one-dimensional wires.

1. Introduction

In the search for the observation of new physical phenomena and the desire to realize revolutionary device designs the ability to fabricate increasingly fine features is a limiting step. Studies of two-dimensional systems have yielded significant results, and it is anticipated that the investigations of lower dimensional systems will also be fruitful. A goal is to obtain complete three-dimensional control over the structure of a material and observe and exploit the effects of lateral quantum confinement. The nature of electrical transport behavior in objects with dimensions less than electron phase coherence lengths and electron elastic scattering lengths differs markedly from the bulk behavior.[1-3] This size regime is now accessible by state-of-the-art technology. At sufficiently small lateral dimensions the finite number of accessible quantum states due to lateral confinement is observable in the transport properties.[4-6] There is also active research on the modification of optical properties in artificially constructed reduced dimensional objects. Potential advantages of enhanced optical nonlinearities and reduced threshold "quantum wire" and "quantum dot" lasers are being pursued.[7] New types of devices based on lateral superlattices and quantum interference have been proposed.[8,9] Advances in these areas are occurring more rapidly as there is increased understanding of how to deal with the problems that arise in working at these reduced dimensions.

Springer Series in Solid-State Sciences Vol. 83: **Physics and Technology of Submicron Structures**
Editors: H. Heinrich • G. Bauer • F. Kuchar

2. Electron Beam Lithography

Scanned electron beams have been used to define the smallest artificially defined objects. Electron beam lithography is an ideal research tool, because of the high quality of electron optics, the availability of high brightness sources and the ability to scan the beam under computer control to define arbitrary patterns. In electron beam lithography a finely focused beam is scanned in the desired pattern over a material that undergoes a chemical change in reaction to electron irradiation (Fig. 1). Even though electron optics and electron sources are of a very high quality, and focused beams less than 1 nm in diameter can be generated, it is not trivially possible to define a structure on the order of the electron beam size. The resolution of the process is limited by the range of scattered electrons and the intrinsic graininess of the beam sensitive material.

The longest range scattering effect is the backscattering of electrons, resulting from large angle scattering with energies near the incident electron energy. The range of these can be many micrometers for typical electron beam energies. The backscattered electrons contribute a diffuse background exposure dose over an area determined by the electron range (Fig. 2). This decreases the process latitude but does not necessarily limit the ultimate lithographic resolution. This backscattered electron range has been determined to have the form $E^{1.7}$ where E is the electron energy, for a wide energy range.[10,11] This

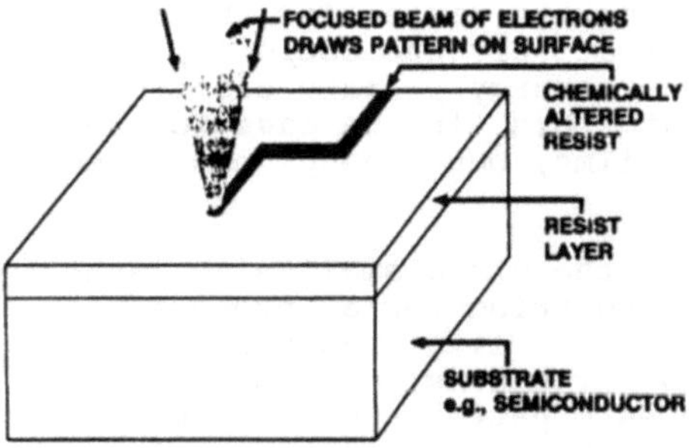

Fig. 1. Schematic of the scanned electron beam lithography process.

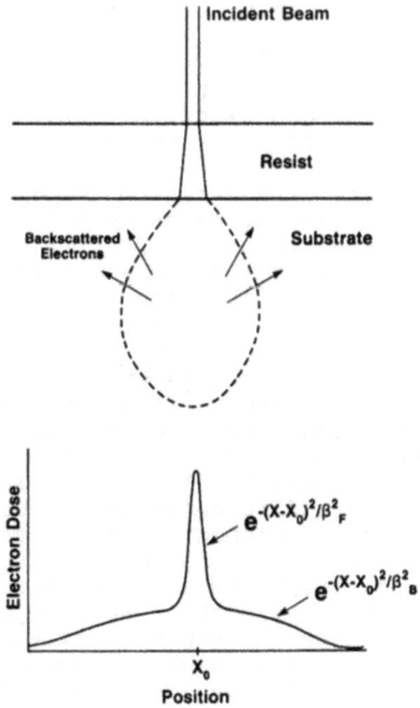

Fig. 2. Electron scattering and qualitative plot of electron dose vs. position as a result of electron scattering.

presents the greatest problem for energies such that the electron range is comparable to the feature spacing.

Much of the early high resolution work was done on thin electron transparent membranes with high resolution scanning transmission electron microscopes.[12-14] In addition to eliminating the backscattered electrons and improving the process latitude, as described above, working on an electron transparent membrane allows the use of high resolution transmission electron microscopes for both exposure and imaging. This is significant because the smallest structures may be difficult to resolve or have little contrast when viewed by conventional scanning electron microscopy. It is with this type of instrument that one obtains the smallest electron beam sizes of down to 0.2 nm. It is more desirable to create robust structures on solid substrates, and this can be done only with care and consideration of the electron scattering effects.[15]

It has been shown both experimentally and by calculation that the resolution limit for exposure of a conventional organic resist (of which polymethyl methacrylate [PMMA] is the prime example) is about 10 nm.[16-18] This is approximately equal to the secondary electron range in the organic resist. To reduce the effects of electron scattering, one can work with the extremes of very high energy electrons, on the order of 100 keV, or very low energy electrons of a few electron volts with a scanning tunneling microscope. The generation of low energy secondary electrons results in a cylindrical exposed volume around a delta function beam. Keyser[17] and others have done Monte Carlo calculations that model this effect. This is a fundamental problem with electron beam exposure that can be reduced only by using resists with shorter ranges for these low energy (≤ 50 eV) electrons. Alternatively one could employ a resist that is exposed only by higher energy primary electrons. There have been suggestions that this is the case for the fluoride and oxide self-developing films, but these processes are not yet fully understood.[20,21]

Another consideration important for the ultra-small size range is the time it takes to write, by a scanning serial process. As a serial process, scanned electron beam lithography is inherently slower than parallel printing processes. This is the price paid for the ability to create patterns without masks. For the scientific investigation of small structures speed is not usually an issue. However, for the creation of large areas of small structures the time required to generate a single pattern for study is a consideration. This time is minimized by using highly beam sensitive, high speed, resists. This is done with consideration of the fact that there is a trade-off between sensitivity and resolution based ultimately on the fact that a sufficiently large number of electrons must expose each resolution element so that statistical fluctuations in the number of electrons is relatively small. For some advanced high speed resists this limit is reached.[23] The smaller the electron beam size the lower the current available for exposure and the longer the exposure time. The highest resolution resists also require higher exposure doses. Fortunately the areas involved for device exposure are small. For experimental devices where a few objects are being written, the time in writing is insignificant, and the benefits and success of scanned electron beam lithography have been great.

Electron beam instruments in the form of analytical electron microscopes are widely available. A conventional scanning electron microscope (SEM), with beam position controlled by a minicomputer, can be used to generate simple patterns over small fields. Control of the mechanical positioning of the sample by laser interferometry allows for the stitching together of electron beam scan fields to cover large areas. This feature is available on commercially produced electron beam writers. For the study of fabrication processes, the analytical features and imaging capabilities of a transmission electron microscope with scanning capability has been effective. The trend in all of this instrumentation for high resolution work is toward higher electron energies, with 100 keV becoming more available and research being done at energies of 250 keV or higher. The process latitude and effects of proximity exposure can be minimized by working at high electron energies. With presently available technology it is routinely possible to create patterns in resist with linewidths of about 25 nm. With greater effort it is possible to make finer features in some materials. Fairly thin layers of conventional polymeric resists such as PMMA are adequate for this size range. For particular applications and in cases where large areas and therefore speed are important other resist systems can be advantageously employed. These procedures can be used to define arbitrary geometries and are being employed for numerous experimental structures and devices.

3. Pattern Transfer

The creation of high resolution patterns in a resist layer is only part of the fabrication process. In fact, pattern transfer is frequently the resolution-limiting step. While holes in thin films as small as 2 nm have been generated by high voltage electron beam lithography,[20,21] they have not been transferred into a solid material for study. Also the effects of process-induced damage can be devastating on ultra-small structures, and this must be taken into consideration in order to create useful samples for study.

The general pattern transfer technique requires an additive or subtractive transfer process, typically an etching or deposition and lift-off step, as illustrated in Fig. 3. The practical limitation in the lift-off procedure is on the aspect ratio and thickness of the film to be patterned in the lift-off step. For a successful removal of the deposited film, the deposited film should be discontinuous over the resist edge. If the deposited film is thick compared to the resist, this will not in general be the case and the lift-off is impossible. For best results a resist profile with an under-cut shape is desirable, but even so the resist thickness must be greater than the deposited film. As noted above the best resolution is obtained with thin resist. The multi-layer resists have been used to

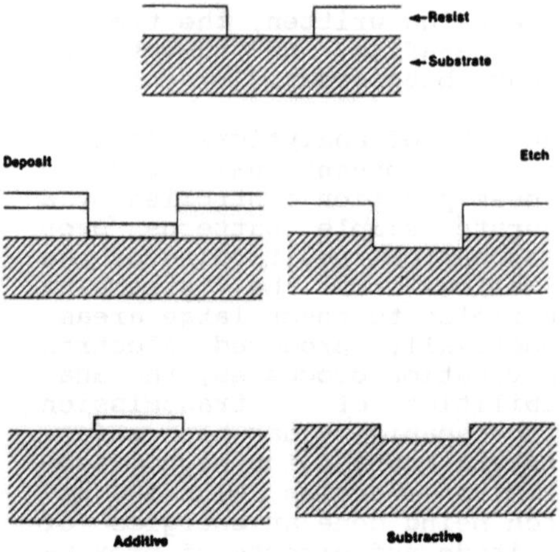

Fig. 3. Schematic of pattern transfer processes.

improve the resist performance by using a thin imaging layer on a thicker under-layer.[22,23] The pattern exposed and developed in the thin imaging layer can be transferred into the subbing layer by a high resolution process such as reactive ion etching. By appropriately choosing the etching chemistry this can greatly increase the resist utility. As an alternative additive transfer process, electroplating has been used to create 0.8 nm diameter gold wires in narrow cavities.[24]

Directed ion etching methods of ion milling, reactive ion etching, reactive ion beam etching and ion beam assisted etching have all been used as high resolution etching processes. These processes are anisotropic because they use a directed beam of ions. In ion milling the removed atoms will deposit on surfaces in the line of sight of the sputtered surface. In the reactive process the reaction products are volatile and do not deposit on exposed surfaces. In etching GaAs for example a relatively non-reactive metal can be used as a mask to prevent the underlying material from being removed in a Cl-containing plasma.[25] Etching narrow grooves can still be a problem because of the required transport into long narrow cavities. The highest aspect ratio structures in semiconductors have been made by ion beam assisted etching.[26]

Let us consider some examples of nanofabrication that illustrate the possibilities and how the techniques build on each other. In one example of a nanofabrication process using high resolution lithography and a lift-off step, arrays of 20 nm gold particles were fabricated on sapphire substrates for optical measurements.[27] The creation of regular arrays of known geometry particles allowed comparison to effective medium theories for the optical response of inhomogeneous media without adjustable parameters. For the creation of these structures a single 60 nm thick layer of PMMA was exposed by a ~2 nm diameter 120 keV electron beam in a scanning electron microscope. The beam was focused by imaging a scratch in the resist using the secondary electron signal. The exposed resist dot areas were removed in a solution of cellosolve in methanol. Pure gold was deposited on the resist by vacuum evaporation. The remaining resist was dissolved in acetone to perform the lift-off and leave the gold disks. Annealing the sample at 125°C caused the gold to consolidate to reduce the surface energy and become more spheroidal. The annealing step made significantly rounder and more regular particles. An electron micrograph of a 50 nm period array of gold particles is shown in Fig. 4. These particles were sufficiently small that the particle size had to be considered as a limit to the electron mean free path.

In another example a technique similar to the previous example was used to define etch masks for creating small GaAs columns.[25] In this case NiCr particles were defined similarly to the gold, as described above, by electron beam exposure and lift-off. The NiCr was resistant to anisotropic reactive ion etching in a $SiCl_4$ plasma that anisotropically etches GaAs. By this method 20 nm diameter GaAs columns were formed in square arrays as seen in Fig. 5.

The next level of complication involves layered GaAs/AlGaAs materials that require more consideration for etching, because of differing chemistry of the alternating layers. Studies of etching such layers were done[28], and the reactive ion etching in the correct BCl_3 and Ar gas mixture was used to create arrays of quantum dots and wires for optical studies of dimensionally reduced structures.[29] Fig. 6 shows a transmission electron micrograph of etched multiple quantum well structures.

Fig. 4. Scanning electron micrograph of gold dots on sapphire

Fig. 5. Scanning electron micrograph of etched GaAs columns.

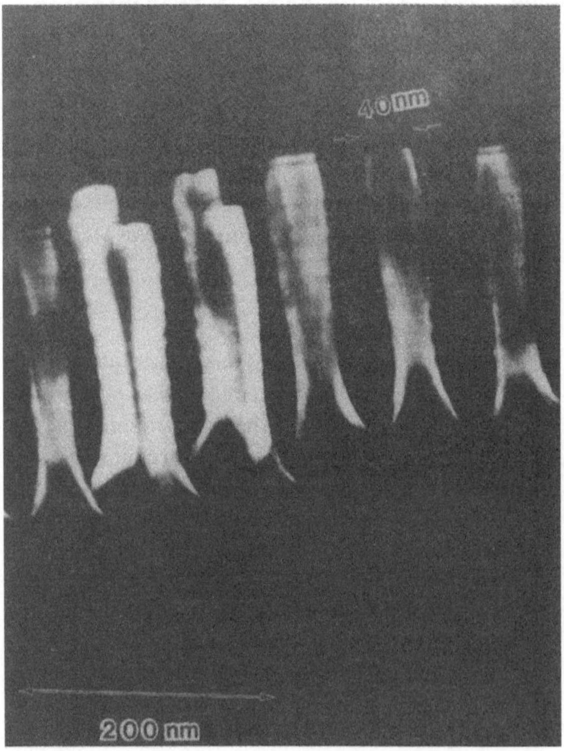

Fig. 6. Cross-sectional electron micrograph of etched GaAs-AlGaAs quantum well samples.

4. Combined Processes

The distinction between lithography and pattern transfer is not clear when growth and patterning are combined. This is a fruitful direction of research and may eliminate many of the problems of processing-induced damage and the exposure of free surface. Growth on patterned substrates by chemical vapor deposition can be used to make narrow wires. [30-32] In situ MBE growth on etched surfaces may be able to preserve the high resolution possibilities of EBL and ion etching with atomic resolution growth.[33]

The ability to etch high aspect ratio structures in GaAs has been used to create high resolution ion masks for selected area disordering of GaAs/AlGaAs heterostructures. In the technologically important case of compound heterostructure materials the effect of impurity and defect induced compositional mixing can be used to define very small lateral geometries. In this process the presence of certain impurities or sufficient amounts of lattice damage substantially reduces the temperature at which the layers of material will interdiffuse.

Much of the work in this area has been motivated by the possibility of fabricating improved heterostructure lasers and waveguides. In the area of nanofabrication, lattice damage induced by Ga or Al ion implantation has been used to locally enhance interdiffusion and modify the mini-band structure to achieve lateral carrier confinement in patterned dots and wires.[34-37] With the implantation of Si impurities even greater enhancement of interdiffusion can be achieved to laterally modify the materials band structure. Si acts as an amphoteric dopant, however, and can complicate the electrical modification control. The patterning resolution limit of selected implant induced disorder has been explicitly studied by cross-sectional transmission electron microscopy,[37] and the indication is that resolution easily better than 30 nm can be achieved. This holds even greater future promise with focused ion beam systems.

5. One-Dimensional Wires

The techniques described above have been employed to create narrow conducting channels in a GaAs/AlGaAs modulation doped two-dimensional electron gas.[38] We employed high voltage electron beam lithography to create a range of channel geometries with widths as small as 75 nm. Using ion beam assisted etching by Cl_2 gas and Ar ions with energies as low as 150 eV, conducting channels were defined by etching only through the thin GaAs cap layer. This slight etching is sufficient to entirely deplete the underlying material without necessitating exposure of the sidewalls that results in long lateral depletion lengths. At 4.2 K, without illumination, our narrowest wires retain a carrier density and mobility at least as high as that of the bulk 2DEG and exhibit quantized Hall effects.

Modulation doped GaAs-AlGaAs heterostructure two-dimensional electron gas systems open a new regime for the investigation of low temperature transport phenomena in structures of reduced dimensionality. This is possible because of the high electron mobility, long scattering lengths and small electron effective mass in this material. The ability to fabricate small structures in III-V compound semiconductors by electron beam lithography and dry etching processes has been described above. The creation of conducting structures with widths less than one micrometer however has been difficult.[39] Fermi level pinning by states at the exposed side walls of narrow etched structures causes depletion of the carriers within the small structure. A method has recently been described that significantly reduced this carrier depletion problem. By patterning only the doped AlGaAs layer of the two-dimensional electron gas (2DEG) structure, conducting paths were defined without exposing the electron gas interface at the sidewalls of the wire.[6] There exists a strong incentive to make even narrower wires, since quantum confinement effects become pronounced when the wire width becomes comparable to λ_f, the Fermi wavelength, which characterizes the spatial extent of the electron wavefunction. For a typical 2DEG with a electron density in the 10^{11} cm^{-2} range, $\lambda_f=2\pi/k_f$ is on the order of 100 nm.

We describe a method of defining narrow conducting channels in a 2DEG structure by removing only a small amount of material from the surface of the material by ion etching. It appears that electrical damage, induced by 150-500 eV ions, is sufficient to destroy the conductivity of the 2DEG. We have demonstrated this by controlled etching of large areas of material by ion beam assisted etching (IBAE) with Ar ions in the presence of Cl_2 gas. We observe a progressive deterioration of carrier density and mobility with increasing etch time. The mobility is clearly reduced as a result of ion damage since similar amounts of material removed by chemical etching do not significantly alter the mobility of the 2DEG. Using this technique, we have patterned wires with mask widths as narrow as 75 nm. At liquid helium temperatures, in the dark, these conducting channels exhibit no degradation in the original mobility and carrier density of the 2DEG.

The electron beam lithography was done in a computer controlled scanning transmission electron microscope (STEM) at an electron energy of 250 keV to expose a single layer of PMMA as described above. The high energy electron beam eliminates concerns of proximity correction in this type of isolated device, and we demonstrated that there was no measurable effect upon the electrical properties of the 2DEG at the electron doses used for exposure. A mask of 120 nm of SrF_2 was formed by deposition and lift-off. The fluoride was used because of its high resistance to etching by halogen-containing plasmas. This was important for our original work where we etched entirely through the 2DEG interface to create mesas. It was also significant in subsequent work where, as have others, we etched only the doped AlGaAs layer by reactive ion etching. With our present technique the strong fluoride etch resistance is less important since only shallow etching is done. It is convenient, however, that the fluoride etch mask is nonconducting and can remain on the surface during the measurements, simplifying the processing steps.

The etching was done by ion beam assisted etching with Ar ions in the presence of Cl_2. A 150 eV ion beam with a specimen current density of 30 μA cm^{-2} was directed at the sample with Cl_2 gas introduced near the etched surface. The total system pressure was $3x10^{-4}$ Torr. These conditions lead to a GaAs etch rate of about 5 nm/minute. This system has been developed for high aspect ratio etching of GaAs. This high aspect ratio is not needed for this method of patterning, but we find that the available range and control of ion energies is appropriate for inducing damage to the necessary depths.

The introduction of electrical damage in GaAs by low energy ion bombardment has been observed previously. Capacitance-voltage (C-V) measurements show that ion milling with 100 eV ions electrically damaged the material to a depth of 90 nm.[40] These measurements are consistent with our sheet resistivity data, where similar ion energies were used. The ion penetration depth at 100 eV is much less than this 90 nm and gross structural changes as obtained by transmission electron microscopy are on the order of a few nm for this type of ion species and energy. The exact nature and mechanism of electrical damage introduction is not entirely understood. The case of ion beam assisted etching of 2DEG structures has been even less widely studied. We have done measurements that establish how to exploit the effect of ion damage for patterning of nanometer scale 2DEG structures, but we have not established the exact nature of the changes introduced by the processing.

Conducting wires and other channel geometries were created by the above conditions. Fig. 7 shows a scanning electron micrograph of a 5 μm long wire with Hall voltage probes along the length. The electron micrograph shows the shallow etch into the GaAs cap layer with the 120 nm fluoride mask. The fine electron beam written pattern connects to the larger gold leads seen as the larger features in this figure.

Some of the considerations of this technique are similar to those in the high resolution definition of GaAs structures by implant-induced disordering of heterostructures. Much narrower conducting channels can be defined by locally changing the electrical properties than by etching and exposing untreated surface. Until surface passivation and regrowth techniques are perfected these selected area modification techniques such as that described here, appear the most promising for creating ultra-small quantum structures. Our present size limit of 75 nm is not a fundamental one. It is limited by the difficulty of fabricating high aspect ratio and continuous ion masks. The narrowest wire we have been able to create has conducted without illumination. We see no reason why, with further work and advances in the mask fabrication, even narrower wires should not be possible with this selective low energy ion damaging.

We have demonstrated that it is possible to define conducting channels in GaAs-AlGaAs 2DEG systems with widths as small as 75 nm. Magneto-transport measurements on these fine wires demonstrate that the electron mobility is not reduced in the masked area and that the geometry of the conducting channels is well defined by the mask geometry.

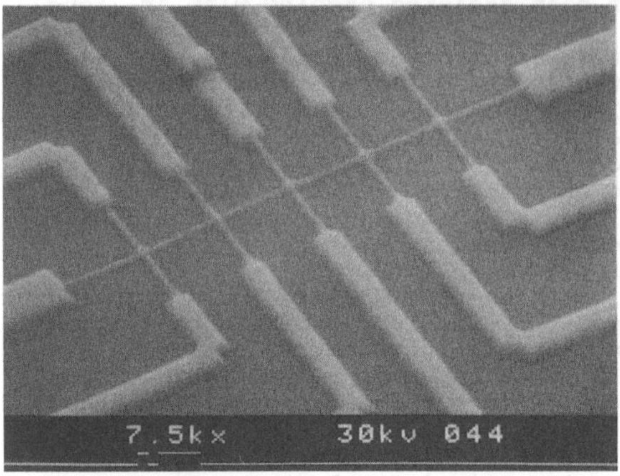

Fig. 7. SEM of a quasi-one-dimensional wire, where the main contrast is obtained from the fluoride mask.

6. Conclusion

The fabrication technology exists to allow fabrication of structures with dimensions on the order of 10-100 nm in a variety of materials. With advanced electron, ion and x-ray techniques the lateral control of semiconductor materials is constantly advancing. With the possibilities of epitaxial growth combined with these techniques complete three-dimensional control of semiconductor structure can be contemplated. This is allowing the exploration of new phenomena and consideration of quantum interference devices, quantum memories, and lateral hot electron devices.

References

1. S. Washburn, H. Schmid, D. Kern, and R. A. Webb, Phys. Rev. Lett. 59, 1791 (1987)
2. G. Timp, A. M. Chang, P. Mankiewich, R. Behringer, J. E. Cunningham, T. Y. Chang, and R. E. Howard, Phys. Rev. Lett. 59, 732 (1987)
3. H. van Houten, C. W. J. Beenakker, M. E. I. Broekaart, M. G. J. Heijman. B. J. van Wees, J. E. Mooij, and J. P. Andre (to appear in Phys. Rev. Lett.)
4. M. L. Roukes, A. S. Scherer, S. J. Allen Jr., H. G. Craighead, R. M. Ruthen, E. D. Beebe, and J. P. Harbison, Phys. Rev. Lett. 59, 3011 (1987)
5. T. P. Smith et al. (to appear in Phys. Rev. Lett)
6. H. van Houten, B. J. van Wees, M. G. J. Heijman, and J. P. Andre, Appl. Phys. Lett. 49, 1781 (1987)
7. Y. Arakawa and H. Sakaki, Appl. Phys. Lett. 40, 939 (1982)
8. G. J. Iafrate, D. K. Ferry and R. K. Reich, Surface Science, 113, 485 (1982)
9. S. Washburn, H. Schmidt, D. Kern, R. A. Webb, Phys. Rev. Lett. 59, 1791 (1987)
10. T. Everhart and Hoff, J. Appl. Phys. 42, 5837 (1971)

11. L. D. Jackel, R. E. Howard, P. M. Mankiewich, H. G. Craighead, and R. W. Epworth, Appl. Phys. Lett. 45, 698 (1984)
12. A. N. Broers, W. W. Molzen, J. J. Cuomo, and N. D. Wittels, Appl. Phys. Lett. 29, 7188 (1976)
13. S. Mackie and S. P. Beaumont, Solid State Communications 28, 117 (1985)
14. H. G. Craighead and P. M. Mankiewich, Appl. Phys. Lett. 44, 468 (1984)
15. H. G. Craighead, J. Appl. Phys. 55, 4430 (1984)
16. A. N. Broers, J. Electrochem. Soc. 128, 166 (1981)
17. D. F. Keyser, J. Vac. Sci. Technol. B 1, 1391, (1983)
18. H. G. Craighead, R. E. Howard, L. D. Jackel, and P. M. Mankiewich, Appl. Phys. Lett. 42, 38 (1983)
19. D. Joy (unpublished)
20. I. G. Sailisbury, R. S. Timsit, S. D. Berger, and J. C. Humphries, Appl. Phys. Lett. 45, 1289 (1984)
21. A. Muray, M. Isaacson, and I. Adesida, Appl. Phys., Lett. 45, 1289 (1984)
22. D. M. Tennant, L. D. Jackel, R. E. Howard, E. L. Hu, P. Grabbe, R. Capic, and B. S. Schneider, J. Vac. Sci. Technol. B 1, 1291 (1983)
23. A. G. Gozdz and P. S. D. Lin, Elec. Lett. 24, 123 (1988).
24. W. D. Willaims and N. Giordano, Rev. Sci. Instrum. 55, 410 (1984)
25. M. B. Stern, H. G. Craighead, P. F. Liao, and P. M. Mankiewich, Appl. Phys. Lett. 45, 410 (1984)
26. M. W. Geiss, G. A. Lincoln, N. Efremow, and W. J. Piacentini, J. Vac. Sci. Technol. 19, 1390 (1981)
27. H. G. Craighead and G. A. Niklasson, Appl. Phys. Lett. 44, 1134 (1984)
28. A. Scherer, H. G. Craighead and E. D. Beebe, J. Vac. Sci. Technol. B, 5, 1599 (1987)
29. A. Scherer and H. G. Craighead, Appl. Phys. Lett. 49, 1284 (1986)
30. R. Bhat,to be published
31. E. Kapon,to be published
32. E. Kapon, M. C. Tamargo, D. Hwang, Appl. Phys. Lett. 50, 347 (1987)
33. A. Scherer, J. Harbison, and E. Beebe, (to appear in Proc. SPIE)
34. J. Cibert et al., Appl. Phys. Lett. 49, 1275 (1986)
35. E. A. Dobisz et al., Proc. Mat. Res. Soc. (1986)
36. P. Petroff, to be published
37. E. A. Dobisz, B. Tell, H. G. Craighead, and M. C. Tamargo, Proc. SPIE (1987)
38. A. Scherer, M. L. Roukes, H. G. Craighead, R. M. Ruthen, E. D. Beebe and J. P. Harbison, Appl. Phys. Lett. 51, 2133 (1987)
39. K. K. Choi, D. C. Tsui, and K. Alavi, Appl. Phys. Lett. 50, 110 (1987)
40. S. W. Pang, G. A. Lincoln, P. W. McClelland, P. D. DeGraff, M. W. Geis, and W. J. Piaconti, J. Vac. Sci. Technol. B 3, 398 (1985)

Electron Beam Lithography and Dry Etching Techniques for the Fabrication of Quantum Wires in GaAs and AlGaAs Epilayer Systems

S.P. Beaumont[1], *C.D.W. Wilkinson*[1], *S. Thoms*[1], *R. Cheung*[1], *I. McIntyre*[1], *R.P. Taylor*[2], *M.L. Leadbeater*[2], *P.C. Main*[2], and *L. Eaves*[2]

[1]Nanoelectronics Research Centre,
Department of Electronics and Electrical Engineering,
University of Glasgow, Glasgow G12 8QQ, Scotland, UK
[2]Department of Physics, University of Nottingham,
Nottingham, NG7 2RD, UK

Introduction

Research on transport in low dimensional semiconductor systems has recently progressed from simple layered structures to patterned devices of much greater complexity. The development of fabrication techniques capable of machining materials to quantum dimensions has been partly responsible for this interest. Many fascinating phenomena have been investigated in such structures, including localisation and trapping effects, universal conductance fluctuations[1], Aharanov-Bohm oscillations[2] and quenching of the Hall effect[3]. In this paper the technology for fabricating quantum wire and related structures in GaAs and GaAs-AlGaAs is described and some measurements of quantum interference phenomena in n^+ GaAs wires are presented.

Fabrication routes

Three routes to the definition of quantum structures have been described in the literature, each with its own detailed variations. These are shown schematically in Fig.1.

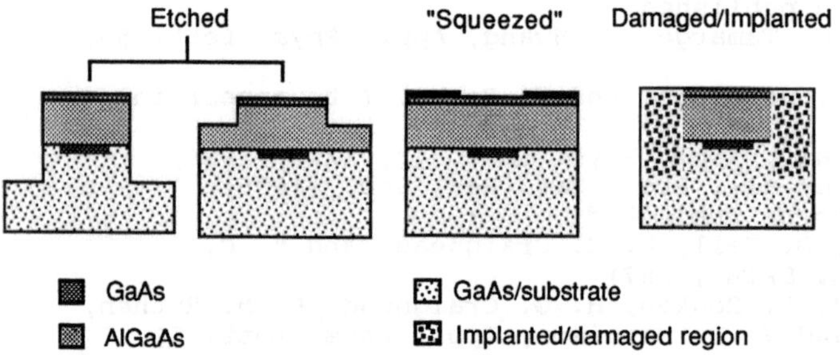

Fig.1 Alternative methods of fabricating quantum wires.

The first route involves etching the semiconductor using wet or dry techniques: the active layers may be removed completely[4] or partially[5]. In the latter case surface depletion is relied upon to pinch off active material outside the required structure. This variant is believed to reduce the influence of surface effects on transport in the underlying channel. The second, highly successful method requires no etching at all: instead, a pattern of Schottky barriers defined on the surface of the material is used to deplete the underlying carriers leaving only a narrow channel for occupation[6]. By varying

Springer Series in Solid-State Sciences Vol. 83: **Physics and Technology of Submicron Structures**
Editors: H. Heinrich • G. Bauer • F. Kuchar

the external potential the width of the quantum channel can be altered and the effects on transport explored. Recent results demonstrate excellent lateral quantisation in such structures. Obviously this method causes minimum damage to the semiconductor and the orginal surface remains intact. A third route employs ion bombardment either to damage the underlying material and bring about a substantial reduction in conductivity[7] or to mix heterojunctions so that the bandgap is locally increased and the electrons confined to the unimplanted regions[8]. Quantum dots and wires for optical studies have been successfully fabricated by this route and have exhibited quantum confinement effects in their photoluminescence spectra. It should be emphasised that these three methods are not entirely distinct: for example, the etched route probably involves some damage to material outside the wire, although this has not been thoroughly investigated.

This lecture concentrates on the first method of fabrication using etching techniques but much of the information is generally applicable to all three.

Dimensional requirements

The external dimensions of a 'quantum wire' vary considerably and depend on the experiments to be performed, the material and the fabrication technique employed. The most stringent constraint would be to ensure that all transport in the wire occurs in the lowest lateral sub-band. Assuming this as a worthwhile goal, we can estimate the width of the channel by calculating the number of available states per unit length in the lowest subband, assuming a suitable well potential, and equate this to the number of electrons which need to be accommodated. If we assume also that the sheet carrier concentration remains constant irrespective of width then we need to fabricate a channel ~40nm wide for typical 2-DEG sheet carrier concentrations. A more refined calculation would have to be carried out self-consistently as the shape of the potential forming the wire determines and depends on the electron concentration. Thus we have to make wire about 40nm wider than cutoff, which occurs at low temperatures when the Fermi level falls below the lowest available sub-band. The condition is determined by the applied potential in "squeezed" wires or by the dimensions and surface depletion of etched wires. It is possible to calculate the cutoff widths of both etched and squeezed wires analytically by assuming that the potential around the surface of the wire is pinned by the surface states[9]. Measured cutoff widths for 2-DEG wires range from 0.25µm for lightly etched structures to 0.5µm for the deep-etch method. Thus to attain our goal of single-subband operation we need to define a submicron wire with precision and uniformity better than 40nm. The only adequate lithographic techniques are electron beam direct writing and x-ray contact printing. Since the latter is suited only to stamping out masks generated by some other means, I shall concentrate on the former.

Electron Beam Nanolithography.

The basic principles of electron beam lithography are described in many textbooks on fabrication technology and there is not space to cover them in this lecture. It is relevant, however, to discuss the techniques of e-beam *nanolithography* and the machine requirements. It is generally accepted that in the best polymeric resist materials the limit to resolution is of the order of 10nm linewidth and ~50nm centre-to-centre spacing (pitch) under ideal conditions. There is, however, some controversy over the underlying reasons for this limit. Of course in any given machine the minimum beam diameter may determine resolution. The probe diameter is determined by contributions from the demagnified Gaussian source, spherical aberration in the final lens and electron diffraction from the beam-limiting aperture. It is commonly accepted that these components can be added in quadrature to give the overall probe diameter via the following equation:

$$d_p^2 = \frac{4I_p}{\pi^2 \alpha^2 \beta} + \frac{1}{4} C_s^2 \alpha^6 + \left(\frac{1.22\lambda}{\alpha}\right)^2 . \tag{1}$$

By comparison with glass optics, electron optical lens aberration coefficients are very large (typically many centimetres and often many metres) and therefore the optimum aperture is very small, of the order of a few tens of mrad. Thus in order to obtain significant beam current in very small probes it is necessary to use sources of high brightness. Most modern sources, typically those used in conventional scanning electron microscopes, have sufficient brightness to form probes of 5nm diameter: but the development of Field emission and Lanthanum Hexaboride cathodes allows beams of 0.5nm diameter to be formed with sufficient current for the operator to be able propertly to set up the probe. By using systems such as these for lithography it has proved possible to demonstrate that resolution limits in useful resists such as PMMA are not determined by the quality of the electron optics[10]. Electron scattering can, however, expand the beam in its passage through the resist into the substrate but such effects can be reduced to negligible proportions by using thin resist films and high energy electrons. Resist can be spin-coated into films less than 50nm thick with acceptable pinhole densities and, whereas commercial electron beam lithography systems operate at 25-30kV, most nanolithography is now carried out in the 50-120keV range: indeed, some recent work has been carried out at even higher energies[11]. The effect of high electron energy is to narrow the forward-scattering distribution and to smear out the backscattering so that the exposure of a given feature is less influenced by the exposure of its neighbours - the proximity effect. These trends are illustrated by considering the variation with energy of the parameters of the two-gaussian model of resist exposure distribution. This expresses the energy density ε at a distance r from the point of incidence of an infinitesimal beam in the form:

$$\varepsilon(r) = k\exp\left(-\frac{r^2}{\beta_f^2}\right) + \eta_e\exp\left(-\frac{r^2}{\beta_e^2}\right) \tag{2}$$

where the first term represents the forward scattering component and the last the wash of electrons backscattered from the substrate. For a 0.5μm film of PMMA on a solid silicon substrate the following parameters have been measured:

kV	β_f	β_e	η_e	[Ref]
20	0.08μm	2μm	0.78	[12],[13]
50	0.04μm	9μm		[12],[13]
60		13.1μm	0.70	[13]
120		43μm	0.76	[13]

Different values result (especially for β_e) on different substrates. The dramatic influence of electron energy on the backscattered distribution is obvious. Simple two-gaussian models based on primary scattering do not adequately describe the exposure of isolated features less than 0.1μm in size, however. Subtle variations in the exposure profile are seen on different substrates close to the point of impact of the beam[14] and at the resolution limit the primary scattering model fails completely to explain the experimental data. Secondary electrons are believed to play some role in determining ultimate resolution. Since polymeric resists are relatively insensitive to fast electrons it is accepted that secondary electrons are largely responsible for initiating the physical and chemical processes that 'expose' the resist. These have a finite range not accounted for in simple models. We have attempted to model completely the secondary electron generation, transport and exposure process using a Monte Carlo model based on empirical data[15]. Secondaries were shown to contribute only 3-5nm additional linewidth, not 10nm as required by the measured resolution limit. In our view the resolution of polymeric resists is determined by long-range molecular chemistry, but the issue is still open. Its solution may be important in the search for higher resolution resist processes to fabricate yet smaller structures.

High beam energy is therefore one important factor in electron beam nanolithography. Other requirements are: adequate beam current, mainly for alignment and beam setup rather than throughput, which is not usually a concern in research applications of this nature; a spot size smaller than the minimum feature size to that multiple passes of the beam can be used to fill in the pattern, giving better edge definition than is possible with a single pass of a larger beam; a fast blanking system to avoid overwriting large areas of high definition structure without loss of resolution or serious distortion; and high electrical and mechanical stability to avoid drifts and vibration during long exposures typical of this work. The engineering of nanolithography machines is as yet in its infancy; the bulk of the work so far having been executed on modified Scanning Electron Microscopes and Scanning Transmission Electron Microscopes which usually have appropriate optics but few of the mechanical and electrical refinements needed for optimum results.

Resist processes.
To fabricate quantum wires by etching we can adopt two alternative patterning routes involving resists of opposite (positive or negative) tone. Our aim is to lay down a stripe of material capable of masking the ion etching process.

The choice of mask is important in this application. Metals are excellent but must be removed after processing to avoid shorting out the wire: it is often difficult the find a suitable metal as common etchants tend to attack the III-V's. Liftoff of insulators is more satisfactory as the mask need not be removed: strontium fluoride and alumina masks have both been used in this way [7],[5]. A more direct route is to employ a negative e-beam resist. Negative resists generally exhibit poorer resolution than their positive counterparts. This is because of their poorer contrast which results in the formation of bridges by backscattered electron exposure, and their tendency to swell during development as the solvent penetrates the cross-linked polymer matrix to form a gel. Consequently negative resists have been neglected for nanolithography. Excellent results are nevertheless obtainable.

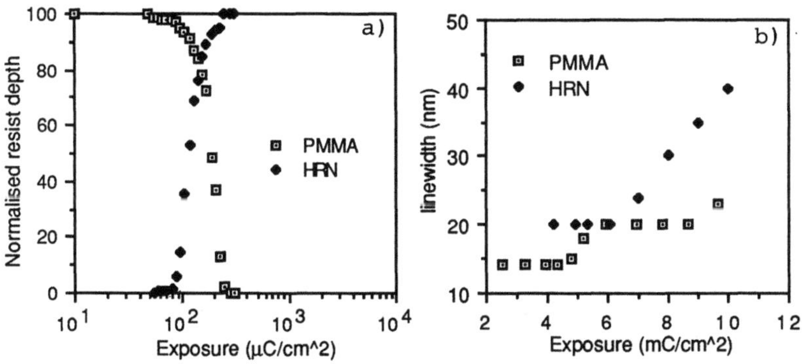

Fig.2a Contrast curves for PMMA (positive resist) and HRN (negative resist)
Fig.2b Resolution vs exposure for single lines in PMMA and HRN on solid GaAs at 50kV

Fig.2a compares the contrast curves of PMMA and a standard high resolution negative resist HRN and Fig.2b shows the linewidth/exposure curves for lines written in the same materials. The lower resolution of the negative resist is obvious but the performance degradation is not severe. 20nm lines can be defined in isolated structures and 50nm features in close-packed (180nm pitch) patterns. The densest pattern we have written in HRN on GaAs consists of lines on 120nm pitch, below which severe bridging causes loss of fidelity. Note that the resolution of the image in PMMA is hardly affected by the change in pattern density.

Reactive ion etching
Having defined a resist pattern it is necessary to preserve its resolution and edge definition in transferring it onto the substrate to define the quantum wire. Wet etch techniques are able to define wires below cutoff dimensions, i.e. < 50nm for n^+ GaAs wires, but not without difficulty. All etchants can be classified as anisotropic or isotropic. Wet anisotropic etches tend preferentially to etch specific crystal planes leading to wire cross-sections which may be suitable in one direction but not in others. For example in GaAs wet etched wires vary from overcut to undercut in orthogonal directions. Isotropic etches always result in undercutting of the mask because of their roughly equal etch rates in the vertical and lateral directions. This effect has been used to reduce the width of photolithographically defined wires but, without extreme care, leads to loss of structure. Wet etches are also very susceptible to surface contamination which leads to nonuniformity across even very small samples and, in particular, makes the light surface etching technique very difficult to control. Most workers have therefore turned to reactive ion etching (RIE) for the patterning of wires.

The apparatus used for RIE is shown in Fig.3. Samples to be etched are placed on a platen which is capacitively coupled to an rf generator at 13.6MHz. The chamber is back-filled with a gas which, when dissociated, produces ions and radicals at least some of which react with the semiconductor to form volatile products. A plasma is struck in this gas at low pressure (10mtorr - 100mtorr) whereupon the driven electrode attains a negative potential by receiving a nett flux of electrons during the first few cycles because of the higher mobility of these particles than the positive ions. The negative charge serves to equalise the electron and ion target currents on each rf half-cycle. The magnitude of the voltage dropped across the dark space just above the driven electrode increases with the ratio of the grounded electrode area to the driven electrode area. RIE systems are designed with a large ratio of areas so that the sample to be etched is bombarded with high energy positive ions. At the same time radicals and atoms diffuse to the surface of the sample. The reactions at the surface are complex, but generally involve the enhancement of the chemical reaction by ion bombardment, a process which usually results in anisotropic etching of the structure.

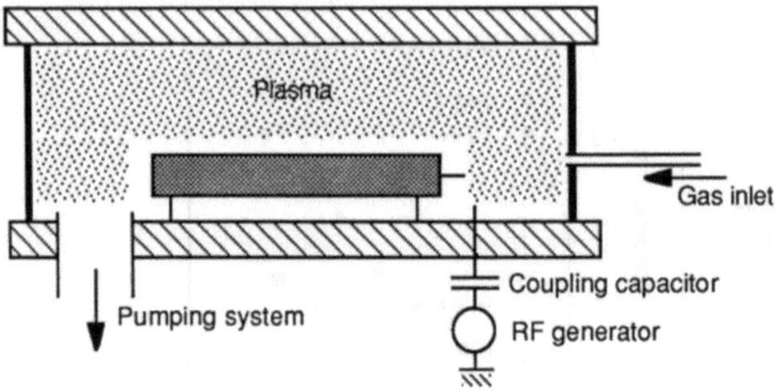

Fig.3 Schematic of reactive ion etching system. The sample to be etched is placed on the electrode coupled to the rf power source.

Chlorine-containing gases are often used for etching GaAs in RIE systems. Pure chlorine and silicon tetrachloride are two of the most common choices. We have used silicon tetrachloride extensively and many of the following results were obtained with this gas. For quantum wire fabrication optimisation of rf power and gas pressure is necessary to obtain a rectangular wire cross-section and to avoid the formation of grassy spikes due to polymer formation and redeposition on the sample. There is a relatively small parameter space in which the correct conditions prevail[16]. Fig.4 shows a typical GaAs quantum wire, 0.15µm wide defined using an HRN directly written e-beam mask as described above. The mask is still in place.

Wires fabricated by dry etching may appear to be of high quality but the bombardment of the substrate by high energy ions inevitably causes surface damage. This is to be expected on horizontal surfaces which receive the highest dose but quantum wire fabrication has revealed significant damage to the vertical walls of etched structures too. We have shown that n^+ GaAs quantum wires fabricated by $SiCl_4$ dry etching cut off at roughly twice the width of those wet etched from the same material[16]. The difference in cutoff width was found to increase with rf power and was interpreted to result from dry-etch damage to the sidewalls.

Damage to dry etched surfaces can be caused in a number of ways: physical disruption of the lattice, radiation damage, redeposition and implantation have all been detected. A variety of physical, electrical and optical techniques can be employed to characterise the damage qualitatively and quantitatively. We have attempted to quantify the damage to GaAs caused by $SiCl_4$ etching and to find alternatively etchants which produce less damage. One candidate is the methane/hydrogen mixture which has been shown to etch InP successfully[17]. The etching mechanism for III-V's is believed to be a reverse MOCVD reaction in which Ga-containing metallorganic species and arsenic hydrides are formed.

Fig.4 Dry etched quantum wire 0.15µm wide running between large contact pads

Fig.5 compares the rate of removal of GaAs by $SiCl_4$ with that of CH_4/H_2 and Fig 6 the etch rate and wire profile as a function of gas composition. Peak etch rate and rectangular wire profiles are obtained simulataneously at a 5:1 H_2/CH_4 composition. The peak rate of 20nm/min is very much smaller than $SiCl_4$ (250nm/min): it is therefore very easy to control the etching of small quantities of material, especially for the fabrication of quantum wires by the shallow etching technique.

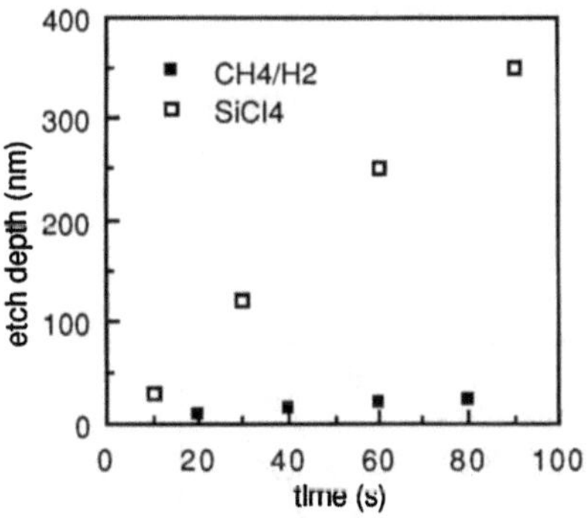

Fig. 5

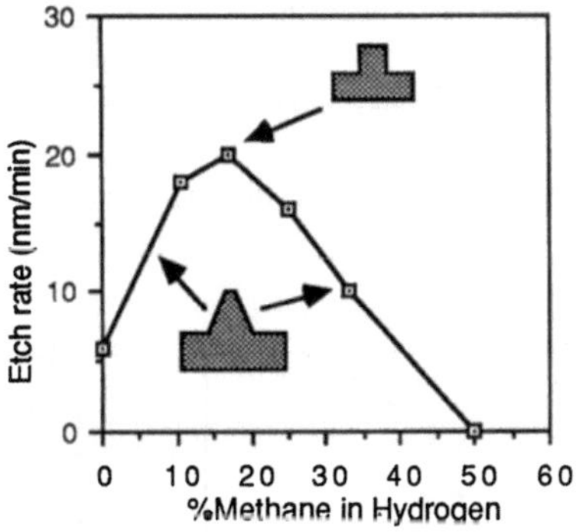

Fig. 6

Fig.5 Removal of GaAs using $SiCl_4$ or CH_4/H_2 with time.
Fig.6 Dependence oftch rate of CH_4/H_2 on gas composition

Figs.7 and 8 demonstrate that CH_4/H_2 causes much less damage to GaAs than $SiCl_4$ as assessed by integrated photoluminescence measurements and the ideality factors of Shottky barrier diodes fabricated on tin-doped material. In both experiments the measurand is believed to be sensitive to the generation of recombination centres (traps) by the etch. CH_4/H_2 appears either to generate fewer traps or to passivate them: the exact mechanism is as yet poorly understood.

One practical difficulty connected with the CH_4/H_2 process has recently been highlighted by fabricating quantum wires from silicon doped material. Fig.9 compares the conductance-width characteristics of $SiCl_4$ and CH_4/H_2 processed 2-DEG wires and shows that the latter, as-etched, have much larger cutoff widths than the former. This is due to the passivation of Si donors by hydrogen and can be eradicated by annealing. Fig.10 shows the recovery of the conductance after annealing CH_4/H_2 processed wires at different temperatures.

In summary, electron beam nanolithography is not seriously challenged by the problem of fabricating useable quantum-scale wires in semiconductor materials. Difficulties can arise in subsequent processing steps when the integrity of the original material may be disrupted by factors which are as yet poorly understood but which can be mitigated by careful processing.

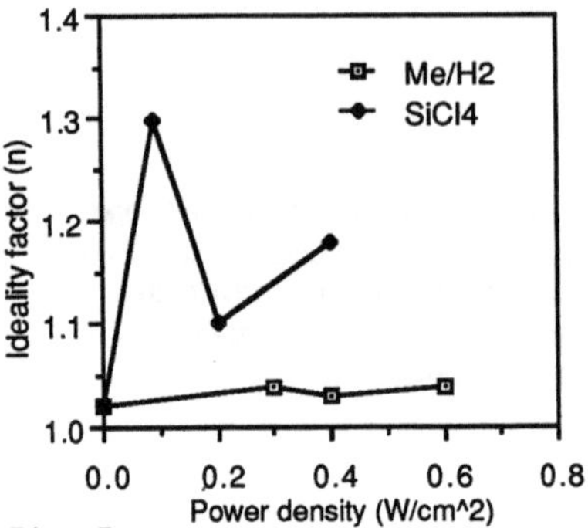

Fig. 7

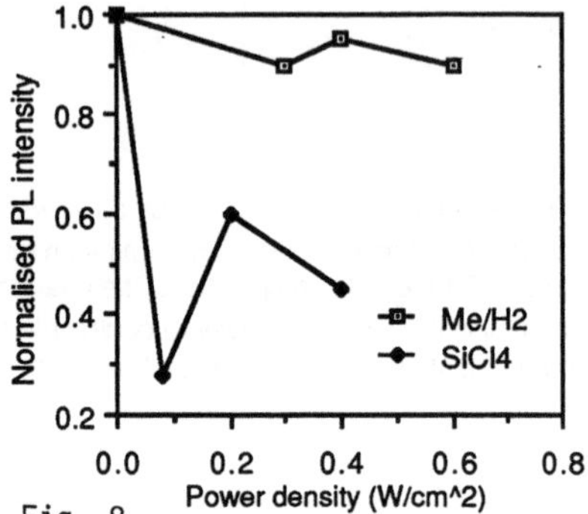

Fig. 8

Fig.7 Ideality factors of Schottky barrier diodes fabricated on as-etched surface of tin-doped GaAs ($5x10^{17} cm^{-3}$)

Fig.8. Integrated photoluminescence from as-etched GaAs surface. Laser wavelength 633nm, temperature 15K.

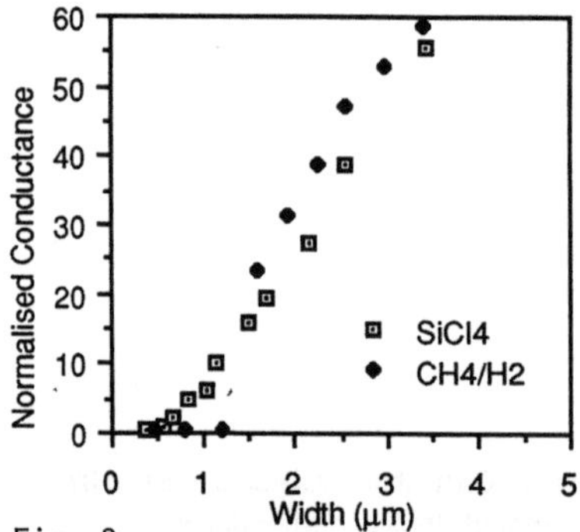

Fig. 9

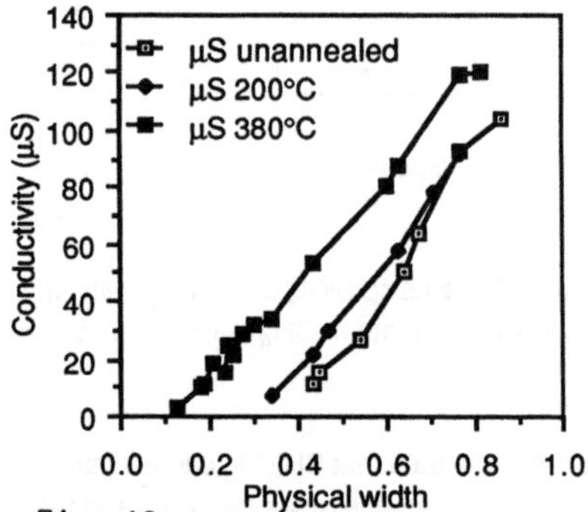

Fig. 10

Fig.9 Cutoff characteristics of 2-DEG quantum wires fabricated by the shallow etching technique. As-etched wires processed with CH_4/H_2 cut off at larger widths than the $SiCl_4$ counterparts because of hydrogen passivation of the donors.

Fig.10. Development of the cutoff characteristics of CH_4/H_2 processed wires after annealing at 200°C and 380°C.

Aperiodic Magnetoresistance fluctuations and weak localisation in n^+ and 2DEG wires

The low temperature transverse magnetoresistance signatures of narrow n^+ GaAswires show a large decrease in resistance as B is increased from zero due to the effects of weak localisation[18,19]. This marked weak localisation is expected due to the strong ionised impurity scattering. Reproducible though aperiodic fluctuations are observed as the field is swept to higher values. Both these phenomena are attributed to the quantum interference of coherent electron partial waves scattered around loops by a sequence of elastic scattering events. In the case of weak localisation it is recognised that these partial waves follow time-reversed paths which scatter the electron from k to -k. Constructive interference round these special loops contributes an additional component to the resistance. The application of a magnetic field **B** to such a loop of area **S** will introduce a phase change $\Delta\phi = 2e\mathbf{B}.\mathbf{S}/\hbar$ between the partial waves, initially destroying the constructive interference and resulting in the sharp drop in resistance observed close to **B**=0. At higher fields the resistance contributed by one of the loops will vary as $0.5(1+\cos(2e\mathbf{B}.\mathbf{S}/\hbar))$ thus introducing a fluctuation of a specific period in **B**. However, it is generally assumed in theories of aperiodic resistance fluctuations in "metallic" systems, that oscillatory effects due to a weak localisation loops are suppressed at fields sufficiently large to thread several (~10) flux quanta through the loop[20]. The aperiodic magnetoresistance fluctuations at higher field are thought to arise from an h/e oscillatory effect.

Fig.11 shows the transverse magnetoresistance of a wire 90nm wide etched in 50nm thick , 5×10^{18} Si-doped GaAs epilayers grown by MBE. The low field negative magnetoresistance and strong conductance fluctuations are readily apparent. The detailed structure in these measurements is perfectly reproduced on each sweep of the field. To assess the dimensionality of the electron trajectories, measurements of the transverse magnetoresistance up to 1T were repeated with the sample tilted in the field about an axis along the length of the wire. If the trajectories are perfectly two-dimensional we would expect the positions of the peaks to shift by $(\cos\theta)^{-1}$ where θ is the angle between the normal to the plane of the wire and the field. Fig.12 shows that instead the peaks shift by $(\cos(\theta+\delta))^{-1}$ where δ is a small angle measuring the angle the plane of the loop makes with the plane of the wire. In n^+ wires, therefore, the trajectories meander in three-dimensional paths[18].

As the temperature is raised the positions of the peaks are unchanged but their amplitudes diminish. The higher-frequency fluctuations disappear most rapidly but the slowly varying fluctuations persist at temperatures up to 100K. Since high-frequency detail results from trajectories enclosing a large area, we would expect the temperature-dependent phase breaking rate to wipe out these features first, as is observed.

For microstructures in which the current and voltage probes are separated by lengths greater than the phase-breaking length l_ϕ it is known that the magnetoresistance is an even function of B[20]. As this is the case in our structures, the observed magnetoresistance can be represented mathematically by a cosine Fourier series

$$R = R_o + \sum_n a_n(T)\cos f_n B \ . \tag{3}$$

Such a representation includes <u>both</u> the pronounced low field negative magnetoresistance associated with weak localisation <u>and</u> the aperiodic magnetoresistance fluctuations up to 12T. The FT spectra of the data in Fig.11 are shown in Fig.13. The peaks in the spectra have well-defined values of f_n up to temperatures of 100K and correspond almost entirely to $a_n>0$. Note that the <u>amplitude and frequency</u> spectra of the Fourier components necessary to produce the shape and amplitude of the negative magnetoresistance are the same at all temperatures as those giving rise to the aperiodic magnetoresistance fluctuations. This leads us to the conclusion that, in these heavily doped n^+-GaAs wires, the flux-enclosing loops which give rise to the aperiodic resistance fluctuations have very similar areas to those associated with the backscattering that causes weak localisation. Note also that the constancy of the peak positions in the Fourier spectra up to temperatures of over 60K indicates the thermal[21] and temporal (>~20 hours at 4K) stability of the loop geometries. Our observation strongly suggests that a significant proportion of the aperiodic structure is arising from non-self averaging interference effects associated with the backscattering loops. This seems plausible since at 10T a

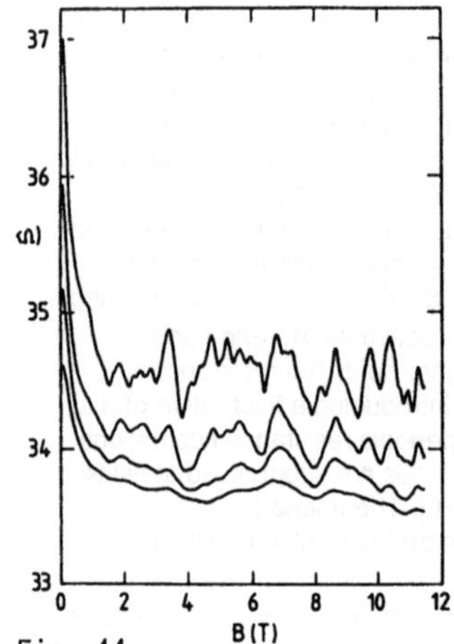

Fig. 11

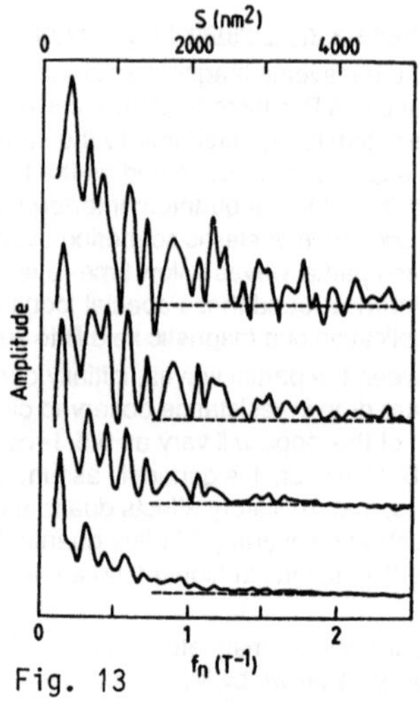

Fig. 13

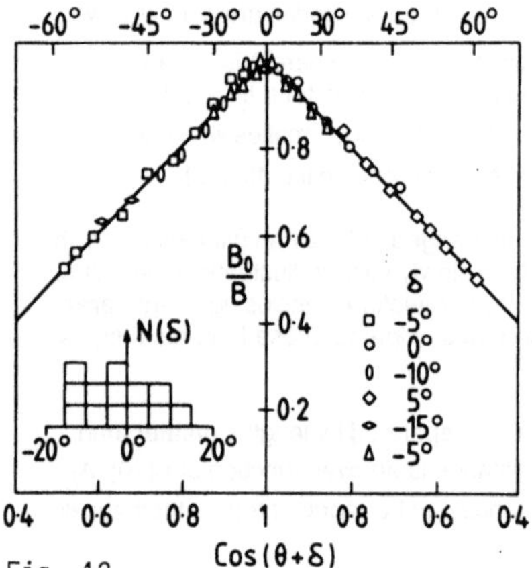

Fig. 12

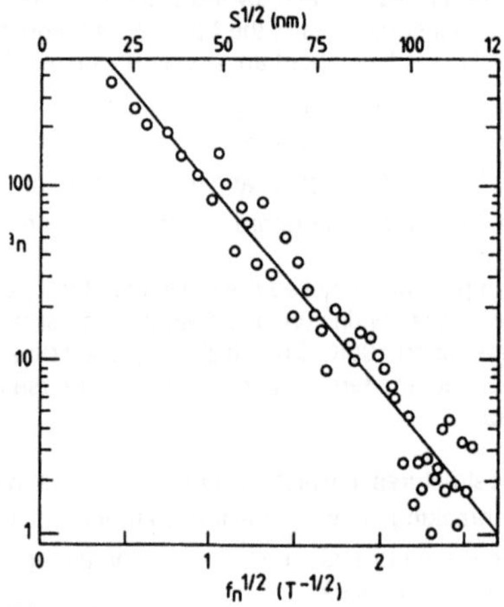

Fig. 14

Fig.11 (Top left) Transverse magnetoresistance of $0.09\mu m$ wide n^+GaAs wire at temperatures of 4.2K (top), 13.4K, 26.0K and 51.5K (bottom). Vertical axis in $k\Omega$

Fig.12 (Bottom left) Relative field position of various features in the transverse magnetoresistance plotted against cost(θ+δ). The inset histogram shows the distribution of the parameter δ.

Fig.13) (Top right) Fourier transforms of the magnetoresistance signatures of Fig (). Curves are displaced for clarity: dashed lines indicate the zeroes. Upper horizontal axis shows loop areas corresponding to f_n.

Fig.14 (Bottom right) Logarithmic plot of Fourier amplitude a_n against $f_n^{0.5}$ for 90nm wide wire at 4.2K. Each circle represents the amplitude of a peak in the Fourier Transform

loop area of $2000nm^2$ has only about five h/e flux quanta threaded through it. In this case, $f_n=2e\mathbf{S_n}.\mathbf{B}/B\hbar$, where $\mathbf{S_n}$ is the area of the nth backscattering loop. The loop associated with the peaks a_n in the FT spectra are of small area $S_n \sim l_e^2$ where $l_e \sim 30nm$ is the elastic scattering length. Hence we treat the electron motion governing the weak localisation as ballistic rather than diffusive and assume that the time taken to complete a loop is l/v_f where v_f is the Fermi velocity and l, the perimeter of the loop, is related to its area by $l=\alpha S^{0.5}$; an analysis of a variety of trajectory shapes involving a small number of elastic scattering events suggests $\alpha=4.5(\pm 1)$. An electron travelling for time τ has a probability of not undergoing a phase-breaking event proportional to $\exp(-\tau/\tau_\phi)$ where $1/\tau_\phi$ $(=v_f/l_\phi)$ is the phase-breaking rate. Assuming that the variation of a_n with f_n is governed by this probability, we write

$$\alpha_n(T) = \beta \exp\left(- \frac{2\alpha S_n^{1/2}}{l_\phi(T)}\right) \tag{4}$$

since the effective path length is twice the loop perimeter in the case of backscattering loops. Fig.14 plots $\log(a_n)$ vs $f_n^{0.5}$ (or $S_n^{0.5}$) for peaks in the FT of the 90nm wire. The fit is excellent over three orders of magnitude of a_n. From this data l_ϕ and τ_ϕ can be extracted. In Fig.15 we show the phase-breaking rate $1/\tau_\phi$ as a function of temperature for 90nm and 260nm wide wires. The data for the wider wire satisfies $1/\tau_\phi=AT+C$ over the full temperature range but data for the 90nm wire shows significant deviations below 30K. We believe that our simple analysis is invalid for lower temperatures because many of the loops involved in the calculation are larger than the wire width and must therefore elongate along the wire violating the assumptions made in deriving the relation between l and S. Note that for T>30K the value of A is the same for both wire-widths but that C is largest for the smaller wire.

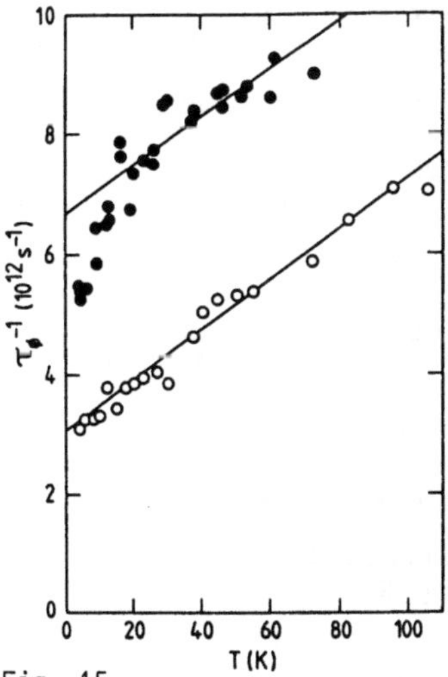

Fig. 15

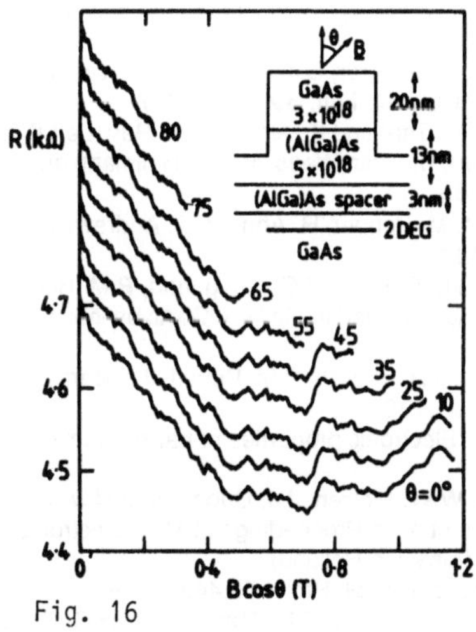

Fig. 16

Fig.15 Phase breaking scattering rate τ_ϕ^{-1} plotted against T for 0.09μm (upper) and 0.26μm (lower) wide wires.
Fig.16 Transverse magnetoresistance of a 0.54μm wide GaAs/AlGaAs heterostructure wire at 4.2 plotted against Bcosθ for various values of θ, the angle between **B** and the normal to the 2DEG plane.

In fact we find that C corresponds to a zero temperature phase breaking length of the order of the wire width suggesting that an inelastic surface scattering process is present in these wires.

To conclude this section we note that most previous workers have treated aperiodic magnetoresistance fluctuations (universal conductance fluctuations) in "metallic" systems as arising from h/e quantum interference. In structures such as ours, which exhibit marked weak localisation due to ionised impurity scattering, each back-scattering loop involves a small number of scattering centres whose separations are random but correlated. We suggest, therefore, that aperiodic magnetoresistance structure can arise from non-self averaging of weak localisation effects due to small (~2000nm^2) backscattering loops at least up to fields (~10T) for which only a small number (~5) flux quanta are threaded through the characteristic loop sizes.

2-DEG wires have also been investigated. These were fabricated by the light-etching technique from MBE-grown layers of composition shown in Fig.16. Wires were 10μm long and had widths between 0.18 and 0.54μm. Aperiodic structure was observed in the transverse magnetoresistance curves. By rotating B in the plane perpendicular to the current axis the magnetoresistance and aperiodic structure varied precisely as Bcosθ where θ is the angle between **B** and the 2DEG plane. This behaviour is shown in Fig.16 and demonstrates the exact 2D nature of the magnetoresistance effects in modulation-doped wires[19].

Acknowledgments
This research has arisen from a continuing close association between the University of Glasgow and the University of Nottingham. In Glasgow, Andy Stark, Dave Gourlay and Victor Law directly contributed to the work. Thanks are also due to John Davies for many illuminating discussions on the properties of wires close to cutoff. The work is supported by grants from the Science and Engineering Research Council.

References

1. W.S. Skocpol in Physics and Fabrication of Microstructures and Microdevices, ed. Kelly M.J. and Weisbuch C. (Springer, Berlin, 1986)
2. K. Ishibashi, Y. Takagoki, K. Gamo, S. Namba, S. Ishida, K. Murase, Y Aogaki and M. Kawabe Proc. Univ.Tokyo Int. Symp on Anderson Localisation 1987 & Solid State Commun. **64**,885-8,1987
3. M.L. Roukes, A. Scherer, S.J. Allen Jr., H.G. Craighead, R.M. Ruthen, E.D. Beebe and J.P. Herbison. Phys. Rev. Lett. **59**,3011,1987
4. K.K. Choi, D.C. Tsui and K. Alavi. Phys. Rev. Lett **B32**,5540,1985
5. H. van Houten, B.J. van Wees, M.G.J. Heijmann and J.P. Andre. Appl. Phys. Lett. **49**,1781,1986
6. T.J. Thornton, M. Pepper, H. Ahmed, D. Andrews and G.J. Davies Phys. Rev. Lett. **56**,1198,1986
7. A. Scherer, M.L. Roukes, H.G. Craighead, R.M. Ruthen, E.D. Beebe and J.P. Herbison "Ulta-narrow channels defined in GaAs-AlGaAs by low energy ion damage", submitted to Appl.Phys. Lett.
8. A.C. Gossard, J.H. English, P.M. Petroff, J. Cibert, G.J. Dolan and S.J. Pearton J. Cryst. Growth **81**,101,1987
9. J.H. Davies, "Electronic properties of narrow semiconducting wires near threshold", to be published.
10. A.N. Broers, W.W. Molzen, J.J. Cuomo and N.D. Wittels Appl. Phys. Lett. **29**,596,1976
11. G.A.C. Jones et al. in Proceedings of Microelectronic Engineering 1986, Interlaken, Switzerland (Elsevier, Holland)
12. H.G. Craighead J. Appl. Phys. **55**,4430,1984
13. L.D. Jackel, R.E. Howard, P.M. Mankiewich, H.G. Craighead and R.W. Epworth Appl. Phys. Lett. **45**,698,1984
14. S.A. Rishton and D.P. Kern J. Vac. Sci. Technol **B5**,135-141,1987
15. S.A. Rishton, PhD thesis, Glasgow University 1984
16. S.Thoms, S.P. Beaumont, C.D.W. Wilkinson, J. Frost and C.R. Stanley "Ultrasmall device fabrication using dry etching of GaAs" Proceedings of Microelectronic Engineering 1986, Interlaken, Switzerland (Elsevier, Holland)

17. R. Cheung, S. Thoms, S.P. Beaumont, G. Doughty, V.J. Law and C.D.W. Wilkinson Electronics Letters **23**,857,1987
18. P.C. Main, L. Eaves, R.P. Taylor, G.P. Whittington, S. Thoms, S.P. Beaumont and C.D.W. Wilkinson Proc. 18th Int. Conf. on the Physics of Semiconductors, Stockholm 1986 p 1591 (World Scientific Press)
19. R.P. Taylor, M.L. Leadbeater, G.P. Whittington, P.C. Main, L. Eaves, S.P. Beaumont, I. McIntyre, S. Thoms and C.D.W. Wilkinson Proc. 2D Conf. Santa Fe 1987 (to be published in Surf. Sci. 1988)
20. R.A. Webb, these proceedings.
21. M.L. Leadbeater, R.P. Taylor, P.C. Main, L. Eaves, S.P. Beaumont, I. McIntyre, S. Thoms and C.D.W. Wilkinson Proc 14th Int. Conf. GaAs and Related Compounds, Crete 1987. (IoP Conf. Ser. 1988)

Fabrication and Optical Characterization of Semiconductor Quantum Wires

A. Forchel, B.E. Maile, H. Leier, and R. Germann

4. Physikalisches Institut, Universität Stuttgart,
Pfaffenwaldring 57, D-7000 Stuttgart 80, Fed. Rep. of Germany

Abstract. We have investigated different approaches to effectively one dimensional III-V semiconductor structures (quantum wires) with dimensionality dependent optical properties. The quantum wires are defined by high resolution electron beam lithography. By dry etching or selective implantation induced interdiffusion the patterns are transferred into the semiconductor material. The nanometer structures are analyzed by photoluminescence spectroscopy. In etched GaAs/GaAlAs wires we observe a strong decay of the quantum efficiency as the lateral width is reduced. This can be related to the high surface recombination velocity and surface depletion in GaAs. In InGaAs/ InP quantum wires, in contrast, the quantum efficiency is only weakly affected by surface effects. Using implantation induced interdiffusion we have defined buried quantum wire structures in GaAs/GaAlAs. The photoluminescence spectra show clear lateral quantization effects.

1. Introduction

Stimulated by the predictions of Esaki and Tsu [1] and the progress of epitaxial technologies the investigation of dimensionality dependent physical properties of semiconductors has become one of the most intensively studied topics of semiconductor physics [2,3]. By molecular beam epitaxy, for example, heterostructures with abrupt interfaces and thicknesses well below the de Broglie wavelength of electrons and holes can be prepared from alternating layers of III-V semiconductors. Due to the different energy gaps of the layer materials, potential wells are formed in real space. In these quantum wells the electrons and holes are confined in one direction and are free in the two dimensions parallel to the interface.

The one dimensional confinement in epitaxial quantum well structures leads to strong changes of the optical [4] and transport properties [5] compared to bulk material. Notably the energy band structure [6,7], the density of states [4], the excitonic properties [8], recombination mechanisms [9] and many body effects [10] show a strong dimensionality dependence.

As an example for the dimensionality dependence observed in going from a three to a two dimensional system, Fig. 1 depicts the variation of the energy gaps at the Γ-point (direct gap, circles) and between the L-point of the conduction band and the top of the valence band (indirect gap, squares) in GaSb/AlSb quantum wells as a function of the GaSb thickness [7]. As expected, both transition energies increase as the well width is reduced. Due to the much smaller electron mass, the increase of the Γ-point transition energy is much faster than the variation of the transition at the indirect gap. Bulk GaSb is a direct gap material. The band gap of the indirect transition is higher by about 90meV (2K) than the direct gap. Due to the different shifts of the quantized conduction band levels, however, the indirect gap defines the band edge in two dimensional GaSb/AlSb layers for thicknesses below ~ 4nm. As a consequence of the size induced bandstructure change the carrier lifetime, the quantum efficiency etc. change drastically.

Strong changes of the physical properties are also expected if the effective dimensions of semiconductor structures can be reduced further. This can be illustrated by the energy variation

Springer Series in Solid-State Sciences Vol. 83: **Physics and Technology of Submicron Structures**
Editors: H. Heinrich · G. Bauer · F. Kuchar

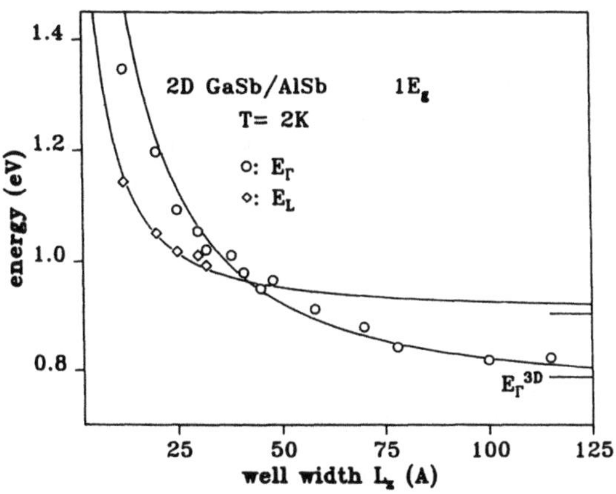

Fig. 1
Size induced direct to indirect bandstructure change in GaSb/AlSb quantum wells of different well widths. The points are experimental data for the position of the emission maxima due to direct gap recombination (circles) and indirect gap recombination (squares). The curves represent calculations of the well width dependence of the direct and the indirect band gap.

of the density of states (DOS). In going from a three dimensional to an effectively zero dimensional structure - in which the carrier motion is inhibited in all directions - the density of states varies between a $E^{1/2}$ - dependence (3D), a step function (2D), an $E^{-1/2}$ - relation (1D) and a discrete series of level (0D) [11]. These changes of the DOS are expected to lead to strong modifications of radiative processes [12], transport properties [13] and e.g. laser performance [11].

In order to fabricate effectively one and zero dimensional structures ("quantum wires" (1D), "quantum dots" (0D)) for the investigation of quantum effects the two dimensional layers have to be patterned with structures of lateral dimensions well below 0.1 μm. The most advanced technology for a pattern definition in the nanometer range is electron beam lithography (EBL). X-ray lithography and ion beam lithography are capable of sub-100nm-resolution as well, but in spite of some significant advances of these processes (e.g. no background exposure due to the proximity effect) they are still limited to somewhat larger dimensions compared to EBL [14,15]. By ultrahigh resolution EBL lines and dots with widths below 30nm can reproducibly be defined. On thick substrates, where the lithography is complicated by backscatter effects, linewidths as small as 8nm have been achieved [16].

In order to obtain quantum wires and dots the lithographic patterns have to be transferred into the semiconductors. This is typically done by dry etching [17] or by other technologies which induce lateral modulations in the energy bandstructure of the quantum wells.

In the following sections of this paper we first describe the basic technological processes for the fabrication of ultra small semiconductor structures for optical studies. We then report results of photoluminescence experiments on GaAs/GaAlAs- and InGaAs/InP- wires defined by dry etching. The last section presents our data on effectively one dimensional GaAs/GaAlAs wires defined by implantation induced interdiffusion.

2. Fabrication of Microstructures for Optical Spectroscopy

The requirements of laterally patterned semiconductor structures for optical investigations of dimensionality dependent phenomena differ to a certain degree from those of nanometer patterns for transport studies: For optical spectroscopy one uses in general undoped material and the intrinsic properties are probed by optically excited electron-hole pairs. This implies that any band bending will drastically affect the number of electron-hole pairs and may determine the properties of the structure (quantum efficiency, energetic position, lifetimes) to a large degree.

Furthermore, the spatial resolution in optical experiments is on the order of 1 μm. Individual quantized objects can only be addressed if the distance between them exceeds the resolution. The number of photons which are absorbed in a nanometer structure is quite small and limits the emission intensity: For a rather strong laser excitation with 10ns pulses with an excitation intensity of $1kW/cm^2$ only about 10^3 photons/pulse will be absorbed in an area of 100nmx100nm. The number of detected photons strongly depends on the external quantum efficiency and the detection efficiency and quickly approaches the noise level as the dimensions are reduced.

In order to increase the sensitivity of optical experiments on nanometer wires one generally defines arrays in which the same structure is repeated over an area larger than the excitation spot. This implies a very good control of the dimensions of the structure.

We have fabricated wire structures from GaAs/GaAlAs- and InGaAs/InP-quantum well wafers by using high resolution electron beam lithography [18,19]. Our commercial high resolution system provides a minimum beam diameter of 8nm. The beam can be scanned over a field of 80μmx80μm size using an address grid with 2.5nm increments. With coarser address structures field sizes up to 1.6mm x 1.6mm can be used.

If a pattern exceeds the field size the sample table is moved by stepping motors. The table position is controlled by a laser interferometer with 5nm resolution. If the table position deviates by more than the interferometer resolution from the expected position, the beam position is automatically corrected. This allows to reduce the "stitching error", i.e. the positional error which occurs in large patterns due to displacements of the table, to values below 50nm. The good positional control is essential for many electronic and optoelectronic devices with submicron dimensions [20].

For the exposure two acceleration voltages can be used (25kV, 50kV). High voltages are generally favourable for electron beam lithography because the proximity effect is modified and the process latitude increases for increasing voltage [21]. The resolution which can be achieved with electron beam lithography depends on the beam diameter, the positional stability and the electron beam resist in which the pattern is exposed. For high resolution lithography polymethylmethacrylate (PMMA) is used mostly. In PMMA 8nm wide lines have been defined with good edge control by Namba and coworkers [16].

PMMA acts as a positive resist for typical electron doses of $400\mu C/cm^2$ at 50keV, i.e. the exposed areas are removed in the developer. If subsequent processes like e.g. dry etching are considered, the positive tone is a certain drawback. In order to obtain a quantum wire a ridge shaped etch mask is necessary, whereas the wire exposure provides a gap in the resist. The gap can be transformed into a wire mask by a "lift-off" process. In this process a thin metal layer is deposited on the resist structures e.g. by evaporation. If the metal layers on top of the resist and in the gaps have no interconnection, the resist and the covering metal can be removed in a solvent, whereas the metal lines deposited directly on the semiconductor surface remain. The metal structures can then be used as masks for the etching of quantum wires and dots.

In order to transfer the mask pattern into the semiconductor dry etching processes are generally used. In these processes the unprotected parts of the semiconductor are removed by an ion beam or in a plasma. If chemically reactive etch gases (e.g. Cl_2, CCl_2F_2) are used the semiconductor and the mask are attacked by physical and chemical effects. Reactive dry etching allows to minimize surface damage and redeposition as well as to optimize the etch rate ratio between the mask and the semiconductor. The GaAs/GaAlAs structures studied in the present paper were prepared by reactive ion etching with an Ar/CCl_2F_2 mixture, whereas the InGaAs/InP structures were obtained by ion milling with an Ar/O_2 mixture.

The lift-off process can be avoided if a negative resist is used. We have investigated the suitability of chloromethylated poly-α-methylstyrene (αM-CMS) for high resolution lithography. Using proper development conditions in combination with a suitable descumming step to remove resist fragments between the exposed structures 20nm wide lines can reproducibly be defined [22]. Figure 2 shows a dot pattern etched into InP. The structure was defined by electron beam lithography in CMS. The InP columns have a separation of 200nm and a top diameter of about 50nm.

Negative resists are especially attractive for the fabrication of structures for optical studies. The resists are transparent at the excitation and the emission wavelengths. Therefore even resist covered structures may be characterized with high sensitivity. Furthermore the resist mask can be removed very easily in an O_2-plasma. In contrast, the metal masks used for the lift-off strongly attenuate the excitation and emission intensity from the wire or dot structures. The metal mask generally remains on top of the structure. Due to the different thermal expansion coefficients of the metal and the semiconductor the mask may significantly stress the structures, particularly in low temperature experiments. This can lead to shifts of the emission energy which are very difficult to distinguish from quantization effects.

Figure 3 shows etched GaAs wires defined by a CMS etch mask. The width in the top section is about 30nm. Together with the high reproducibility this clearly demonstrates the suitability of negative resists for the fabrication of quantum wire and dot arrays for optical studies.

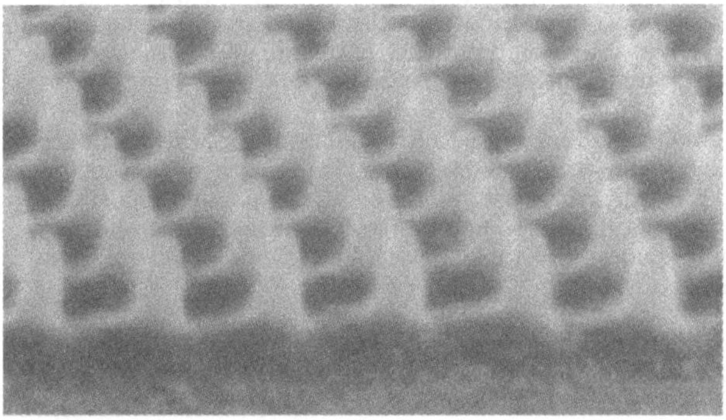

Fig. 2
Etched dot structure in InP.

Fig. 3
Etched GaAs wires with a width of about 30nm in the top regions. αM-CMS was used as etch mask.

3. Luminescence Studies of Etched III-V Semiconductor Structures

We have fabricated wire structures from GaAs/GaAlAs- and InGaAs/InP quantum wells. For the optical investigation the samples contained a number of grids, in which the distance between the wires and the wire width were kept constant. Typically the grids extend over 600 μm x 600 μm fields. Wire widths between 5 μm and about 0.05 μm were studied. In addition to the patterned areas the samples contain reference fields which were completely etched or completely masked.

For the luminescence measurements a special set-up was developed which permits a low temperature characterization of the patterns with approximately 5 μm spatial resolution. The high spatial resolution is obtained by using a specially designed cryostat and a reflective objective with a working distance of 24mm for the excitation of the sample and the collection of the luminescence. The achromatic reflection optics allows to use the same objective for excitation and detection. For spatial scans the cryostate can be moved by a stepping motor driven stage. The samples are excited by an Ar - Laser. The emission is spectrally dispersed in a 0.3m monochromator and detected by a GaAs photomultiplier tube (GaAs samples) or a Ge-detector (InGaAs samples).

We have used the free exciton emission of a 4nm GaAs/GaAlAs quantum to monitor the wire etching. The high emission intensity and the narrow linewidth of the as-grown structure indicates the excellent quality of the epitaxial material. The quantum well is located about 45nm below the surface of the sample. We have prepared wires from this heterostructure by etching through the top and the quantum well layers. Therefore the carrier confinement in the wires is due to the GaAlAs (z-direction) and due to the surface (x-direction).

The strongest changes of the photoluminescence due to a variation of the wire width are observed for the quantum efficiency. Figure 4 depicts the width dependence of the exciton emission intensity from etched GaAs/GaAlAs wires with widths between 5 μm and 100nm in a double logarithmic plot. Over the entire width range we observe a decrease of the emission intensity by about four orders of magnitude with shrinking wire dimensions. The decrease is particularly strong below about 1 μm.

The strong dependence of the quantum efficiency on the wire width is a particular feature of GaAs wires. As shown in Fig. 5 the emission intensity from etched InGaAs wires varies by about 50 percent only if the wire width is reduced from 5μm to 500nm.

The different behaviour of GaAs- and InGaAs-wires can be attributed to a large degree to the different influence of the surfaces for the two materials. GaAs is known to have a strong surface recombination (surface recombination velocity ~ 10^6 cm/s) [23]. The surface

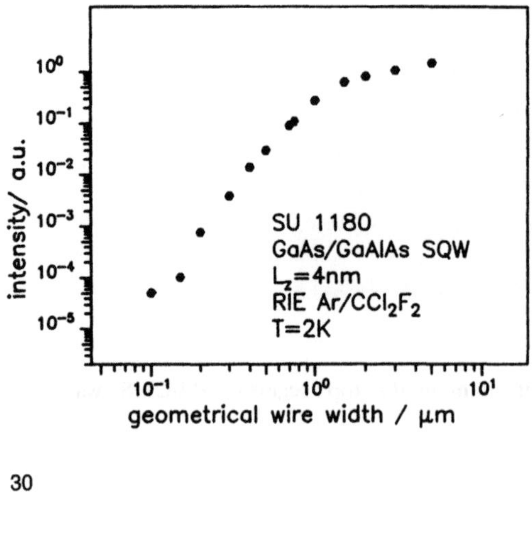

Fig. 4
Width dependence of the band edge emission intensity of etched GaAs wires (T=2K). The emission intensity has been normalized in order to account for the different amount of material in the wires.

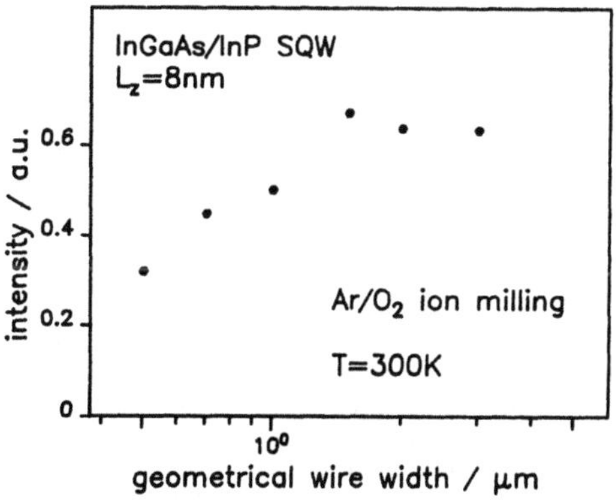

Fig. 5
Width dependence of the exciton emission intensity of etched InGaAs/InP quantum well wires (L_z= 10nm, T= 300K).

recombination velocity of InGaAs, on the other hand, is of the order of 10^5 cm/s only [24]. To a first approximation the carrier lifetime due to surface recombination can be estimated as 100ps and 1ns for a 2μm wide GaAs and InGaAs wire, respectively. Because the carrier lifetime in the unprocessed quantum wells amounts to about 500ps for both materials, surface effects will dominate the lifetime in micrometer width GaAs stripes, whereas they are negligible in InGaAs.

It should be noted here, however, that the decrease of the GaAs emission intensity in the submicrometer range is much too strong to be explained by nonradiative recombination at the surface alone. For small wire widths one has to consider the nonzero penetration depth of the etching particles used in the dry etch process. The surface layers are expected to contain a high concentration of defects and therefore they will not contribute to the luminescence ("dead layers"). The thickness of the dead layers has to be subtracted from the geometrical wire width. Using a model calculation for the surface limited lifetime in the GaAs wires we estimate the thickness of the damaged layers to about 100nm [19].

In view of the smaller surface recombination the InGaAs-system is a promising candidate for the fabrication of quantum wires and dots by dry etching. First experiments on structures with a geometrical width of 30nm have been reported by Temkin and coworkers [25]. Nanometer structures on the basis of GaAs, on the other hand, require the fabrication of buried structures to avoid surface effects [26]. Recently submicrometer wires for transport studies have been reported [27], which were defined by etching only part of the GaAlAs-barrier layer. The lateral confinement is then obtained by band bending, similar to the technology of Hansen et al. [28]. As will be discussed in the following section, we have used the implantation induced interdiffusion to define buried GaAs quantum wires.

4. Quantum Wire Fabrication by Implantation Induced Interdiffusion

Intrinsic quantum wells exhibit a remarkable thermal stability. This stability is particularly surprising if one considers the strong concentration gradients of the matrix atoms which occur at the interfaces between the barrier and the quantum well layer. Exposure to 875°C for one hour, for example, results only in an interdiffusion by 1 lattice constant. This corresponds to interdiffusion coefficients of $1x10^{-17}$ cm^2/s.

Laidig and coworkers [29,30] have noticed that the activation energy for interdiffusion can be reduced drastically, if the quantum wells are subject to an impurity diffusion (e.g. by Zn or Si) or to ion implantation followed by a low temperature annealing. Due to the defects produced in the implantation the diffusion coefficients of the matrix atoms increase by about three orders of magnitude at 850°C.

The interdiffusion significantly changes the energy band structure of the quantum wells. As shown in Fig. 6 the effective width at the bottom of the GaAs/GaAlAs-quantum well is reduced by the Al-diffusion. This leads to an increase of the subband energy for the lowest levels of the quantum well (left hand side, Fig. 6). Furthermore, for sufficiently large interdiffusion lengths the Al-concentration at the center of the well increases. Especially for the case of comparable quantum well widths and interdiffusion lengths this mechanism dominates the shift of the quantum well band edge to higher energies.

Because of the strong bandstructure changes, the interdiffusion can be studied by luminescence spectroscopy. Figure 7 displays the dose dependence of the band edge emission from Ga-implanted quantum wells. The quantum well is located 47nm below the surface of the sample. The implantation energy (100keV) was choosen appropriately to observe strong interdiffusion effects. All samples were annealed for 30min at 800°C. During this time the defects themselves are annealed and therefore longer annealing times do not change the spectra further. As shown in fig. 7, the emission shifts strongly to higher energies for increasing Ga-dose. The maximum energy shifts which can be realized are of the order of 100meV. By studies of the energy dependence of the interdiffusion effects we can show that the enhanced interdiffusion after Ga-implantation is due to implantation induced defects [31].

The implantation induced interdiffusion may be used for lateral patterning of quantum well devices. The increase of the band gap in the implanted areas gives rise to electrical and optical confinement. By implantation or diffusion induced disorder quantum well buried heterostructure lasers and waveguides have been developed [32]. The major advantage of the interdiffusion technology for lateral patterning is the possibility to create buried

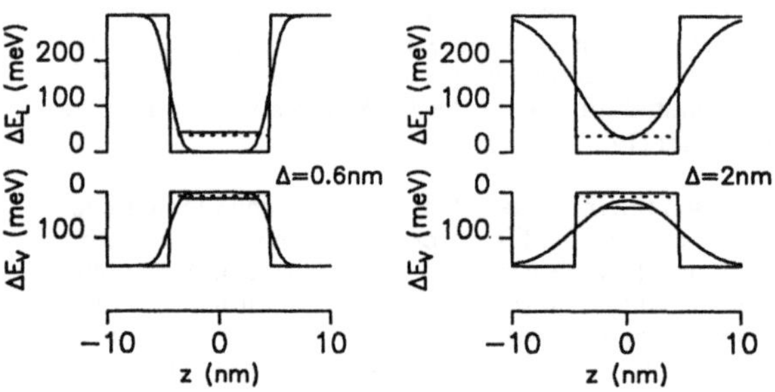

Fig. 6
Model calculation of interdiffusion effects on the lowest energy levels of a quantum well with a width L_z =8nm. For the calculation an error-function shaped diffusion profile characterized by an interdiffusion length of 2nm (left hand side) and of 5nm (right hand side) has been assumed.

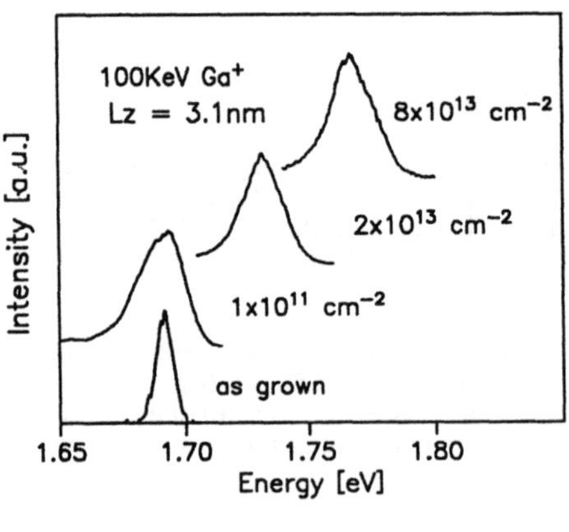

Fig. 7
Photoluminescence spectra (T= 2K) from Ga implanted GaAs/ GaAlAs quantum wells (L_z = 3.1nm) for different implantation doses.

heterostructures by relatively small changes of the quantum well layer and without the need of a multilevel epitaxy. Petroff et al. [33] and others [19], [34] have used this process to define quasi one- and zerodimensional GaAs/GaAlAs-structures.

For the fabrication of quantum wires by implantation induced disorder we have developed Au-masks with widths of about 100nm. The Au-wires are defined in PMMA by a lift-off process. Because the wires act as absorbers for the implanted ions a large thickness is required (in our case 40nm). Due to the larger atomic number of Au the range of the Ga-ions in the mask is much shorter than in GaAs and GaAlAs. Neglecting edge effects for the masks and assuming an ion energy of 100keV the maximum of the Ga distribution is located in the center of the gold layer and the Ga concentration decreases by a factor of 30 until the quantum well is reached. The defect concentration which is directly related to the interdiffusion shows a similar behaviour. In the uncovered areas of the sample, on the other hand, the maximum of the defect concentration occurs in the quantum well. Therefore very strong interdiffusion effects are expected in a subsequent annealing step for the uncovered sections of the samples, whereas only rather small interdiffusion should occur beneath the masked wires.

In order to estimate the optimum mask width, one has to consider the lateral distribution of the defects. The lateral defect distribution is given by the width of the mask, the straggling of defects beneath the mask and the defect diffusion parallel to the quantum well in the annealing. The contribution of ion straggling can be calculated fairly accurately. For the present case the straggling profile can be characterized by an error-function with a width of 9nm. Only coarse estimates are available for the magnitude of the effective defect diffusion length. From studies of the energy dependence of the interdiffusion, we estimate a value of 15nm to 20nm. This implies that the influence of the mask on the defect distribution can be described using the geometrical width and an error-function with a characteristic width of about 20-25nm.

The actual lateral potential is given by changes of the quantum well width and composition. According to the defect distribution in our case, changes of the width will dominate under the mask whereas in the unmasked sections the energy gap is mainly increased due to an increasing Al-content in the well.

Photoluminescence spectroscopy gives a simple access to the height of the lateral potential barrier. Figure 8 displays the emission of a GaAs/GaAlAs-sample with an original quantum well width of 3.1nm after a Ga implantation with a wide mask (130nm) and annealing. The emission of the masked wire sections occurs at 1.7eV. The interdiffused areas (distance between Au-wires 800nm) give rise to an emission band peaked at 1.76eV. The total lateral potential barrier height in the conduction and the valence band is simply given by the energetic difference of the two bands and amounts to about 60meV in the present case.

In contrast to the GaAs-wires fabricated using dry-etching, the wires defined by implantation induced disorder emit luminescence very efficiently. Firstly, this is due to the small number of defects at the interface between wire and barrier. Secondly, the wires may capture carrier pairs very efficiently from the barrier regions. Both effects lead to a quantum efficiency of the wires which is comparable to the efficiency of the original quantum wells.

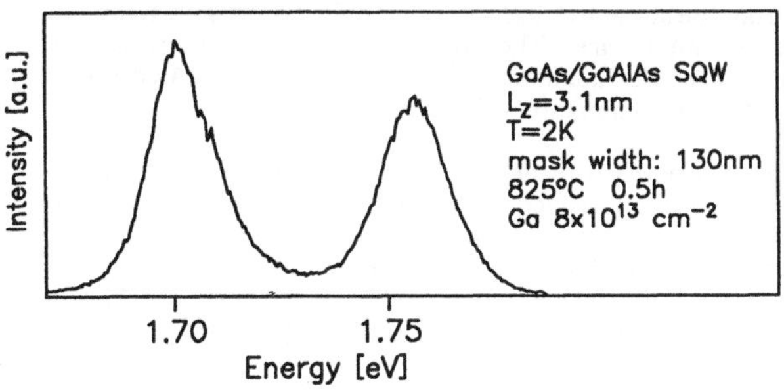

Fig. 8
Luminescence spectrum (T=2K) of a GaAs/GaAlAs quantum well (L_z=3.1 nm) after Ga implantation using Au-masks (L_x=130nm) and annealing.

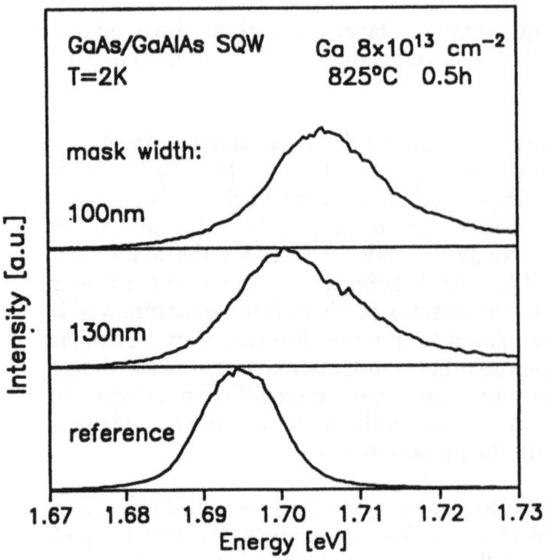

Fig. 9
Luminescence spectra of GaAs/GaAlAs-quantum wires (T=2K) fabricated by implantation induced disorder. Top trace: mask width L_x = 100nm, Center trace: L_x = 130nm, Bottom: reference field.

We have investigated the emission from quantum wires defined by implantation induced disorder as a function of the mask width L_x between 250nm and 70nm. Figure 9 shows luminescence spectra from GaAs/GaAlAs-quantum wires (L_z=3.1nm) obtained with a mask width of 130nm and 100nm in comparison with the emission of a part of the sample which was only annealed (reference spectrum in Fig. 9). With decreasing wire width the emission shifts significantly to higher energy and the lines broaden slightly.

The shift of the emission lines to higher energy is due to two effects. Firstly, a lateral confinement will induce subband shifts which increase if the lateral dimensions shrink. Secondly, lateral straggling and the diffusion of the defects enhance the interdiffusion even at the center of the wire mask. Both effects can be separated by a calculation of the lateral subband ladders [19]. For the calculation we assume that the lateral potential can be described by an error function with a characteristic length of 25nm. The quasi parabolic shape of the potential in the valence and conduction band leads to approximately equidistant subbands at the bottom of the bands. For a 100nm mask the energetic spacing between subbands amounts to about 4meV in the conduction band and to about 1meV in the heavy hole valence band. At the center of the lateral potential well significant interdiffusion effects lead to an increase of the energy gap by about 60 percent of the observed energy shift (L_x= 130nm).

5. Summary

We have discussed the fabrication and optical characterization of narrow III-V-semiconductor wires. By using high resolution electron beam lithography and dry etching GaAs/GaAlAs- and InGaAs/InP- wires with lateral dimensions down to 30nm have been fabricated. The evaluation of the photoluminescence intensity from the structures indicates that only the InGaAs-system is suitable for the fabrication of etched quantum wires. The properties of the GaAs - wires are dominated by surface effects. In particular, the quantum efficiency of etched GaAs-wires is drastically reduced due to the high surface recombination velocity.

Ion implantation induced disorder can be used to define GaAs quantum wires which are buried in the quantum well substrates. Therefore surface effects play no role for the properties of these structures. We have fabricated quantum wires by this technology using Au masks with widths between 80nm and 200nm and Ga-implantation for the wire definition. The emission spectra of the wires shift to higher energy with reducing mask width. In the case of a 100nm wire these shifts are given by comparable contributions of the lateral quantization and of changes of the quantum well under the mask.

In summary, lateral patterning techniques now start to be able to provide effectively one and zero dimensional structures for physical investigations. In further improved structures a broad variety of new optical and transport effects are expected to show up. This makes the field of nanometer device development and characterization one of the most promising areas of semiconductor physics.

Acknowledgements

We are grateful to M.H. Pilkuhn for many helpful discussions. The experimental assistance by A. Menschig and H. Rothfritz in different steps of the fabrication and characterization of the structures has been very valuable. We especially thank G. Weimann, Forschungszentrum der Deutschen Bundespost, Darmstadt and H. Meier, IBM Research Center, Zürich for the high quality GaAs/GaAlAs samples, and F. Scholz, Universität Stuttgart for the good InGaAs/InP heterostructures. The financial support by the Stiftung Volkswagenwerk, the Land Baden-Württemberg and the Deutsche Forschungsgemeinschaft was essential for our activities and is gratefully appreciated.

References

1. L. Esaki, R. Tsu, IBM Research Note RC-2418 (1969);
2. G. Weimann, W. Schlapp, in Springer Series in Solid State Physics 53, 88 (1984).
3. N.T. Linh, Advances in Solid State Physics 23, 227 (1983), Vieweg-Verlag, Braunschweig.
4. R. Dingle, Advances in Solid State Physics 15, 21 (1975), Vieweg-Verlag, Braunschweig.
5. R. Dingle, W. Wiegmann, C.H. Henry, Phys. Rev. Lett. 13, 827 (1974).
6. R.C. Miller, D.A. Kleinmann, J. Lum. 30, 520 (1985).
7. A. Forchel, U. Cebulla, G. Tränkle, H. Kroemer, S. Subbanna, G. Griffiths Surf. Science 174, 143 (1986).
8. J.C. Maan, G. Belle, A. Fasolino, M. Altarelli, K. Ploog, Phys. Rev. B 30, 2253 (1984).
9. J. Feldmann, G. Peter, E.O. Göbel, P. Dawson, K. Moore, C. Foxon, R. Elliot, Phys. Rev. Lett. 59, 2337 (1987).
10. G. Tränkle, H. Leier, A. Forchel, C. Ell, H. Haug, G. Weimann, Phys. Rev. Lett. 58, 419 (1987).
11. Y. Arakawa, H. Sakaki, Appl. Phys. Lett. 40, 939 (1982).
12. H. Hassan, H. Spector, J. Vac. Sci. Tech. A 3, 22 (1985).
13. H. Sakaki, Jap. J. Appl. Phys. 19, L735 (1980).
14. A.C. Warren, I. Plotnik, E.H. Anderson, M.L. Schattenburg D.A. Antoniadis, H.I. Smith, J. Vac. Sci. Tech. B 4, 365 (1986).
15. S. Namba, Microelectronic Eng. 6, 315 (1987).
16. F. Emoto, K. Gamo, S. Namba, N. Samoto, R. Shimizu, Jap. J. Appl. Phys. 24, L809 (1985).
17. A.R. Reinberg, in "VLSI Electronics - Microstructure Science", Vol. 2 Academic Press, New York, 1981.
18. B.E. Maile, A. Forchel, R. Germann, A. Menschig, K. Streubel, F. Scholz, G. Weimann, W. Schlapp, Microcircuit Eng. 6, 163 (1987).
19. A. Forchel, H. Leier, B.E. Maile, R. Germann, Advances in Solid State Physics, Vieweg Verlag, 1988, in press.
20. M. Korn, A. Forchel, M. Möhrle, R. Germann, K. Streubel, F. Scholz, Microcircuit Eng. 6, 551 (1987).
21. R.E. Howard, L.D. Jackel, W.J. Skocpol, Microelectronic Eng. 3, 3 (1985).
22. B.E. Maile, A. Forchel, A. Menschig, R. Germann, H. Meier, to be published.
23. H.C. Casey, E. Buehler, Appl. Phys. Lett. 30, 247 (1977).
24. The value of the surface recombination velocity in InGaAs can be estimated from ref. 23 in conjunction with ref. 25.
25. H. Temkin, G.J. Dolan, M.B. Panish, S.N.G. Chu, Appl. Phys. Lett. 50, 413 (1987).

26. P.M. Petroff, A.C. Gossard, R.A. Logan, W. Wiegmann, Appl. Phys. Lett. 41, 635 (1982)
27. R.E. Behringer, P.M. Mankiewich, R.E. Howard, J. Vac. Sci. Technol. B 5 326 (1987).
28. W. Hansen, M. Horst, J.P. Kotthaus, U. Merkt, Ch. Sikorski, Phys. Rev. Lett. 58, 2586 (1987).
29. W.D. Laidig, N. Holonyak, M.D. Camras, K. Hess, J.J. Coleman, P.D. Dapkus, J. Bardeen, Appl. Phys. Lett. 38, 776 (1981)
30. J.J. Coleman, P.D. Dapkus, C.G. Kirkpatrick, M.D. Camras, N. Holonyak, Appl. Phys. Lett. 40, 904 (1982).
31. H. Leier, H. Rothfritz, A. Forchel, G. Weimann, to be published.
32. M. Komuro, H. Hiroshima, H. Tanoue, T. Kanayama, J. Vac. Sci. Technol. B 4, 985 (1983).
33. J. Cibert, P.M. Petroff, G.J. Dolan, S.J. Pearton, A.C. Gossard, J.H. English, Appl. Phys. Lett. 49, 1275 (1988).
34. Y. Hirayama, S. Tarucha, Y. Suzuki, H. Okamoto, Phys. Rev. B 37, 2774 (1988).

Electron-Beam Lithography of Ultra-Submicron Devices

D.K. Ferry, G. Bernstein, and Wen-Ping Liu

Center for Solid State Electronics Research,
Arizona State University, Tempe, AZ 85287, USA

Abstract. In the ultra-submicron device area, in which individual feature sizes can be as small as a few tens of nanometers and comparable to the inelastic mean free path of the electrons, many novel new concepts arise. These can include ballistic transport and cooperative effects such as Bloch oscillations. In this paper, we discuss the use of high resolution electron-beam lithography for the fabrication of such ultra-submicron devices. The appearance of ballistic transport in MESFETs and of the possible Bloch oscillations in suitably prepared surface superlattices will be discussed.

1. Introduction

The advent of electron-beam lithography (EBL) has allowed the investigation of physical phenomena which occur in devices whose characteristic dimensions are of the order of a few tens of nanometers. In particular, the transport of electrons undergoing reduced scattering due to fewer degrees of dimensional freedom can be better investigated in structures whose scales are on the order of 50 nm, the expected inelastic mean free path length for electrons in GaAs at room temperature.

We have used a converted scanning electron microscope as an EBL system to fabricate both ultra-submicron GaAs MESFETs [1-3] and novel quantum devices [3-5]. The MESFETs have gate lengths ranging from 25 to 90 nm, and we have obtained preliminary evidence of the effects of very small dimensions in both the MESFETs and the quantum devices. In the case of the former, we observe an increase in the transconductance as the gate length is decreased below 50 nm (at room temperature). We believe this is due to velocity overshoot of the electrons passing through the high field region just under the gate.

The quantum devices, called BlochFETs, are based upon a two-dimensional lateral surface superlattice (LSSL), which is induced by a grid replacing the normal metal gate. The grid of the LSSL typically is formed by patterning 40 nm metal lines on a pitch of 160 nm, with the lines being formed by normal "lift-off" processing. When this grid forms the gate of a large, but conventional, high electron mobility transistor (HEMT), we can take advantage of the high electron mobility, and the resulting long inelastic mean free path. The grid gate forces the electrons to accumulate in an array of small quantum boxes, formed within the quasi-two dimensional inversion channel, which remain coupled due to the long

Springer Series in Solid-State Sciences Vol. 83: **Physics and Technology of Submicron Structures**
Editors: H. Heinrich · G. Bauer · F. Kuchar

inelastic mean free path. These quantum boxes form the LSSL in the quasi-two dimensional electron layer, especially near the pinch-off point of the channel. The BlochFET characteristic curves show stron negative differential conductivity for gate voltages near pinch-off. We believe this may be due to the onset of Bloch oscillations within the mini-bands formed by the LSSL. If so, this would be the first direct observations of such Bloch oscillations.

2. Fabrication

We utilize an ISI-100B scanning electron microscope, which has been modified to interface to a desk-top computer, operating in a vector-scanned mode. The computer provides both the control mechanism to drive x and y position and deblank times, but also the high-resolution graphics pattern generator with which the e-beam masks are generated. All fine geometry exposures are carried out at 40 keV, using the finest (5 nm) beam diameter, even though resultant exposures have larger dimensions. Larger areas are exposed using 20 keV and larger spot sizes. While most of the devices are made using the positive resist PMMA, we have also experimented with a negative resist that also provides fine line capabilities.

The normal approach to writing nanometer lines in electron-beam lithography is to use the positive resist PMMA, or some combination of resists for a multilevel approach. Here, we use entirely single level resists, and negative resists are desired for use as a mask in dry etching or implantation. However, high resolution in negative resists has been hard to achieve, so that the normal approach is to use the positive resist and subsequent metal lift-off to create the mask. We have used the new negative resist SAL 601-ER7 by Shipley to achieve high contrast resist structures. For 40 keV electrons, we measure the contrast factor γ=3 and

Fig. 1: Examples of exposed resist structures attainable in negative resist. The grid shown in (a) is composed of lines as small as 70 nm (on 100 nm pitch), while in (b) we can see the extremely high aspect ratio attainable.

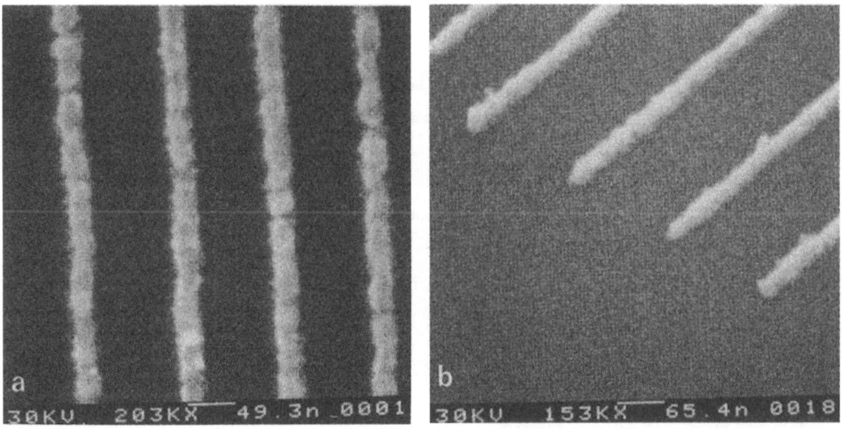

Fig. 2: Lifted-off Au lines produced with the PMMA photoresist and a new high contrast developer solution.

D_0=8.6x10^{-6}, D_1=1.9x10^{-7} C/cm^2. Typical grids and test structures measured in such resists are shown in Fig. 1. In preliminary work on such a thick layer, we have achieved exposed lines of the order of 70 nm thick (on pitches as small as 100 nm), with developed aspect ratios of approximately 5.7:1. The resist profiles exhibit extremely steep side walls, which are necessary to utilize the resist's high dry etch resistance.

We have also achieved a new developer solution for PMMA, which has demonstrated an improvement in the contrast factor and increased the sensitivity by lowering the minimum required dose, as compared to the normal developers cellosolve or MIBK. This new combination of normal developers has achieved a contrast factor near 9, which is considerably larger than that normally achieved for 40 keV electrons. With this new developer, we have achieved a much more uniform development and less feature loss, which has allowed the routine fabrication of extremely narrow lines of very high pitch. In Fig. 2, we show Au lines which have been fabricated by lift-off processing. By these techniques, we have been able to achieve 30 nm lines on a 95 nm pitch in the lifted-off metal, which are higher density lines than we have been able to previously achieve.

3. MESFET Devices

We have made a series of ultra-submicron MESFET devices using the EBL techniques discussed above. The purpose of studying these structures lies in achieving devices which actually show the effects of the inelastic mean free path. People have discussed the appearance of ballistic transport and the velocity overshoot effect in devices for some years. These two effects are in fact equivalent statements of the role which non-equilibrium transport plays in these devices. However, there has been no clear evidence of the appearance of these effects in

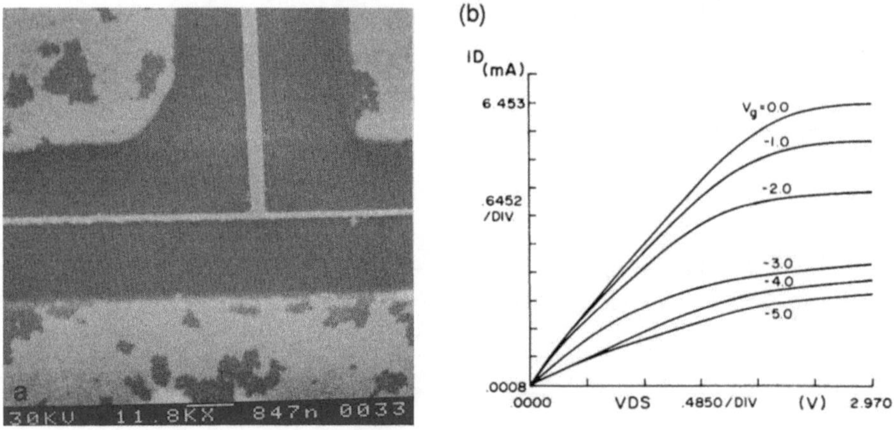

Fig. 3: (a) A blowup of the gate region and active device area of the MESFET, which is fabricated in a "source-cutaway" configuration. (b) The characteristic curves of a device with a gate length of 35 nm.

normal devices, primarily because the gate lengths have been too long. It is to be expected that for a clear observation of velocity overshoot, it will be necessary for the entire high electric field region, over which the velocity overshoot occurs, to be completely under the gate and to be reduced in length to below the inelastic mean free path. The overshoot length is that over which the energy relaxes [6], or the carriers lose memory of the initial energy (phase). This is just the quantum mechanical inelastic mean free path, which has been estimated to be only about 50-75 nm in GaAs at 300 K [7]. If the gate region is longer than this, charge is redistributed within the device, leading to inhomogenous distribution of the charge under the gate. Where the velocity is high, the density will be low. Where the velocity is low, the density will be high, which is in keeping with Kirchoff's current law. Thus, truly short gate length devices are required in order to explore this effect.

The devices we have made have the gate length as short as 25 nm. In Fig. 3, such a gate structure, and the resulting *i-v* curves for a 35 nm gate length device, are shown. For gate lengths below about 50 nm, we see an increase in the transconductance, when all other parameters (doping, channel thickness and width, etc.) are maintained at the same values. Normally, with the doping we use, $n=2x10^{17}\,cm^{-3}$, we expect (and model incorporating the complete energy transport process also predict [8]) that the transconductance will peak, as the gate length is reduced, near that length which gives an aspect ratio between 2 and 3 (gate length to channel thickness). For smaller gate lengths, the transconductance drops and can only increase again if some new physical process causes the carriers to transit the gate region faster. In our case the aspect rato is below 0.16 for a gate length of 100 nm, so that the rise we observe can reasonably be attributed to the onset of velocity overshoot in these devices.

4. BlochFET devices

The use of a LSSL in a BlochFET was first suggested by Reich *et al.* [9]. In this structure, we attempt to create a superlattice in the two dimensional plane of the inversion layer, so that the carriers see a type of quantum confinement in all three dimensions. Then, carriers accelerated by the source-drain electric field can be accelerated into the upper half of the narrow energy band and undergo Bloch oscillations (oscillations in **k**-space which causes their localization in real space). Although, the existence of Bloch oscillations, and the resultant Stark ladder in energy, has been controversial [10], the treatment of quantum high field transport in an extended superlattice has clarified the ideas greatly [11]. Indeed, numerical simulations of transport in a LSSL device have clearly indicated the nature of the negative differential conductivity that will result [12].

We have fabricated BlochFETs in a device using the HEMT structure. In a HEMT, the use of modulation doping produces an inversion layer at the GaAlAs/GaAs interface in which the electrons have quite high mobility due to their physical separation from the donor atoms. The details of the structures used has been given previously [3-5], and the gate grids were produced by lift-off techniques using PMMA. Improved structures, which we are currently fabricating, however, use dry processing techniques to produce a grid with much smaller dimensions. In Fig. 4, the basic energy structure of the HEMT, and a micrograph of the gate grid, are shown. Here, the grid is composed of 40 nm lines on a 160 nm pitch. The devices are typically measured at 4 K, as it is only at the low temperatures that the inelastic mean free path is sufficiently long to produce coherence from one quantum well to the next. A measured set of *i-v* curves and a photograph of the completed device are shown in Fig. 5.

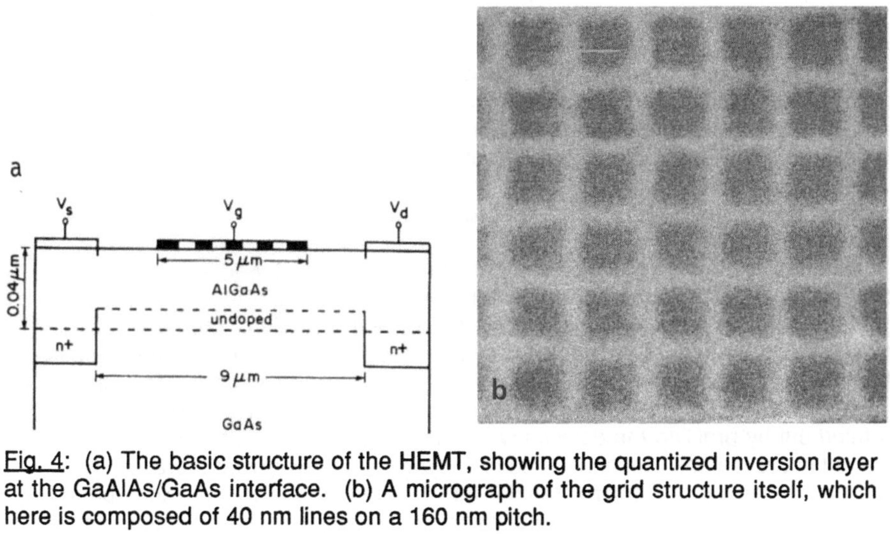

Fig. 4: (a) The basic structure of the HEMT, showing the quantized inversion layer at the GaAlAs/GaAs interface. (b) A micrograph of the grid structure itself, which here is composed of 40 nm lines on a 160 nm pitch.

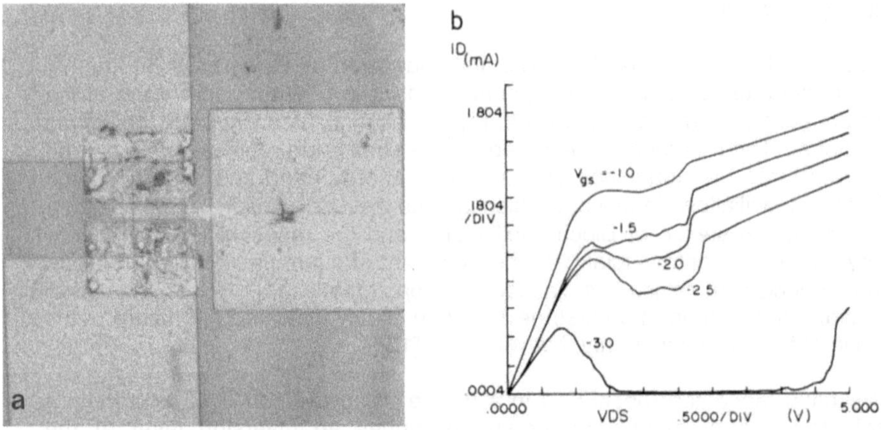

Fig. 5: (a) A micrograph of the completed HEMT structure, and (b) the resulting characteristic curves.

The striking feature of the characteristic curves is the strong NDC present in the current at large values of the gate reverse bias. These features can be explained as follows. Near pinchoff (largest negative bias), the electrons in the quantum wells are fully localized in their wells, and the electron's self energy actually works to deepen the wells. However, the inelastic mean free path is still larger than the interwell separation, so that a degree of quantum coherence is maintained from one well to the next. Thus, a superlattice band structure exists for the two dimensional reciprocal space. Under the source-drain bias, electrons are thought to be accelerated into the upper parts of the band, and to thereby undergo Bloch oscillations. When this occurs, those electrons experiencing the oscillations are localized and removed from the conduction process so that the current drops. A subsequent rise at very high values of the drain potential is due to field breakup of the localization (or equivalently excitation to a higher subband in the superlattice band structure). As the channel is turned more on by a less negative bias, a background of electrons exists which is not localized to the quantum wells. This background charge acts to screen out the periodic potential, thus weakening its effect, and reducing the effect of the NDC. Thus the interwell coupling (tunneling) is eventually reduced to near zero for sufficiently positive gate voltage.

In addition to Bloch oscillations, other phenomena exist which could lead to such strong NDC regions in the characteristic curves. One possibility is degradation of the contacts. This is not likely to be the cause of the NDC, as the curves are consistant in their behavior away from the NDC regions, and devices with solid gates do not show this behavior. Another is trapping and detrapping of carriers in a buried quantum well layer, such as occurs in a NERFET device. This was ruled out by profiling the devices with secondary mass spectroscopy (SI MS), which did not reveal the presence of such a layer. If the layer were there, the NDC would be caused by real-space transfer into it. Real space transfer to a buried layer, and not to the doped AlGaAs layer, would be required as the effect is stronger with more negative bias, which pushes the carriers deeper into the structure. If real-space transfer into the upper AlGaAs layer were the cause, one would expect it to be more pronounced for more positive values of gate bias, which is opposite to the experimental observation. Moreover, we would expect it to also occur with a solid gate, which is not the case, as even most superlattice gate devices do not show the effect.

Another possibility is sequential tunneling, in which the tilting of the bands by the drain potential allows electrons to tunnel from one quantum well to the next. The values of the potential for which the peak and the valley currents are observed, and for which the next rise is observed, suggest that in this case the carrier density (for the observed band widths and gaps) is somewhat higher than that expected to exist in this devices, but not completely out of reason. However, if this were the cause, we note that it also requires the presence of the superlattice band structure in keeping with the assertions for the Bloch oscillations.

In this light, it is difficult to separate the latter mechanism from Bloch oscillations by the use of simple d.c. measurements. Conclusive evidence must come from the a.c. measurements of the negative conductivity, or from radiation at the Bloch frequency, both of which should be strong linear functions of the drain potential. On the other hand, sequential tunneling is expected to be a weaker function of the drain potential. In any case, definitive and conclusive confirmation of the presence of the Bloch oscillations has not been obtained so far, but the search is still underway. On the other hand, the NDC observed in Fig. 5b is consistent with the theoretical expectations [12] for Bloch oscillations. If we used the simple relation expected for the *i-v* curves, the relaxation time inferred from the peak of the current near 0.75 V on the drain is compatible with the transconductance of the transistor itself. In addition, when we calculate the expected scattering by the dominant polar optical phonons, we achieve a product of $\omega\tau=1$ near the observed peak field in this figure. Thus both of these observations reinforce the suggestion that Bloch oscillations are the cause of the NDC.

5. Acknowledgements

The authors would like to thank Tom Aucoin, of the U.S. Army's Electronics Device and Technology Laboratory, and Jim Comas, of the Naval Research Laboratory, for providing the epitaxial layers used in these studies, and to W. Porod, W. Poetz, R. O. Grondin, M. A. Reed, F. Capasso, and G. J. Iafrate for many helpful discussions. This work was supported in part by the Office of Naval Research.

6. References

1. G. Bernstein and D. K. Ferry: Superlatt. and Microstruc. **2**, 147 (1986)
2. G. Bernstein and D. K. Ferry: IEEE Trans. Electron Dev., in press
3. G. Bernstein and D. K. Ferry: Superlatt. and Microstruc. **2**, 373 (1986)
4. G. Bernstein and D. K. Ferry: J. Vac. Sci. Technol. B **5**, 964 (1987)
5. G. Bernstein and D. K. Ferry: Z. Phys. B **67**, 449 (1987)
6. D. K. Ferry: in *Handbook of Semiconductors*, Ed. by W. Paul, Vol. 1 (North-Holland, Amsterdam, 1978) p. 1130

7. M. Heiblum, I. M. Anderson, and C. M. Knoedler: Appl. Phys. Letters **49**, 207 (1986)
8. W. Curtice, private communication.
9. R. K. Reich, R. O. Grondin, D. K. Ferry, and G. J. Iafrate: IEEE Electron Dev. Lett. **EDL-3**, 381 (1982)
10. J. N. Churchill and F. E. Holstrom: Phys. Lett. **85A**, 453 (1981)
11. J. B. Krieger and G. J. Iafrate: Phys. Rev. B **33**, 5494 (1986)
12. R. K. Reich, R. O. Grondin, and D. K. Ferry: Phys. Rev. B **27**, 3483 (1983)

GaAs Quantum Well Wire Structures: Their Fabrication by Focused Ga Ion Beam Implantation and Their Optical Properties

Y. Hirayama and H. Okamoto

NTT Basic Research Laboratories, 9–11 Midor-icho, 3-Chome, Musashino-shi, Tokyo 180, Japan

Abstract. Interdiffusion of Al and Ga at the GaAs/AlGaAs heterointerface is enhanced by focused Ga ion beam implantation and subsequent annealing. Quantum well wire structures are fabricated using this interdiffusion enhancement resulting from focused Ga ion beam raster scanning. Fabricated structures reveal multiple fine structures in their low temperature photoluminescence and its excitation spectra. These multiple fine structures are considered to originate from the density of states specific to one-dimensional carrier systems.

1. Introduction

Quantum wells and superlattices fabricated by well-defined epitaxial growth technology have stimulated many studies about two-dimensional carrier systems. Quantum well wires (QWWs), in which carriers are confined not only in the epitaxial growth direction but also in the lateral direction, have received growing interest in proportion to the progress of in-plane fine pattern fabrication technology.

Electron conduction in QWWs has been reported by many authors[1]. The unique characteristics specific to the small wire structure, such as universal conductance fluctuation and Aharanov-Bohm oscillation, have been obtained in Si MOS wires[2] and AlGaAs/GaAs modulation doped wires[3]. However, it is difficult to observe the density of states specific to a one-dimensional carrier system by conduction experiments relating to such interference effects[4].

The QWW structures are expected to have many attractive features also in the optical characteristics. By optical measurement we can detect the density of states independently of carrier interference effects. Enhanced non-linearity and an increase of exciton binding energy[5] are expected. Application of QWW structures to present optical devices are also anticipated[6,7]. Furthermore, these features can be extended to zero-dimensional carrier systems. However, optical studies of QWW structures are less intensive than those of electron conduction. This is due to the difficulty of fabricating quantum well wires for optical measurement. For measuring optical properties, it is necessary to confine both electrons and holes in the same space. It requires bandgap variation in the lateral direction as well as in the growth direction. Some microstructure fabrication procedures, such as reactive ion etching (RIE)[8–10], chemical etching[11], and implantation-enhanced compositional disordering[12–14] have been proposed and tried to fabricate QWWs for optical characterization. The combination of electron beam (EB) lithography and RIE

Springer Series in Solid-State Sciences Vol. 83: **Physics and Technology of Submicron Structures**
Editors: H. Heinrich · G. Bauer · F. Kuchar

was studied by Kash et al.[9]. They realized small structures less than 30 nm in their minimum dimension and reported the enhancement of luminescence intensity. However, the sides of these structures are exposed to the atmosphere, and no features characterizing a one-dimensional carrier system other than the blue shift of luminescence energy were observed. On the other hand, with implantation enhanced-compositional disordering, the fabricated wires are completely buried in the crystal. Though this process does not result in an abrupt lateral interface or a large lateral confinement potential depth, multiple fine structures were observed in the luminescence spectra measured at low temperature[13,14]. These fine structures are considered to be related to the discrete energy levels resulting from a one-dimensional carrier system.

We use Ga ion implantation for enhancing interdiffusion. This is because Ga is the most stable ion source in the focused ion beam (FIB) system[15]. Further, Ga implantation makes a highly resistive region in GaAs and does not make shallow donors or acceptors[16,17]. This is an important advantage in measuring optical properties of the fabricated strucutures. Fabrication of QWW structures by Ga implantation-enhanced compositional disordering was first reported by the present authors[12] and later applied to wire and box structures by Cibert et al.[14]. In this paper, we describe fabrication of a QWW structure by Ga FIB and its optical properties.

2. FIB scanning and annealing conditions

The acceleration voltage and beam current of Ga FIB were 100 kV and about 20 pA, respectively. The ion dose of the implanted region was controlled by changing either the scanning duration or the number of scans. The beam diameter of Ga FIB was 80–100 nm in this experiment. Post implantation annealing was carried out to remove the damage resulting from implantation. During annealing the sample surface was covered with another fresh GaAs wafer (face-to-face condition) in an H_2 ambient.

3. Interdiffusion of GaAs-AlGaAs heterointerface enhanced by Ga implantation

Since the first demonstration of impurity-enhanced compositional disordering (IECD) of a superlattice by Zn diffusion[18], much effort has been devoted to studying it[19]. It is becoming apparent that compositional disordering is almost universal. It has been induced by a wide variety of impurity species, including constituent atom implantation[12,20,21]. The compositional disordering enhancement is considered to be driven by the diffusion of impurities[21,22]. For example, Si pair diffusion is considered to enhance the interdiffusion in case of Si implantation. In general, a greater diffusion of impurities brings about greater interdiffusion of Al and Ga at the heterointerface. In the case of constituent atom implantation, so-called defects, such as Ga interstitials, As vacancies or their complexes, play the same role as Si-Si pairs in Si implantation. Diffusion of such defects is supposed to enhance the interdiffusion[21,23]. We will refer to such a defect as an interdiffusion enhancing defect (IED).

The degreé of interdiffusion of a GaAs-AlGaAs heterointerface is expressed in terms of the interdiffusion length (Δ_{int}). Δ_{int} was measured as a function of Ga ion dose and is shown in Fig.1. Though interdiffusion

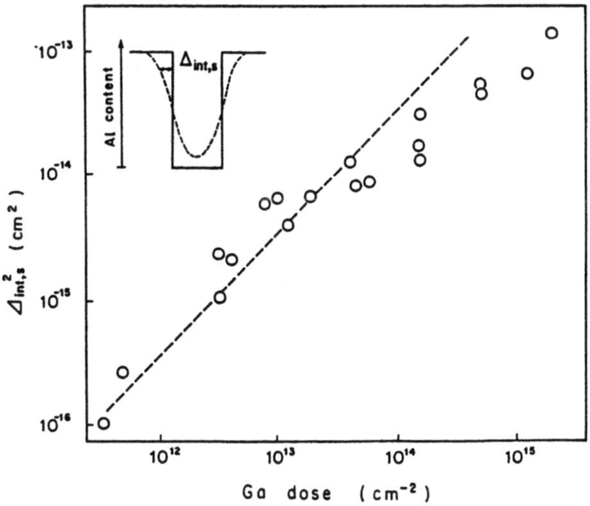

Fig.1 : The saturated interdiffusion length ($\Delta_{int,s}$) as a function of Ga dose. Broken line indicates the linear relationship between $\Delta_{int,s}^2$ and Ga dose.

gradually proceeded with prolonged annealing, most interdiffusion occurred within the first few minutes of the annealing under the face-to-face annealing condition. Δ_{int} values in Fig.1 were obtained after 5-60 min annealing. Therefore they represent saturated values ($\Delta_{int,s}$).

In general, Δ_{int} is given by

$$\Delta_{int}^2 = \eta N_t \Delta^2 \qquad (1)$$

where η and N_t are the proportionality constant and implanted ion density, respectively[21,23,24]. As shown in Fig.1, $\Delta_{int,s}^2$ has an approximately linear relationship with the Ga dose in the low dose region, and tends to saturate above 10^{14}cm^{-2}. N_t is considered to be proportional to the density of IEDs in the case of Ga implantation. The linear relationship can be explained by simply assuming that the diffusion of only one kind of IED enhances the interdiffusion and that the diffusion length indicated by Δ is independent of the ion dose. This simple linear relationship between Δ_{int}^2 and Ga dose is a characteristic peculiar to Ga implantation. In the case of Si implantation, the relation between Δ_{int}^2 and Si density is not simple. This is because the diffusion length strongly depends on the Si density. It is considered that the increase of Si pair formation rate brings about a larger Δ value in the higher Si dose region[22].

The saturation characteristics of $\Delta_{int,s}^2$ vs. Ga ion dose above 10^{14}cm^{-2} may be due to the decrease of Δ_s, i.e. the saturated value of Δ. The large density of damage induced by high dose implantation is considered to prevent the diffusion of IEDs. Such reduction of diffusion under high ion dose implantation has also been reported for Si implantation[19]. The characterization of IEDs in more detail is a subject for future study.

The crystalline property of the Ga implantation-enhanced compositionally disordered region was observed by TEM images. These images indicate an existence of dislocation loops after Ga implantation and subsequent annealing. However, the density of such defects is less than that observed in Si- or Ar-implanted samples [25].

4. Fabrication of QWW structures

The QWW structures were fabricated by raster scanning Ga FIB on a SQW wafer (5.5 nm GaAs layer sandwiched by $Al_{0.5}Ga_{0.5}As$ barriers). The interdiffusion is enhanced only in the Ga-implanted region as shown in a schematic cross sectional view in Fig.2. The lateral broadening of the interdiffused region is governed by the diameter of the Gaussian shaped ion beam, lateral straggling of implanted ions and the lateral diffusion of IED during annealing.

The lateral profile of QWW fabricated by Ga FIB raster scanning is approximately calculated by using (1). When most interdiffusion occurred within the first few minutes of annealing ($t<\tau$), we assume that diffusion of the IED stops at an annealing time of τ:

$$\Delta^2 = Dt \ (t<\tau), \quad \Delta^2 = \Delta_s{}^2 = D\tau (t>\tau) \qquad (2)$$

Under this assumption, the lateral profile of the interdiffused region after long time annealing ($t>\tau$) can be obtained from the two parameters Δ_s and η, independently of τ.

For fabricated QWW structures, a blue shift of the photoluminescence (PL) peak energy was observed. Though a part of the blue shift of the PL peak energy is due to the lateral carrier confinement effect, most of the blue shift of QWW structures is explained by the lateral diffusion of IEDs. Therefore, Δ_s can be approximately determined from this energy shift. Filled circles in Fig.3 show the energy shift observed for various periods of FIB raster scanning. Solid lines 1 to 4 indicate the energy shift obtained from the calculation based on the relation (1). The FIB beam diameter of 100 nm and lateral straggling of implanted Ga of 19 nm at 100 keV implantation are assumed in this calculation. A period of 100 nm approximately corresponds to uniform implantation. η is determined for each Δ_s such that the energy shift of the uniformly implanted region coincides with the broken line in Fig.1. Taking into account the fact that the broken line in Fig.1 overestimates the interdiffusion in high dose regions, Δ_s of 30-54 nm and $\eta=(1.5\text{-}3)\times10^{-22}cm^{-3}$ are obtained in the present process. These values approximately agree with those reported by Cibert and Petroff[24] using conventional Ga implantaion.

Using Δ_s = 50 nm and $\eta=2\times10^{-22}cm^{-3}$ the lateral profile of Al content in QWWs was calculated. The lateral Al content profiles at the center of a GaAs SQW are shown in Fig.4. Only one lateral period was represented in this figure. The important parameter for QWWs is not the period, but the space width. However, in this experiment, the implanted line width was kept

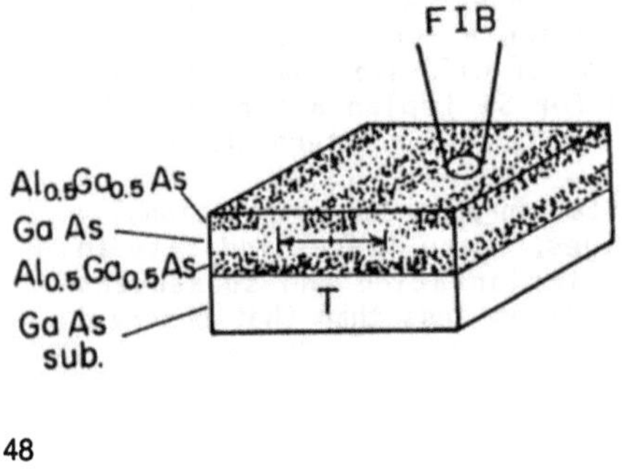

Fig.2 : Schematic cross-sectional view of a structure obtained by GaFIB raster scanning. Dotted regions indicate the interdiffused regions.

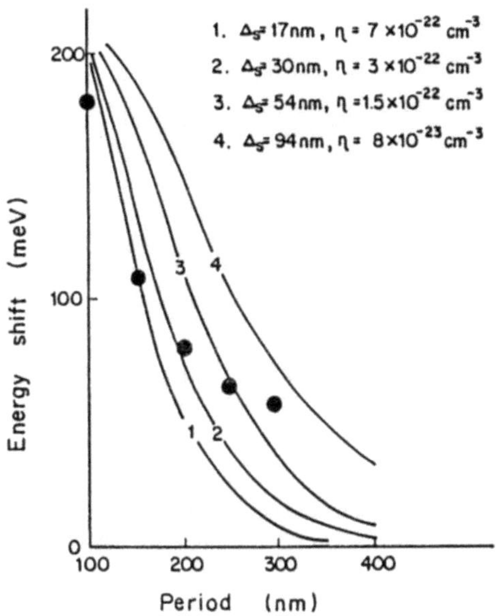

Fig.3 : Blue shifts of the PL peak energy of QWW structures as compared with that of a reference SQW without implantation. Solid lines indicate calculated blue shift for various Δ_s values.

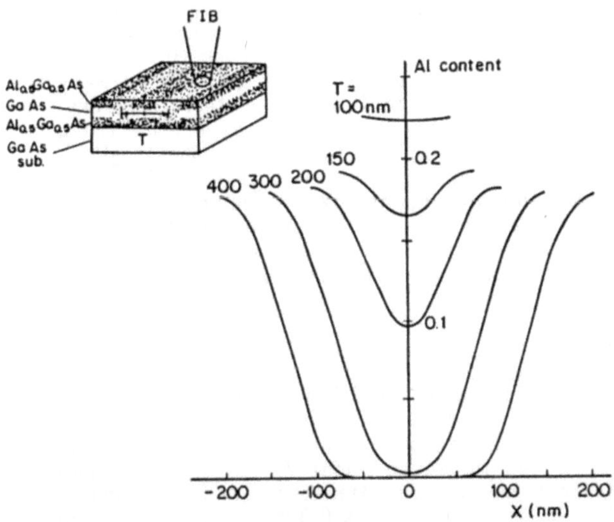

Fig.4 : The lateral profile of Al content at the center of GaAs well. Only one lateral period is represented.

at a constant value of 100 nm. Therefore we use the period T as the parameter. In the large T region, lateral profiles are approximately rectangular in shape. In the small T region, lateral profiles become approximately parabolic or V-shaped[13,14,24]. The effective lateral width for electron and hole confinement is considered to be governed by the curvature at the bottom of the lateral profile. This curvature has its

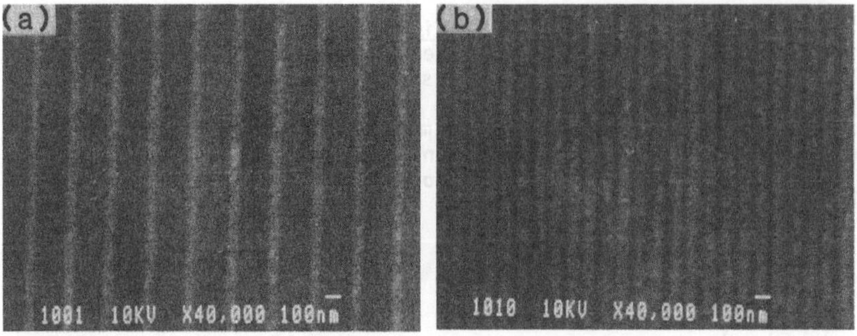

Fig.5 : The in-plane observation of fabricated QWW structures by a backscattered electron (BE) image. The repetition period of FIB scanning was 300 nm (a) and 125 nm (b), respectively.

maximum value at T = 200 – 250 nm and decreases both in the larger T region and in the smaller T region. The maximum curvature corresponds to an effective wire width of 30-50 nm, which is determined by the first and second laterally quantized energy levels.

The in-plane observation of fabricated QWWs was carried out by using the backscattered electron (BE) image mode of a scanning electron microscope (SEM). BE images give higher contrast to compositional differences without suffering from perturbations such as small surface steps or surface potential variations. In addition, a BE image measured at 10kV primary electron beam energy has spacial resolution of 15 nm or better. Therefore, BE images are suitable for nondestructive observation of the periodic structures fabricated by FIB raster scanning and subsequent annealing[26]. Observed images of QWWs are shown in Fig.5. The Ga ion dose was $4\times10^{14}cm^{-2}$ in this experiment. Figure 5 (a) shows an image for a scanning period (T) of 300 nm. The period was reduced to 125 nm in (b). These observations show that lateral compositional modulation with a period as small as 125 nm is certainly realized by GaFIB induced compositional disordering.

5. Optical properties of QWWs

The optical properties of the QWW fabricated by GaFIB scanning with 300 nm period were studied by photoluminescence (PL) and photoluminescence excitation (PLE) spectra at low temperature. The Ga dose was set to $1.6\times10^{14}cm^{-2}$. The PL spectrum was measured under Ar laser (514.5 nm) excitation. The excitation laser power and beam diameter were 500 μW and 5-7 μm, respectively. The spectral resolution was 0.5 meV. In Fig.6, the solid line indicates the PL spectrum measured for the QWW structure and the broken line indicates the reference spectrum measured for the original SQW epilayer which was annealed under the same conditions as in the QWW but without implantation. Most important in the QWW spectrum is the well-defined multiple peak structures denoted by solid bars. They are distinct from the noise level of the measuring system. The energy interval between the adjacent peaks is 3.8-6.2 meV for this QWW.

Figure 7 shows the photoluminescence excitation (PLE) spectra for the QWW and for the reference SQW. The excitation power was between 400 μW and

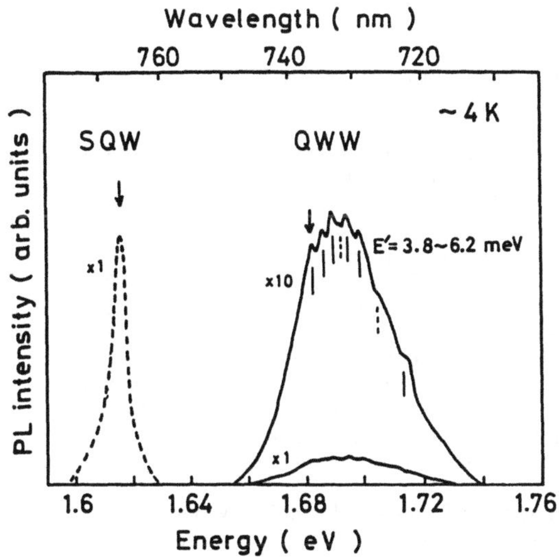

Fig.6 : PL spectra of a QWW structure with T = 300 nm (solid line) and reference SQW (broken line). Downward arrows indicate the energies where excitation spectra shown in Fig.7 were measured.

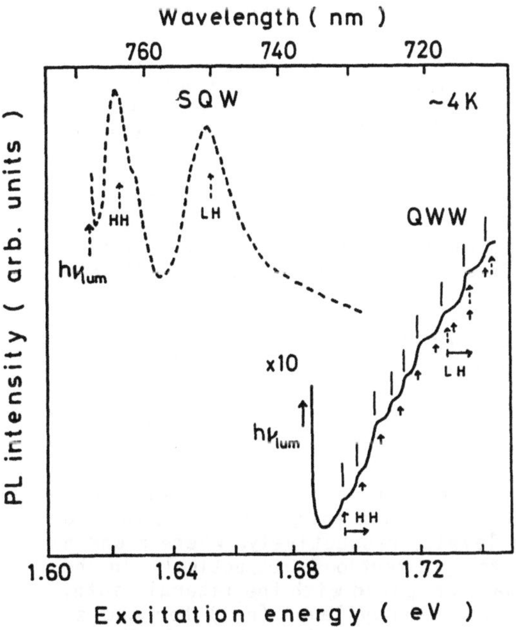

Fig.7 : PLE spectra of a QWW structure with T = 300 nm (solid line) and reference SQW (broken line). The detection energy is denoted by the upward arrow $h\nu_{1um}$. Solid bars above the QWW spectrum indicate the energy position of the observed fine structures. Solid arrows and broken arrows under the spectrum indicate calculated heavy-hole related and light-hole related transitions, respectively.

2 mW, and beam diameter was again about 7 μm. The detection wavelength ($h\nu_{1um}$) was set to 737 nm (1.683 eV) and 768 nm (1.615 eV) for the QWW and for the reference SQW, respectively, as indicated by downward arrows in Fig.6. In the PLE spectrum for the reference SQW, the peaks at 1.623 eV and at 1.654 eV are the n=1 heavy-hole exciton and light-hole exciton, respectively. In the PLE spectra of the fabricated QWW, the exciton peaks are not clear, however, the fine structures are clearly observed again. In

Fig.7, energy positions of fine structures are indicated by solid bars above the spectrum. The energy intervals of the adjacent solid bars are in the range between 5.2 and 7.4 meV, and approximately agree with the energy interval of multiple peaks observed in PL spectrum.

Energy levels of quantum states were calculated from the Al content profiles shown in Fig.4. Energy level separation is approximately constant in this case, because the potential profile is approximately parabolic. The calculated energy level positions of optical transitions are also shown in Fig.7 by upward arrows. Solid arrows and broken arrows under the spectrum indicate the heavy-hole related peaks and light-hole related peaks, respectively. For the convenience of comparison between the calculation and the observation, the lowest calculated transition energy was fitted to the lowest energy fine structure observed in PLE spectra, and the differences of the transition energies from this ground state are indicated. The energy position of observed fine structure agrees reasonably well with the calculated levels. This agreement suggests that observed fine structures originate from the lateral carrier confinement in the fabricated QWW.

Next, we show low-temperature PL spectra of QWWs fabricated with various scanning periods (T). T ranges from 150 to 300 nm in Fig.8. The sample with T=300 nm shows a multiple peak spectrum with an energy interval (E′) of 3.8 to 6.2 meV. This energy interval increases as T decreases and reaches 6 to 7 meV for T=200 nm. Multiple peak structures become obscure for T=150 nm, and the energy separation of these structures seems to be smaller. The calculation mentioned before shows that the effective wire width at the bottom of lateral potential profile has a minimum value at approximately T=250 nm. The minimum effective wire width corresponds to E′ of about 6 meV. Therefore the change in E′ with T, as well as the value of E′, approximately agrees with the calculated value.

Finally, a discussion is given of the influence of the inhomogeneity of lateral width along a QWW upon the optical absorption spectral shape. The optical absorption coefficient α_{QWW} at a photon energy $h\nu$ is given as follows, taking energy broadening ΔE of the sawtooth function[27] into consideration:

$$\alpha_{QWW} \propto \sum_{n,m,i} \sqrt{\mu_i} \left(\frac{h\nu - E_{n,m,i} + [(h\nu - E_{n,m,i})^2 + (\Delta E)^2]^{1/2}}{2[(h\nu - E_{n,m,i})^2 + (\Delta E)^2]} \right)^{1/2} . \qquad (3)$$

$$\Delta E = kT + \frac{\partial E_{n,m,i}}{\partial W} \Delta . \qquad (4)$$

Here, μ_i and $E_{n,m,i}$ indicate the reduced effective mass of the heavy hole (i=h) and the light hole (i=l), and the transition energy between quantized electron levels and quantized hole levels, respectively, where n and m indicate the quantum number in the z and y directions, respectively. In the present case, L_z(=5.5 nm) is very small compared with the lateral width, therefore we consider only the lowest energy level (n=1) for the z-direction quantization. Exciton effects are neglected in this calculation. As an origin of ΔE, we consider two factors. One is thermal broadening (kT) and the other is broadening due to the inhomogeneity of the lateral wire width. Here, W indicates the lateral width of QWW determined from the half maximum of Al content and ΔW indicates the fluctuation of W. Figure 9 shows the calculated absorption spectrum of the QWW for several degrees of inhomogeneity ΔW. Large ΔW makes the absorption peak obscure. In Fig.7, we observed fine structures in PLE spectra, but they are not clear peaks. The reason for this is considered to be the inhomogeneity ΔW as shown in Fig.9.

Fig. 8

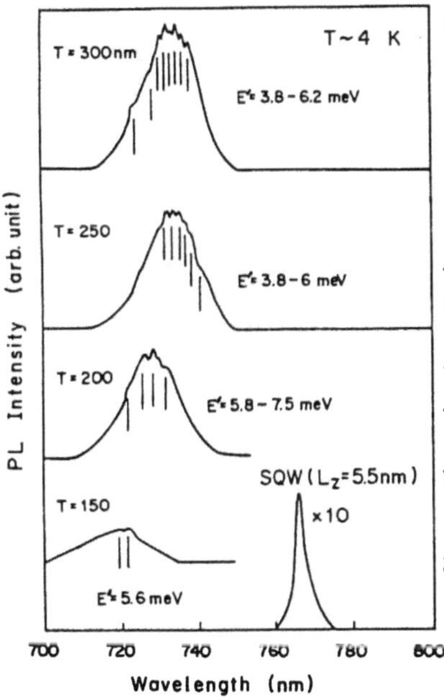

Fig. 9

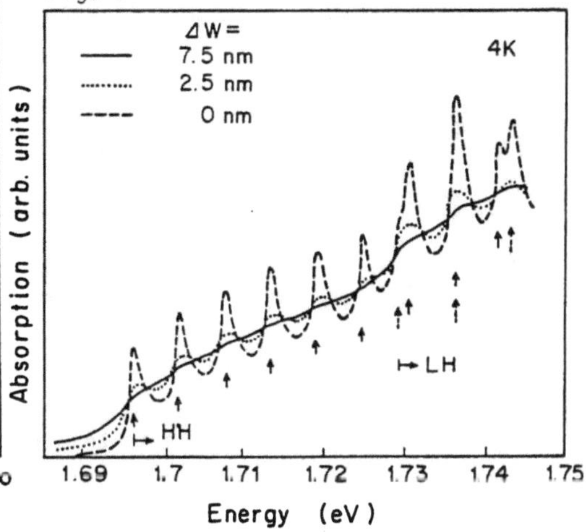

Fig.8 : Low-temperature PL spectra of QWWs fabricated with various scanning periods (T). T ranged from 150 to 300 nm.

Fig.9 : Calculated absorption spectra of QWW structure (T = 300 nm) under several degrees of wire width inhomogeneity ΔW.

The uniformity of FIB scanning was checked by TEM in-plane observation of implantation-induced damage patterns present before annealing. Figure 10(a) shows the line-and-space pattern formed by the raster scanning of Ga FIB with a period of 300 nm. Ion dose was about $10^{14}cm^{-2}$. Figure 10(b) shows an enlarged image of Fig.10 (a). From this damage pattern, we can estimate the inhomogeneity of FIB scanning to be less than 6 nm. This value is fairly good for the submicron technology. However, it is not sufficient for fabricating QWW structures. Further improvement in the uniformity of FIB scanning is expected for observing more clearly low-dimensional optical characteristics. Moreover, inhomogeneity was introduced during the annealing stage after implantation. The precise control of annealing process is then also required.

6. Conclusions

By using Ga FIB raster scanning, we fabricate lateral compositional variation with a period of a few hundred nanometers, i.e. QWW structures. These QWW structures reveal multiple fine structures in their low temperature PL and PLE spectra. Energy separations of these fine structures agree approximately with the calculation based on two-dimensional quantum confinement. Consequently, these multiple peak structures are suggested to originate from the density of states specific to the one-dimensional carrier system. We also discuss briefly the remaining problems which need to be overcome in the future.

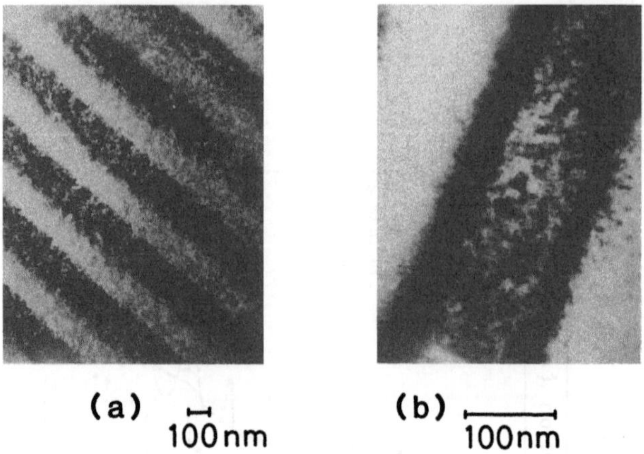

Fig.10 : TEM observation of damage patterns formed by Ga FIB scanning. Post implantation annealing was not carried out for observing damage patterns. (b) shows the enlarged image of (a).

Acknowledgement The authors wish to thank Drs. Y.Suzuki, S.Tarucha, T.Saku and T.Miyazawa for collaboration in this work and valuable discussions. They would also like to thank Drs. Y.Horikoshi, T.Izawa, T.Kimura and Y.Kato for their encouragement throughout this work.

References

1. For example A.B. Fowler, A.Hartstein and R.A. Webb, Phys. Rev. Lett. 48, 196 (1982)
2. For example W.J.Skocpol, P.M.Mankiewich, R.E.Howard, L.D.Jackel, D.M.Tennant and A.D.Stone, Phys. Rev. Lett. 56, 2865 (1986)
3. For example G.Timp, A.M.Chang, J.E.Cunningham, T.Y.Chang, P.Mankiewich, R.Behring and R.E.Howard : Phys. Rev. Lett. 58, 2814 (1987)
4. A.C.Warren, D.A.Antoniadis and H.I.Smith, Phys. Rev Lett. 56, 1858 (1986)
5. M.Matsuura and T.Kamizato, Surf. Sci. 174, 183 (1986)
6. Y.Arakawa and H.Sakaki, Appl. Phys. Lett. 40, 939 (1982)
7. M.Asada, Y.Miyamoto and Y.Suematsu, IEEE J. of Quantum Electron. QE-22, 1915 (1986)
8. P.M.Petroff, A.C.Gossard, R.A.Logan and W.Wiegman, Appl. Phys. Lett. 41, 635 (1982)
9. K.Kash, A.Scherer, J.M.Worlock, H.G.Craighead and M.C.Tamargo, Appl. Phys. Lett. 49, 1043 (1986)
10. H.Temkin, G.J.Dola, M.B.Panish and S.N.G.Chu, Appl. Phys. Lett. 41, 635 (1982)
11. Y.Miyamoto, M.Cao, Y.Shingai, K.Furuya, Y.Suematsu, K.G.Ravikumar and S.Arai, Jpn. J. Appl. Phys. 26, L225 (1987)
12. Y.Hirayama, Y.Suzuki, S.Tarucha and H.Okamoto, Jpn. J. Appl. Phys. 24, L516 (1985)
13. Y.Hirayama, Y.Suzuki, H.Iguchi, S.Tarucha and H.Okamoto, Electronics Materials Conf., Amherst, 1986, N-7, Y.Hirayama, S.Tarucha, Y.Suzuki and H.Okamoto, Phys. Rev.B, B37, 2774 (1988-I)
14. J.Cibert, P.M.Petroff, G.J.Dolan, S.J.Pearton, A.C. Gossard and J.H. English, Appl. Phys. Lett. 49, 1275 (1986)
15. J.Melngaillis : J.Vac. Sci. & Technol. B5, 469 (1987)

16. K.Nakamura, T.Nozaki, T.Shiokawa, K.Toyoda and S.Namba, Jpn. J. Appl. Phys. 24, L903 (1985)
17. Y.Hirayama and H.Okamoto, Jpn. J. Appl. Phys. 24, L965 (1985)
18. W.D.Laiding, N.Holonyak, Jr., M.D.Camras, K.Hess, J.J.Coleman, P.D.Dapkus and J.Bardeen, Appl. Phys. Lett. 38, 776 (1981)
19. R.D.Burnham, R.L.Thornton, N.Holonyak Jr., J.E.Epler and T.L.Paoli, Proc. Int. Symp. GaAs and Related Compounds (Las Vegas, 1986), Inst, Phys. Conf. Ser. No.83, 9 (1987)
20. P.Gavrillovic, D.G.Deppe, K.Meehan, N.Holonyak, Jr, J.J.Coleman and R.D.Burnham, Appl. Phys. Lett. 47, 130 (1985)
21. Y.Hirayama, Y.Suzuki and H. Okamoto, Jpn. J. Appl. Phys. 24, 1498 (1985)
22. P.Mei, H.W.Yoon, T.Venkatesan, S.A.Schwarz and J.P.Harbison, Appl. Phys. Lett. 50, 1823 (1987)
23. J.Cibert, P.M.Petroff, G.J.Dolan, S.J.Pearton, A.C.Gossard and J.H.English, Appl. Phys. Lett. 49, 223 (1986)
24. J.Cibert and P.M.Petroff, Phys. Rev. B, 36, 3234 (1987-II)
25. Y.Suzuki, Y.Hirayama and H.Okamoto, Jpn. J. Appl. Phys. 25, L912 (1986)
26. Y.Hirayama and H.Okamoto : to be published
27. H.H.Hassan and H.N.Spector : J.Vac, Sci, Technol. A3, 22 (1985)

Progress in Ion Projection Lithography

G. Stengl, H. Löschner, and E. Hammel

IMS – Ion Microfabrication Systems GmbH, Schreygasse 3, A-1020 Vienna, Austria

1. Introduction

Ion projection lithography (IPL) uses demagnifying ion optics to project open stencil mask structures onto a substrate with reduced (.. 5x, 10x, ..) scale [1]. The IPL technique offers high throughput potential by using duoplasmatron ion sources with high angular current densities [2] and with virtual source sizes of less than 10 µm. Thus, for organic resists, chip exposure times of ≲ 0.1 sec are easily obtained.

2. IPL resolution and depth of focus

With a new research type ion projection lithography machine (IPLM-01-2, Fig. 1) resolution, pattern transfer characteristics (modulation transfer) and process latitudes were studied using a 5x and a 10x reduction projection lens, respectively.

Using test masks with 1 µm smallest opening width, the IPL exposures on wafer substrates resulted in smallest lines and spaces of 0.2 µm and 0.1 µm, respectively. Pattern transfer in PMMA and AZ5206 organic resist layers was achieved with both positive and image reversal [3] development. Pattern aspect ratios of 4:1 were realized. A detailed description will be published elsewhere [4].

Figures 2 a - d demonstrate the large depth of focus [5] of ion projection lithography. (These results were obtained with the system IPLM-01 [1, 2]). In Si wafers 4 µm deep grooves were etched with isotropic plasma etching resulting in vertical profiles at the top. This topographical wafer surface was covered with a 200 nm thick thermal SiO_2 layer. This oxide layer was used as inorganic ion sensitive positive resist [6] with ion-bombardment enhanced etching in 5 % HF. No etching of course is observed in the vertical parts of the grooves where IPL exposure resulted in negligible dose.

3. Development of defect-free IPL pattern transfer

Exposing ions through resist layers may cause device degradation. For low dose levels annealing may lead to sufficient recovery [7]. In contrast to electron and X-ray lithography, the penetration depth of ions can be adjusted by selecting

Springer Series in Solid-State Sciences Vol. 83: **Physics and Technology of Submicron Structures**
Editors: H. Heinrich · G. Bauer · F. Kuchar

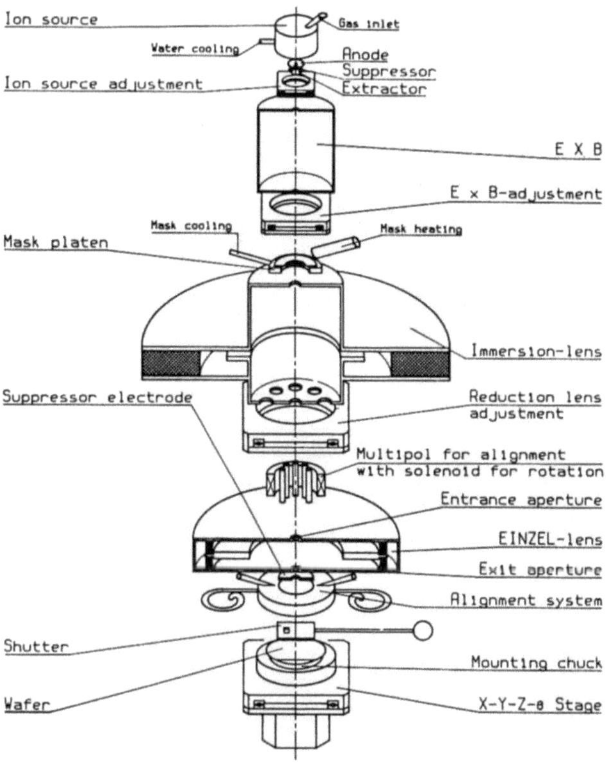

Fig. 1: Schematic of the ion projection lithography machine IPLM-01-2

proper ion energy and ion species (H^+, He^+, N^+, ..). Device damage can also be avoided by the use of a stopping layer beneath the resist layer. This has been demonstrated using oxide layers for the manufacture of GaAs devices [8].

By using surface sensitive organic resists it is also possible to stop the ions within a single layer organic resist using proper surface reactions for pattern transfer [9]. Dry development is possible by excimer laser photoablation of unexposed resist sites as IPL exposed resist surface patterns act as conformal mask for DUV light [10].

4. Intrafield distortion of the ion-optical column

Figure 3 is an optical micrograph of the superposition of two images. The larger structures belong to a shadow image of mask structures obtained at reduced scale by using just the immersion gap lens of the IPLM-01-2. This shadow image clearly shows a pincushion distortion. The picture with the sharp patterns in Fig. 3 is due to the addition of the projective

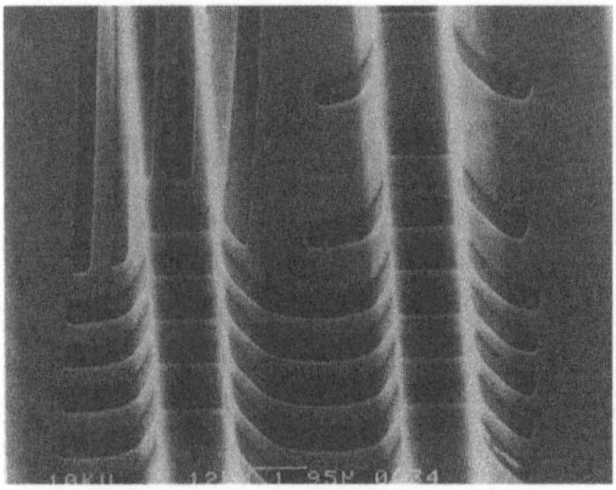

Fig. 2a

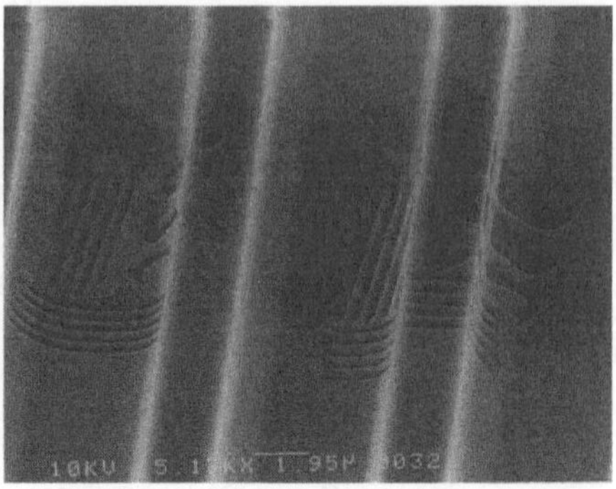

Fig. 2b

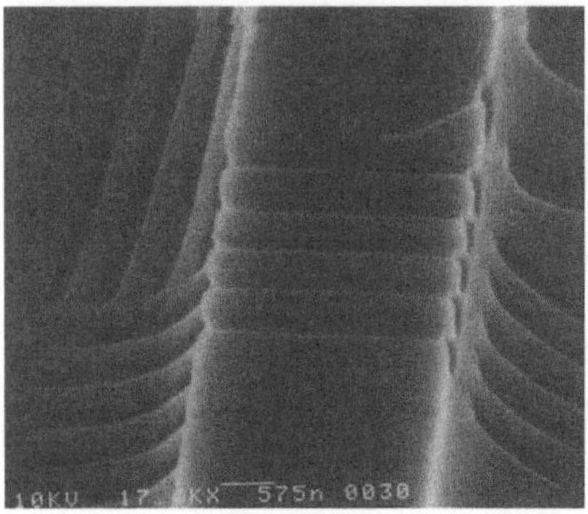

Fig. 2c

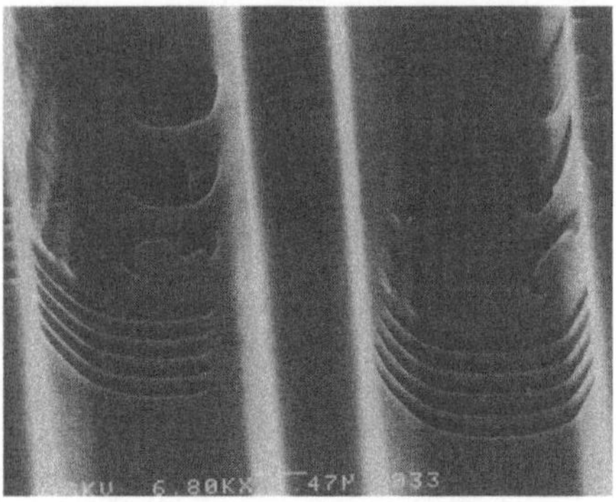

Fig. 2d

Fig. 2: IPL exposure results in a Si wafer substrate with 4 µm deep grooves covered by a thermal SiO_2 layer (initial thickness: 200 nm). IPLM-01 exposure with 80 keV He^+ ions using a 5x reduction lens, ion dose: $3x10^{15}$ ions/cm^2, ion bombardment enhanced etching of SiO_2 layer: 5% HF. The periodicity of the structures (on flat surface) equals 1.8 µm in Fig. 2a; 0.72 µm and 0.62 µm in Fig. 2b; 0.72 µm in Fig. 2c; 0.62 µm and 0.54 µm in Fig. 2d. No etching can be observed in the vertical walls of the Si grooves beyond the range of ion penetration.

lens. No distortion is visible in this real image of the mask openings projected onto the wafer at 5x scale. Distortion values of these projected images have been measured using an electron beam writer as a metrology tool [11]. Distortions of IPLMM-01-2 projected images can be altered from pincushion to barrel by selecting lens voltages and wafer Z-position, as previously reported for the test bench system [3].

In addition to the global pincushion distortion observed in Figure 3 there are irregularities that distort the shape of the images. These defects are caused by local deflections due to charge buildup from dust particles or organic residues at the

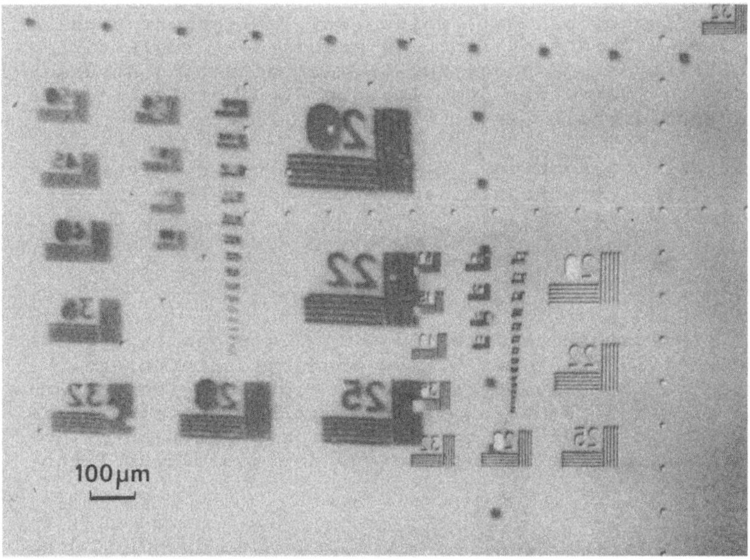

Fig. 3: Optical micrograph of details of two superimposed ion images of resolution test patterns. Exposure was done in self-developing nitrocellulose resist on Si. The coarse structures at the left hand side show a shadow image (using only the immersion lens) of Ni open stencil mask pattern at reduced scale. The sharp, undistorted image at the right hand side is a projected ion image (using immersion lens and projective lens) at 5x demagnification.

mask surface or at the sidewalls of mask openings. These local deflections are not visible in the projected ion image as demonstrated in Fig. 3 unless mask openings are covered by particles.

5. Open stencil mask technology, inspection and repair

The fabrication of Si open stencil masks with beveled opening structures has been reported [12]. Extended stencil supported mask foil structures have been replicated with the IPLM-01-2 at 5x and 10x demagnification [4].

SEM transmission techniques are useful for open stencil mask inspection [9]. Focused ion beam techniques are best suited for the repair of clear and also pinhole defects using ion-assisted metal deposition [13].

6. Electronic alignment and electronic fine adjustment of intrafield distortion

The IPL technique offers the advantage of an electronic fine adjustment of the projected ion image in X, Y, rotation and scale without moving mechanical parts [14]. Alignment signals can be generated by scanning ion beam probes over registration marks. A feasibility study shows sub-0.1-μm detection accuracy using channeltron detectors to measure ion probe induced secondary electron signals [14].

With ion optical correction elements (e.g. electrostatic multipoles) a fine electronic adjustment of intrafield distortion is possible [14, 15].

7. Conclusion

Demonstrated lithographic performance with ion projection shows promising potential as a sub-0.5-μm production technique. However, substantial improvements in ion-optical lens column performance are still needed as well as appropriate electronic alignment results in order to establish ion projection lithography as a cost effective technique for sub-0.5-μm lithography.

References

[1] G. Stengl, H. Löschner, J.J. Muray, Solid State Technol. 29(2), 119 (1986)
[2] G. Stengl, H. Löschner, W. Maurer, P. Wolf, J. Vac. Sci. Technol. 4(1), 194 (1986)
[3] G. Stengl, H. Löschner, E. Hammel, E.D. Wolf, J.J. Muray, NATO Workshop on Emerging Technologies for In Situ Processing, Corsica, May 4-8, 1987, in press (North-Holland)
[4] L.-M. Buchmann, L. Csepregi, K.P. Müller, A. Chalupka, E. Hammel, H. Löschner, G. Stengl, 32nd Int. Symp. Electron, Ion and Photon Beams, May 31 - June 3, 1988, Ft. Lauderdale, FL, t.b.p. J. Vac. Sci. Technol.
[5] R. Fischer, E. Hammel, H. Löschner, G. Stengl, P. Wolf, Microelectronic Engineering 5, 193 (North-Holland, 1986)
[6] G. Stengl, H. Löschner, P. Wolf, Nucl. Instr. Meth. in Phys. Res. B19/20, 987 (1987)
[7] C.W. Slayman, J.L. Bartelt, C.M. McKenna, Proc. SPIE Vol. 333, 68 (1982)
[8] S.W. Pang, T.M. Lyszcarz. C.L. Chen, J.P. Donnelly, J.N. Randall, J. Vac. Sci. Technol. B5(1), 215 (1987)
[9] G. Stengl, H. Löschner, E. Hammel, E.D. Wolf, Proc. 17th European Solid State Device Research Conf., ESSDERC'87, Bologna, Sept. 14-17, 1987, p. 625 (t.b.p. North-Holland)
[10] G. Stangl, E. Cekan, E. Hammel, W. Fallmann, 5th Int. Winterschool on New Developments in Solid State Physics: Physics and Technology of Submicron Structures, Mauterndorf, Salzburg, Austria, Febr. 22-26, 1988, these Proceedings.
[11] E-beam metrology measurements were done by H. Noll, W. Kräuter and M. Strmsek at Austria Micro Systems International GmbH, Unterpremstätten, Austria.
[12] A. Heuberger, L.-M. Buchmann, L. Csepregi, K.P. Müller, Microelectronic Engineering 6, 333 (1987)

[13] J. Melngailis, J. Vac. Sci. Technol. B5(1), 469 (1987)
[14] G. Stengl, H. Löschner, J.J. Muray, Techn. Proc. SEMICON/West 1986, p. 42 (SEMI Inc., Mountain View, CA, 1986)
[15] G. Stengl, H. Löschner, J.J. Muray, Ext. Abstr. 18th Int. Conf. Solid State Devices and Materials, Tokyo, 29 (1986)

Part II

Vertical Transport and Tunneling Phenomena

Vertical Electronic Transport in Semiconductor Nanostructures

M.A. Reed

Central Research Laboratories, Texas Instruments Incorporated, P.O. Box 655936, MS 154, Dallas, TX 75265, USA

Abstract. The investigation of vertical electronic transport in semiconductor heterojunction systems has recently undergone a renaissance due to improved epitaxial techniques in a number of material systems. Presented in this paper are investigations into a number of novel vertical transport systems. Recent advanced in microfabrication techniques have allowed the realization of structures where lateral dimensions approach those achievable with epitaxial technology. Results of transport through structures with reduced dimensionality will be presented.

1. Resonant tunneling spectroscopy

Resonant tunneling in double barrier heterostructures, first investigated in the GaAs/AlGaAs system by Chang, Esaki, and Tsu [1], has recently been the subject of intense investigation. The remarkable submillimeter wave experiments of Sollner *et al* [2] has generated remarkable interest and success due to improvements in GaAs epitaxial growth techniques. Peak-to-valley tunnel current ratios as large as 3.9:1 at 300K, and 21.7:1 at 77K have been demonstrated in the GaAs/AlAs system [3]. Resonant tunneling by holes [4], sequential resonant tunneling through a multiquantum well superlattice [5], double superlattice barrier resonant tunneling structures [6], and resonant tunneling in triple barrier structures [7] are only a few among the intriguing investigations that have been performed.

The ability to engineer peak positions now allows us, in general, to use double barrier structures as spectrometers on suitably designed quantum well or superlattice structures. Specifically, we demonstrate here that we can perform spectroscopy on large quantum wells placed between double barrier structures. The resonant positions of the confining structures are either tunable by changing the quantum well widths or by employing different InGaAs quantum wells [8], where the resonant peak is thus tuned by In composition. Thus, the resonant voltages discriminate which region of the epitaxial structure is being examined. These resonant tunneling regions throughout a complex epitaxial structure gives us a microscopic spectrometer to examine the energy states in the structure.

Figure 1 illustrates an experimental embodiment of such a structure. The structure consists of a 40Å $Al_{.25}Ga_{.75}As$ double barrier / 90Å $In_{.05}Ga_{.95}As$ quantum well / 750Å GaAs quantum well / 40Å $Al_{.25}Ga_{.75}As$ double barrier / 90Å $In_{.05}Ga_{.95}As$ quantum well structure, with a n+ GaAs contact on the other side of the GaAs quantum well resonant tunneling structure. The large GaAs quantum well (750Å) will have a small energy splitting in comparison to the quantum wells of the double barrier structures (90Å). In this specific case, the GaAs quantum well resonant tunneling structure has a n+ contact region so as to provide a "source" or "sink" of available carriers. Thus, when the structure is

Springer Series in Solid-State Sciences Vol. 83: **Physics and Technology of Submicron Structures**
Editors: H. Heinrich · G. Bauer · F. Kuchar

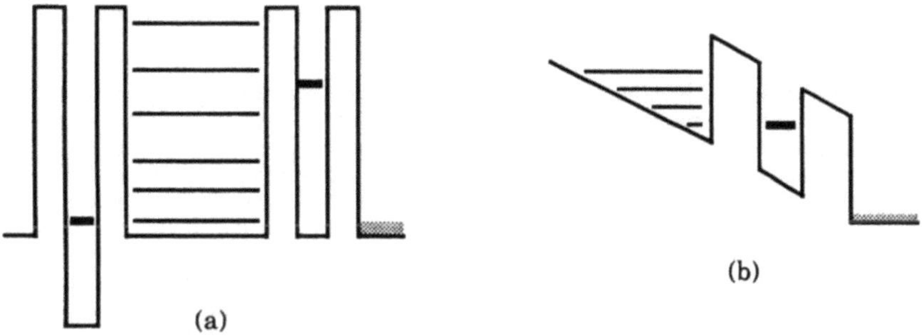

Fig. 1: (a) Schematic conduction band diagram of the large GaAs quantum well / resonant tunneling spectrometer structure. (b) Bias condition for injection from the large GaAs quantum well.

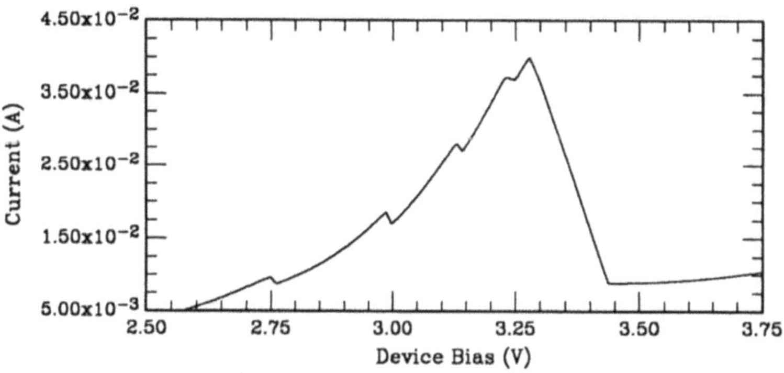

Fig. 2 : I-V characteristic at T=4.2K of the large GaAs quantum well / double barrier structure, corresponding to electron injection from the 750Å GaAs quantum well. The structure on the low bias side of the major peak is due to tunneling from the levels in the 750Å GaAs quantum well.

biased such that carriers tunnel from the large quantum well into the double barrier structure, they will be injected from a ladder of states in the large quantum well. For simplicity, we will discuss results using the GaAs quantum well resonant tunneling structure as the spectrometer, though similar results are also obtained from the InGaAs resonant tunneling structure (at a different bias position).

Figure 2 shows a I-V characteristic at T=4.2K corresponding to electron injection from the large GaAs quantum well. A series of peaks corresponding to electron injection from the states in the 750Å quantum well through the state in the 90Å quantum well are clearly observable. The structure appears on the low bias side of the major peak only. The experimentally observed splittings are 59 meV, 103 meV, 143 meV, and 238 meV. No peaks corresponding to n>4 in the large quantum well are observed, indicative of the position of the Fermi level in this region. The ratios of the splittings to the ground state splitting (1 : 1.74 : 2.42 : 4.03) are in excellent agreement with calculated values (1 : 1.69 : 2.36 : 3.01) except for the n=4 level, presumably due to band bending. Clearly this technique can be generalized to more complex regions, such as parabolic injectors to verify equal splitting in such structures.

Now consider electron injection from the n+ GaAs region, through the 90Å GaAs quantum well double barrier structure, into the 750Å GaAs quantum well region. Below the resonant peak, the tunneling current is a sum of elastic scattering to available states on the other side of the structure, and inelastic tunneling through the entire structure or through the intermediate quantum well state (which should be negligible at these temperatures). Thus, no structure in the I-V characteristic below the resonant peak should be seen due to the large quantum well. However, when the structure is biased beyond the major resonant peak position, elastic scattering can then occur via the 90Å quantum well state; i.e., elastic scattering to the 90Å quantum well ground state subband, relaxation to the bottom of this band, then tunneling out of the structure. Thus, peaks in the I-V characteristic should occur at biases greater than the major resonant peak bias.

Figure 3 shows the I-V characteristic of this structure at T=4.2K corresponding to electron injection from the n+ GaAs region. There are no observable peaks at biases less than the resonant bias. However, a series of oscillations appear at biases greater than the resonant bias in accordance with the mechanism described above. The spacings of these oscillations (~130meV) are approximately constant, corresponding to the asymptotic energy level spacing of a large quantum well, and quantitatively are in good agreement with the size of the splittings of these levels when the asymmetry of the resonant peak positions is taken into account.

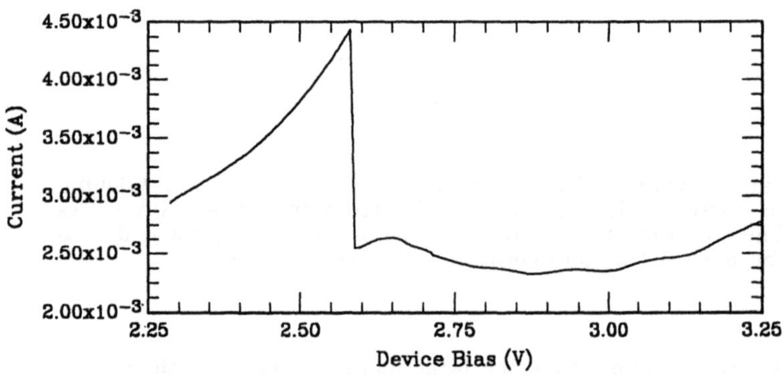

Fig. 3 : I-V characteristic at T=4.2K of the large GaAs quantum well / double barrier structure, corresponding to electron injection from the n+ GaAs contact. The structure on the high bias side of the major peak is due to tunneling into the levels in the 750Å GaAs quantum well.

2. Transport in quantum wires

Recently, there has been a great deal of interest in telegraph noise exhibited in electronic microstructures such as metal-oxide-semiconductor field-effect transistors (MOSFET's) [9,10] and MOS tunnel diodes [11,12]. The discrete resistance fluctuations of the device are thought to be due to the trapping and escape of single electrons at defects in a tunnel barrier or in an inversion layer. In the case of MOSFET wires, electrons in the conduction channel are trapped or detrapped at defects, thus changing the charge state of the defect(s). The traps act as variable scattering centers in the quasi-1D conduction path of the MOSFET wires due to the difference in the scattering cross section of a neutral versus a charged defect. This causes a change in the impedance of the channel, which is manifested as discrete resistance changes in the inversion layer. A conceptually different mechanism has been observed in MOS tunnel junctions, in which trap states in the oxide layer fluctuate around the Fermi level until electron tunneling is favorable. The electrons tunnel through the top of the potential barrier, and as the electrons are trapped and detrapped at defects in the barrier the potential height of the barrier fluctuates. This modulates the tunneling current through the potential barrier, and thus the device impedance fluctuates. The conceptual distinction between the two cases is that, in the tunnel junction case the detailed potential in the region of the barrier is modified due to the trapping of an electron, whereas in the MOSFET wire embodiment the impedance is affected by the change of defect scattering cross sections in the conduction path.

To elucidate the differences between these two effects, we chose an embodiment of a vertical, microfabricated GaAs quantum wire that contains a tunnel junction [13]; specifically a double barrier/single quantum well resonant tunneling structure. The barriers serve as a current limiter and, as will be seen, allows us to distinguish between the two different mechanisms mentioned previously. This is a physical embodiment of a tunnel junction/quantum wire combination. The barriers in the device should give analogous results to the MOS tunnel diodes, and the narrow GaAs wire with the barrier in the conduction path should allow for scattering as in the narrow MOSFET wires. Fabricating the structure this way gives the possibility of simultaneously observing and distinguishing between these mechanisms described above.

The resonant tunneling structure consisted of two 50 Å undoped $Al_{.27}Ga_{.73}As$ barriers enclosing a 50 Å undoped GaAs quantum well. Two undoped 50 Å GaAs spacer layers were grown on either side of the barriers, with two 5000Å Si-doped GaAs buffer layers on top and bottom. An e-beam lithography step and lift-off process followed, leaving a circular metal pad of ~2500Å diameter that served as an etch mask and top Ohmic contact. A highly anisotropic reactive ion etch (RIE) process etched down into the well structure forming columns (vertical wires) with the Ohmic contacts on top. To make contact with the tops of the wires, an insulating and planerizing polyimide layer was spun onto the sample. The tops of the columns were then uncovered by an O_2 RIE. Once the tops of the wires were revealed, a gold bonding pad was defined to provide electrical contact to the tops of the wires. A backside Ohmic contact provided electrical continuity. The samples used in this study are single quantum wires of dimension 2500Å in diameter.

Figure 4 shows the impedance of the device, at fixed bias, as a function of time. Since the resistance fluctuates only between two discrete values, qualitatively identical to what has been previously observed, the behavior indicates the trapping of single electrons onto a given trap in the structure. To verify this, the average rate at which the resistance fluctuates was determined (at fixed bias) as a function of temperature. This rate was clearly activated, with the activation of this specific trap at this bias of 280 meV. Measurement on other traps in this (and other similar) device(s) indicated a wide distribution of trap energies, with switching due to some traps observable to room temperature.

In some instances the simultaneous effects of two different traps are observed. Figure 5 shows an example of a resistance increase to an "up" state (ΔR_1) and larger decrease to a "down" state (ΔR_2) due to different traps. These traps appear to act independently, since the impedance when both traps "switch on" is simply

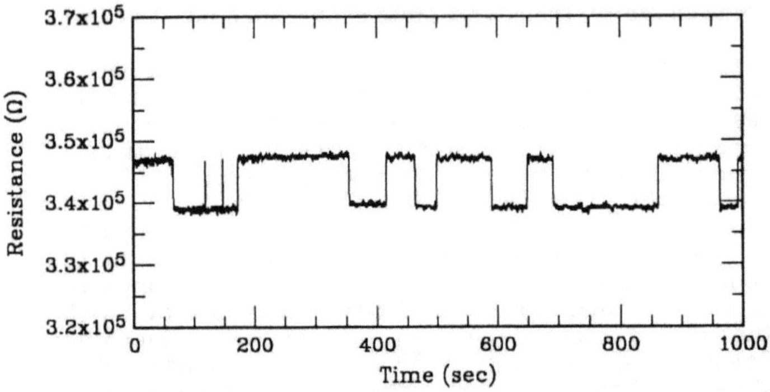

Fig. 4 : Resistance as a function of time for a single heterostructure quantum wire, at fixed bias potential, for T = 180K

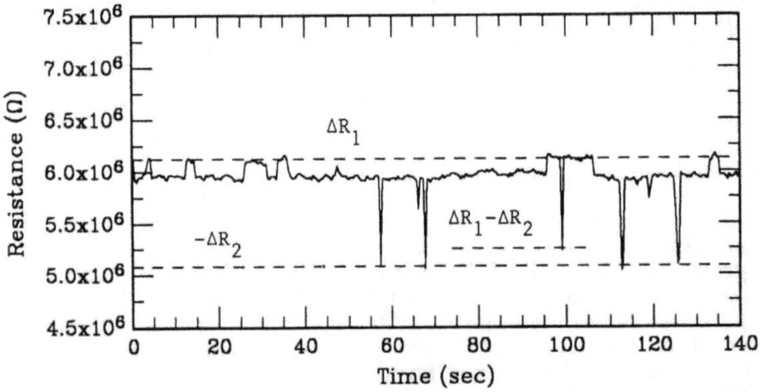

Fig. 5 : Resistance as a function of time for a single heterostructure quantum wire. The effects of two independent traps (ΔR_1 and ΔR_2) and simultaneous switching due to both (ΔR_1-ΔR_2) is observed.

the sum of the impedances (i.e., at $\tau \sim 100$ sec, $\Delta R = \Delta R_1 - \Delta R_2$). These two traps clearly have different activation energies, and appear to be separated sufficiently well in real space to eliminate interaction effects.

The current-voltage characteristics of one of these structures at 100°K is shown in Figure 6. The discrete switching between two impedance states is observed at a number of different bias positions (0.6V, 0.75V-0.85V, 1.0V-1.1V, 1.25V-1.35V). These fluctuations are superimposed onto the wellknown negative differential resistance characteristic of the double barrier/single quantum well heterostructure (due to the quantum well state passing through the conduction band edge of the contact). The overall structure of the characteristics were the same upon repeated bias sweeps, though the detailed structure in the fluctuation regions were not; in particular, the characteristics in the non-fluctuating regions (e.g., 0.65V-0.75V, 0.9V-0.95V, etc.) were exactly repeatable, but the fluctuating regions were not repeatable in detail.

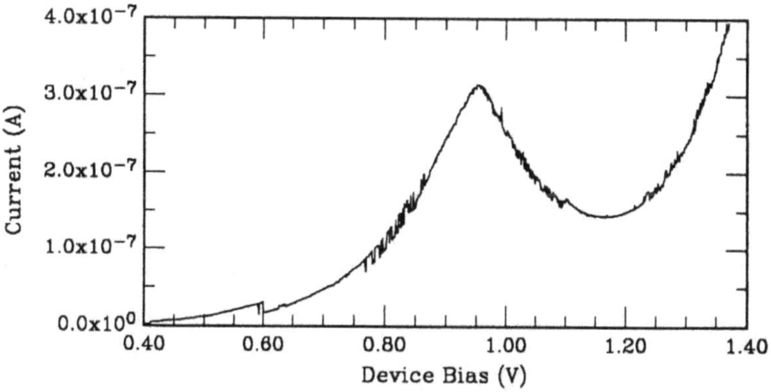

Fig. 6 : Current-voltage characteristic of a single heterostructure quantum wire at 100°K. The physical diameter of the wire is approximately 2500Å.

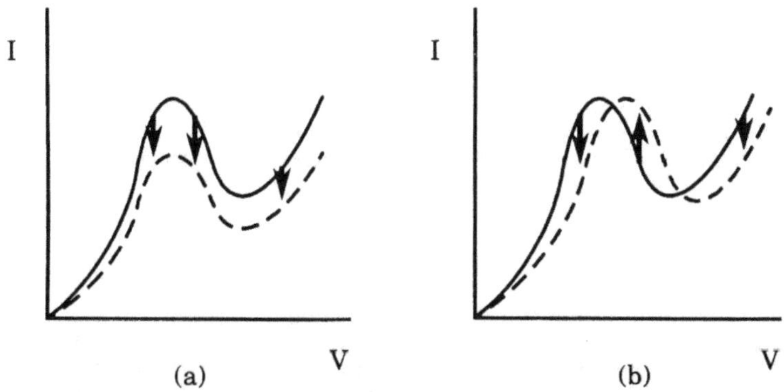

Fig. 7 : Schematic current-voltage characteristics illustrating the sense of switching due to (a) scattering, and (b) potential fluctuations. In (a), the shift to the dashed curve indicates trapping onto a neutral impurity, and in (b) the shift to the dashed curve indicates trapping onto either a neutral or charged impurity. The arrows indicate the asymptotic sense of switching.

The negative differential resistance characteristic of the heterostructure quantum wire enables us to determine which mechanism (scattering or potential fluctuation) is causing the observed switching phenomena. Let us first consider the case of scattering. If the defect in the wire is a positively charged defect, capture of an electron converts it to a neutral, thus causing a decrease in the wire impedance. Thus, as the trap is biased into occupancy, the current state will switch from a low to asymptotically high current state. Let us define this sense of asymptotic current value "switching high". Similarly, a neutral trap capturing an electron to become a negatively charged trap will cause a decrease in the current, which asymptotically "switches low" with occupancy. The situation of a neutral going into a negative is schematically shown in Figure 7(a). The scattering mechanism predicts that the sense of switching ("high" or "low") will be the same no matter where in bias position the trap becomes occupied. Thus, the sense of switching will be the same on either side of the resonant peak position; i.e., "symmetric".

The second case to consider is a fluctuation in the local potential of the double barrier structure. The change in the charge state of a defect located in the accumulation, double barrier, or depletion regions of the structure can modulate the potential distribution across the double barrier structure. This shift in potential will move the position of the resonant tunneling state with respect to the conduction band edge of the injection contact, thus changing the tunneling current through the structure. This can be thought of as an effective shift in the local current-voltage characteristics. Capture of an electron by a positively charged defect to become neutral, or of a neutral defect to become a negative, causes an effective shift to higher biases of the local current-voltage characteristics. In this case, the asymptotic sense of current direction depends on where in bias position the trap becomes occupied. This is schematically illustrated in Figure 7(b). At device biases below the resonant peak position, the asymptotic sense is "switching low" as the trap becomes occupied; in the negative differential resistance region, the sense is "switching high"; and at large biases, the sense will eventually be "switching low". The sense of switching is not the same on either side of the resonant peak position; i.e., this is an "asymmetric" case.

As can be seen in Figure 6, this device exhibits asymmetry in the sense of switching. In the region 0.75V-0.85V, the current switches "low" as the trap becomes occupied. In the region 1.0V-1.1V, the current switches "high", and at very high bias (1.3V-1.35V) the state switches "low". Note that the structure in the range 1.2V-1.275V appears to first switch high, then low. This is consistent with the intersection of the shifted current-voltage characteristics occurring at approximately 1.26V.

3. Transport in quantum dots

Carrier confinement to reduced dimensions in a semiconductor quantum well was first demonstrated in GaAs - AlGaAs by electronic [1] and optical [14] spectroscopy in 1974. This achievement has led to numerous important developments in basic semiconductor physics and device technology. Structures produced by ultra-thin film growth are inherently two-dimensional, and thus investigations have been largely confined to heterostructures where only the carrier momentum normal to the interfaces is quantized. It is expected that the realization of semiconductor heterostructures with quantum confinement to zero dimensions ("quantum dots") will yield physically intriguing phenomena. Attempts to observe such confinement optically have been reported recently [15-17], but the spectra do not show the characteristic structure of a series of isolated peaks expected from a zero-dimensional electron-hole gas. We have therefore studied such structures by electronic transport, and here present evidence for electronic transport through a discrete spectrum of states in a nanostructure confined in all three spatial dimensions.

The approach used to produce quantum dot nanostructures suitable for electronic transport studies was to laterally confine resonant tunneling heterostructures, as above, at a lateral dimension that will quantize lateral momenta. The initial MBE epitaxial structure is a 0.5 μm n^+ GaAs contact (Si-doped at 2×10^{18} cm^{-3}, graded to approximately 10^{16} cm^{-3} over 200Å, followed by a 100Å undoped GaAs spacer layer), a 40Å $Al_{0.25}Ga_{0.75}As$ tunnel barrier, and a 50Å undoped $In_{.08}Ga_{.92}As$ quantum well. The structure was grown to be nominally symmetric about a plane through the center of the quantum well. Large area ($\geqq$2μm square) mesas of a typical structure fabricated by conventional means exhibited two resonant peaks; a ground state at 50 mV with a peak current density of 30 A/cm^2, and an excited state at 700 mV with a peak current density of 8.1×10^3 A/cm^2, both measured at 77°K. Nanofabrication was similar to that described above, with lateral dimensions now at approximately 1000Å.

Figure 8 schematically illustrates the lateral (radial) potential of a column containing a quantum dot, and the spectrum of 3-dimensionally confined electron states under zero and applied bias. A spectrum of discrete states will give rise to a series of resonances in transmitted current as each state drops below the conduction band edge of the injection contact. To observe lateral quantization of quantum well state(s), the physical size of the structure must be sufficiently small that quantization of the lateral momenta produces energy splittings > kT. Concurrently, the lateral dimensions of the structure must be large enough such that pinch-off of the column by the depletion layers formed on the sidewalls of the GaAs column does not occur. Due to the Fermi level pinning of the exposed GaAs surface, the conduction band bends upward (with respect to the Fermi level), and where it intersects the Fermi level determines in real space the edge of the central conduction path core. The radial potential $\phi(r)$ in the column will be parabolic for $(R-W) \leq r \leq R$), where r is the radial coordinate, R the physical radius of the column, W is the depletion depth, and ϕ_T is the height of the potential determined by the Fermi level pinning. When the lateral dimension is reduced to 2W or less, the lateral potential becomes parabolic though conduction through the central conduction path core is pinched off.

Figure 9 shows the current-voltage characteristics of one of these single microstructures as a function of temperature. Assuming that the current density through the structure is approximately the same as a large area device, measurement of the peak resonant current implies a minimum (circular) conduction path core of 130Å for this structure; thus, a lateral parabolic potential approximation is valid. This implies a depletion depth of ~430Å at the double

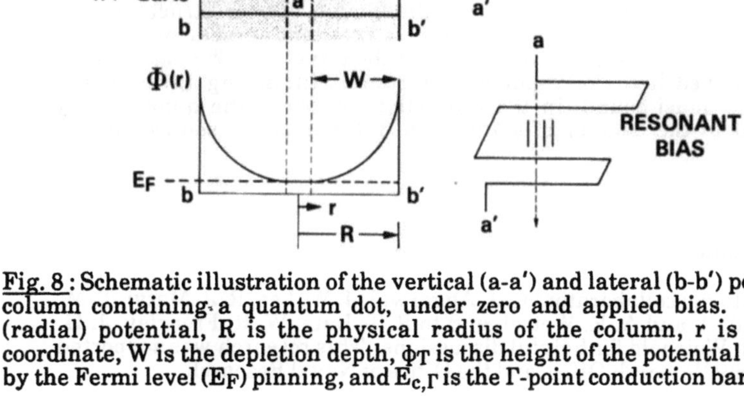

Fig. 8: Schematic illustration of the vertical (a-a') and lateral (b-b') potential of a column containing a quantum dot, under zero and applied bias. $\phi(r)$ is the (radial) potential, R is the physical radius of the column, r is the radial coordinate, W is the depletion depth, ϕ_T is the height of the potential determined by the Fermi level (E_F) pinning, and $E_{c,\Gamma}$ is the Γ-point conduction band energy.

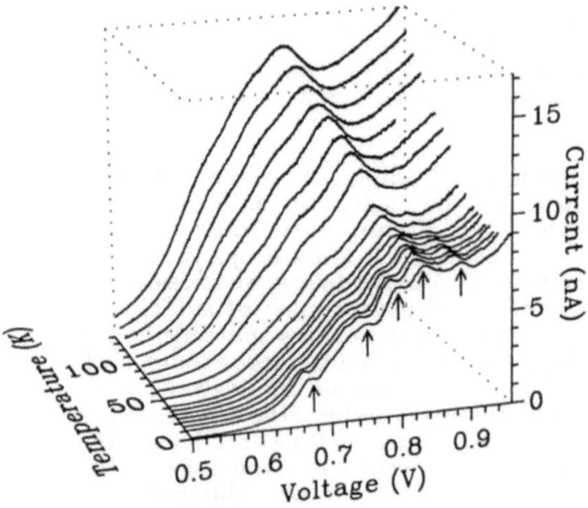

Fig. 9 : Current-voltage characteristics of a single quantum dot nanostructure as a function of temperature, showing resonant tunneling through the discrete states of the n=2 quantum well resonance. The arrows indicate the voltage positions of the discrete states for the T = 1.0 K curve.

barrier structure, in reasonable agreement with that expected from the known doping level (at $2x10^{18}$ cm^{-3}, W=220Å) and realizing that W will enlarge in the undoped double barrier region. The splitting of the discrete electron levels in the quantum dot should be split evenly by 26 meV.

Only the excited state resonance of this structure is observed since the current expected from the ground state resonance is at the detection limits of the test apparatus. At high temperature, the characteristic negative differential resistance of the double barrier structure is evident. As the temperature is lowered, two effects occur. First, the resonant peak shifts slightly higher in voltage (due to the increase of the GaAs contact resistance with temperature) and decreases in current (due to the freezing out of excess leakage current). Secondly, there appears a series of peaks superimposed on the negative differential resistance peak. In the range 0.75V-0.9V the peaks are approximately equally spaced, with a splitting of ~50 mV. Assuming that most of the bias is incrementally dropped across the double barrier structure, the splitting of the equally spaced series is 25 meV, in excellent agreement with the value determined from the physical dimension of the structure. We believe that the structure observed here corresponds to resonant tunneling through the spectra of discrete quasi-bound (in the z-direction) states in the quantum dot which correspond to the density of states of a 3-dimensional semiconductor quantum well.

4. Acknowledgements

I am indebted to my collaborators R. J. Aggarwal, J. W. Lee, R. J. Matyi, T. M. Moore, J. N. Randall, and A. E. Wetsel, and to R. T. Bate, W. R. Frensley, J. H. Luscombe, and M. Weichold for helpful discussions. This research was supported in part by the Army Research Office and the Office of Naval Research.

References

1. L. L. Chang, L. Esaki, and R. Tsu : Appl. Phys. Lett. 24, 593 (1974).
2. T.C.L.G. Sollner, W. D. Goodhue, P. E. Tannenwald, C. D. Parker, and D. D. Peck : Appl. Phys. Lett. 43, 588 (1983).
3. C. I. Huang, M. J. Paulus, C. A. Bozada, S. C. Dudley, K. R. Evans, C. E. Stutz, R. L. Jones, and M. E. Cheney : Appl. Phys. Lett. 51, 121 (1987).
4. E. E. Mendez, W. I. Wang, B. Ricco, and L. Esaki : Appl. Phys. Lett. 47, 415 (1985).
5. F. Capasso, K. Mohammed, and A. Y. Cho : Appl. Phys. Lett. 48, 478 (1986).
6. M. A. Reed, J. W. Lee, and H-L. Tsai : Appl. Phys. Lett. 49, 158 (1986).
7. T. Nakagawa, T. Fujita, Y. Matsumoto, T. Kojima, and K. Ohta : Appl. Phys. Lett. 51, 445 (1987).
8. M. A. Reed and J. W. Lee : Appl. Phys. Lett. 50, 845 (1987).
9. W. J. Skocpol, L. D. Jackel, E. L. Hu, R. E. Howard and L. A. Fetter : Phys. Rev. Lett. 49, 951 (1982).
10. K. S. Ralls, W. J. Skocpol, L. D. Jackel, R. E. Howard, L. A. Fetter, R. W. Epworth and D. M. Tennant : Phys. Rev. Lett. 52, 228 (1984).
11. C. T. Rogers and R. A. Buhrman : Phys. Rev. Lett. 55, 859, (1985).
12. K. R. Farmer, C. T. Rogers, and R. A. Buhrman : Phys. Rev. Lett. 58, 225 (1987).
13. J. N. Randall, M. A. Reed, T. M. Moore, R. J. Matyi, and J. W. Lee : J. Vac. Sci. Technol., to be published.
14. R. Dingle, A. C. Gossard, and W. Wiegmann : Phys. Rev. Lett. 33, 827 (1974).
15. M. A. Reed, R. T. Bate, K. Bradshaw, W. M. Duncan, W. R. Frensley, J. W. Lee, and H.-D. Shih : J. Vac. Sci Technol. B4, 358 (1986).
16. K. Kash, A. Scherer, J. M. Worlock, H. G. Craighead, and M. C. Tamargo : Appl. Phys. Lett. 49, 1043 (1986).
17. J. Cibert, P. M. Petroff, G. J. Dolan, S. J. Pearton, A. C. Gossard, and J. H. English : Appl. Phys. Lett. 49, 1275 (1986).

Magnetic Field Studies of Resonant and Non-resonant Tunnelling in n-(AlGa)As/GaAs Double Barrier Structures

L. Eaves[1], *E.S. Alves*[1], *T.J. Foster*[1], *M. Henini*[1], *O.H. Hughes*[1], *M.L. Leadbeater*[1], *F.W. Sheard*[1], *G.A. Toombs*[1], *K. Chan*[1], *A. Celeste*[2], *J.C. Portal*[2], *G. Hill*[3], *and M.A. Pate*[3]

[1]Department of Physics, University of Nottingham, Nottingham NG7 2RD, UK

[2]Department de Génie Physique, Institut National des Sciences Appliquées, F-31077 Toulouse, France, and Service National des Champs Intenses, Centre National de la Recherche Scientifique, F-38042 Grenoble, France

[3]Department of Electronic and Electrical Engineering, University of Sheffield, Mappin Street, Sheffield S1 3JD, UK

Abstract. The electrical properties of a series of double barrier tunnelling devices with well widths between 5 and 60 nm are investigated. It is shown that the bistability effect in the current-voltage characteristics of a typical resonant tunnelling device can be removed by connecting a suitable capacitance or resistance to the device. These measurements cast serious doubt on the recent interpretation of the bistability as an intrinsic space-charge effect. In the stabilised section of the I(V) curve, at voltages above the main resonant peak, the magnetoquantum oscillations observed with $\underline{B}||\underline{J}$ are used to investigate tunnelling assisted by LO phonon emission and by elastic scattering processes. The resonant tunnelling device with well width 60 nm exhibits sixteen regions of negative differential conductivity. The effect of a transverse magnetic field $\underline{J}\perp\underline{B}$ on the resonances in the I(V) characteristics is investigated. At sufficiently high magnetic field, a transition from tunnelling into hybrid magneto-electric quantised states to tunnelling into magnetically quantised cycloidal skipping states is observed.

1. Introduction

The Resonant Tunnelling Structure has attracted considerable interest recently as a promising high speed device and as a system whose electrical properties are controlled by a fundamental quantum mechanical process. In this article, we report a series of investigations of the electrical properties of these devices, using high magnetic fields. The structures, based on n-(AlGa)As/GaAs were grown at Nottingham by molecular beam epitaxy using a Varian Gen II reactor. Devices exhibiting negative differential conductivity with peak/valley ratio of up to 3.6/1 at room temperature and 25/1 at liquid helium temperatures have been fabricated. We describe three topics of current interest: (1) the question of whether

Springer Series in Solid-State Sciences Vol. 83: **Physics and Technology of Submicron Structures**
Editors: H. Heinrich · G. Bauer · F. Kuchar

or not these devices exhibit intrinsic bistability associated with charge build-up in the well; (2) the contribution of elastic and inelastic scattering processes to the tunnel current as revealed by analysis of magnetoquantum oscillations in the current-voltage characteristics for $\underline{J} \parallel \underline{B}$; (3) the transition from tunnelling into electrically quantised states to tunnelling into magnetically quantised interface states in the presence of a large quantising magnetic field $\underline{J} \perp \underline{B}$.

2. Has Intrinsic Bistability been Observed?

Recently, Goldman et al.[1] reported the observation of bistability in the current-voltage characteristics I(V) of double barrier resonant tunnelling semiconductor heterostructures. This bistability occurs in the region of negative differential conductivity (NDC). It was interpreted as an intrinsic feature arising from the electrostatic effects due to the build-up of negative space-charge in the quantum well between the two barriers.

This conclusion was challenged by Sollner [2] who claimed that the observed bistability in I(V) was a common characteristic of devices exhibiting NDC and was due to current oscillations in the device and in the external circuit. In this case, bistability arises because of the difference between the turn-on and turn-off points for oscillatory behaviour when sweeping the applied DC voltage up or down through the NDC region. The range of applied DC voltage over which the bistability occurs then depends on the external circuit parameters as well as on the intrinsic characteristics of the device itself.

Recently, we investigated the magnetic field dependence of the current-voltage characteristics [3,4] of resonant tunnelling double barrier structures based on n-(AlGa)As/GaAs. Our measurements confirmed the build-up of charge in the well at the resonant tunnelling voltages. In the region of NDC, the devices exhibited a current bistability similar to that originally reported by Goldman et al. [1]. We also reported [4] that the form of the I(V) curves in the bistable region can be simulated quite well by numerical calculation of the current oscillations in an equivalent circuit consisting of a device exhibiting NDC in parallel with a capacitor and connected by resistive and inductive leads to a steady voltage source. This is consistent with Sollner's point of view. Sollner also pointed out that if the magnitude of the negative differential resistance $|dV/dI|$ is sufficiently large, the current oscillations should be suppressed by a capacitor placed in parallel with the device. In order to study this, we have made further measurements on double-barrier devices which have a higher impedance than those used by Goldman et al. [1].

The measured device was a 100 μm-diameter mesa consisting of the following layers (structure A), in order of growth from the n^+GaAs substrate: (i) 1 μm of GaAs, $n = 2 \times 10^{17}$ cm^{-3} buffer layer; (ii) 8.8 nm of undoped (AlGa)As, [Al] = 0.33; (iii) 5.6 nm of undoped GaAs; (iv) 8.8 nm of undoped (AlGa)As, [Al] = 0.33; (v) 1 μm of 2×10^{17} cm^{-3} GaAs, top contact.

The I(V) characteristics shown in Fig. 1 were measured at 77 K with the mesa mounted on a standard transistor header. The upper curve of Fig. 1

was obtained with no external capacitance across the device and a region of hysteresis is clearly observable. An oscilloscope connected across the device indicated the presence of oscillations in the NDC region. The oscillations were not suppressed and the bistability persisted if a capacitor was connected across the leads emerging from the cryostat. When, however, a 0.28 μF chip capacitor was mounted directly on the transistor header so as to minimise the length of the inductive connecting leads, then the smooth I(V) curve shown in the lower curve of Fig. 1 was obtained. In this case, there was no bistability and no oscillations were observed throughout the NDC region.

An alternative method of suppressing the oscillations in the region of NDC is to connect, in parallel with the device, a small chip resistor r (25 Ω) of sufficiently low impedance that $r < |dV/dI|$ throughout the region of NDC. It is necessary to mount the resistor in close proximity with the device with very short connecting leads. A stable I(V) plot of the device and resistor in parallel can then be obtained by applying a ramped voltage V. Even in the region of NDC, the circuit does not break into oscillation. The stable I(V) characteristics of the tunnelling device alone are calculated by subtracting, from the total current, the

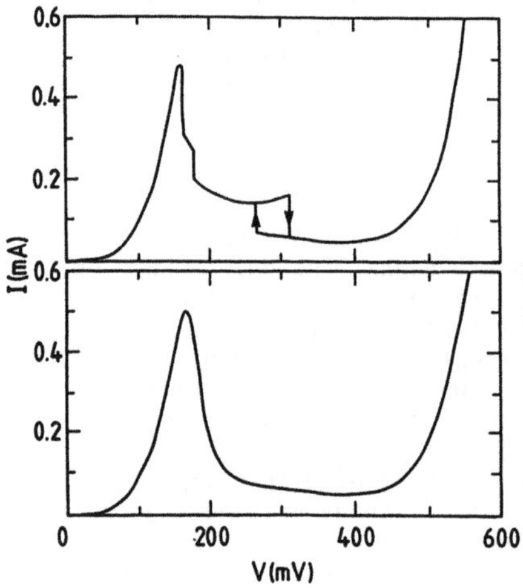

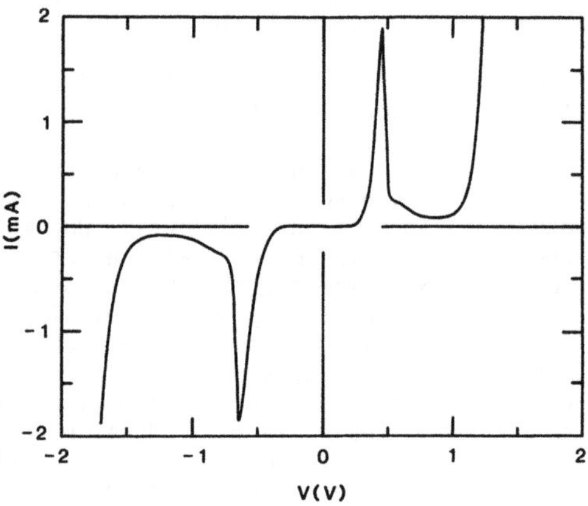

Fig. 1 (left)

The current-voltage characteristics at 77 K of a 100 μm mesa fabricated from the double barrier structure A (5.6 nm well) described in the text. The upper curve shows the d.c. current with no external capacitor across the mesa. In the voltage range corresponding to current bistability, the device is oscillating at a high frequency. In the lower curve, the oscillations and bistability are completely suppressed by placing a 0.28 μF capacitor directly across the mesa.

Fig. 2 (right)

The current-voltage characteristics at 77 K of a 5 μm diameter mesa fabricated from the double barrier structure B (5 nm well) described in the text. The oscillation and circuit bistability are suppressed by connecting a small resistor (25 Ω) in parallel with the device.

current V/r flowing through the parallel resistor. Typical results obtained at 77 K with this simple procedure are shown in Figure 2 for a 5 μm diameter mesa device (structure B) comprising the following layers, in order of growth from the n^+GaAs substrate:

(i) 1.0 μm of $n = 2 \times 10^{18}$ cm^{-3} GaAs, (ii) 50 nm of $n = 2 \times 10^{16}$ cm^{-3} GaAs, (iii) 2.5 nm of undoped GaAs, (iv) 5.6 nm of undoped (AlGa)As, [Al] = 0.4, (v) 5 nm of undoped GaAs, (vi) 5.6 nm of undoped (AlGa)As, [Al] = 0.4, (vii) 2.5 nm of undoped GaAs, (viii) 50 nm of $n = 2 \times 10^{16}$ cm^{-3} GaAs and (ix) 0.5 μm of $n = 2 \times 10^{18}$ cm^{-3} GaAs.

The purpose of the undoped GaAs spacer layers on either side of the barriers was to prevent dopant diffusion into the barrier region and so to reduce ionised impurity scattering which, as discussed later, degrades the peak/valley ratio.

We conclude that our measurements support the objections raised by Sollner concerning the reported observation by Goldman et al. of intrinsic bistability in double barrier resonant tunnelling devices. However, the magneto-oscillation studies [3,4,5] show clearly the existence of space-charge build-up in the quantum well. Thus intrinsic bistability driven by electrostatic feedback remains a theoretical possibility. Theoretical modelling [6] shows that the intrinsic bistability is removed by sufficient inhomogeneous broadening of the bound state level. A carefully designed structure would therefore be required to observe the intrinsic bistability effect.

3. Elastic and Inelastic Scattering and the Valley Current

Either of the above methods of stabilising the devices against oscillations and circuit bistability allow us to investigate in detail the dynamics of tunnelling electrons in the voltage region beyond the main resonant peak in the current. As can be seen from Figures 1 and 2, there is a weak shoulder in the I(V) characteristics of both structures A and B in the voltage region beyond the main peak where the current is a minimum. Goldman et al. [7] have attributed this feature to tunnelling of electrons, assisted by LO phonon emission, from the emitter contact into the quantum well. The emission of the LO phonon breaks the law governing conservation of k-vector component, $k_\perp$, perpendicular to the direction of the tunnel current. $k_\perp$ is generally assumed to be conserved for the majority of tunnelling electrons contributing to the main resonant peak of the I(V) curve. However inter-subband scattering due to LO-phonon emission has been identified in resonant tunnelling structures [8] from an analysis of magnetoquantum oscillations. The LO phonon-related peak and the contribution to the valley current due to elastic scattering can be revealed more clearly by applying a quantising magnetic field B parallel to the direction of the tunnel current. In the absence of a magnetic field, $k_\perp$ is conserved if an electron does not scatter during its transit of the barriers. In the presence of a magnetic field, the energy due to motion in the plane of the barriers is quantised such that the total electron energy in the accumulation layer of the emitter and in the quantum well are given by

$$\text{Accumulation Layer} \quad \varepsilon = \varepsilon_0 + (n + \tfrac{1}{2})\hbar\omega_c$$
$$\text{Quantum Well} \quad \varepsilon = \varepsilon_1 + (n' + \tfrac{1}{2})\hbar\omega_c ,$$

where n(n') is the Landau level number, $\omega_c = eB/m^*$, and ε_0 and ε_1 are lowest bound state energies in the emitter and well respectively. The conservation of k for B = 0 corresponds to the requirement that $p = n' - n = 0$ in the presence of a quantising magnetic field, so that in both cases resonant tunnelling occurs at an applied voltage for which $\varepsilon_0 = \varepsilon_1$.

Figure 3 shows the effect of a magnetic field ($\underline{B}||\underline{J}$) on the I(V) characteristics of a 100 µm diameter double barrier device with a well width of 117 Å (Structure C). At zero magnetic field the oscillations in the NDC region were suppressed by connecting a 25 Ω chip resistor across the device. This structure comprises the following layers in order of growth from the n^+ substrate:
(i) 2 µm, 2 x 10^{18} cm^{-3}, n^+GaAs buffer layer; (ii) 50 nm, 2 x 10^{16} cm^{-3} GaAs; (iii) 2.5 nm undoped GaAs; (iv) 5.6 nm undoped (AlGa)As, [Al] = 0.4; (v) 11.7 nm undoped GaAs; (vi) 5.6 nm undoped (AlGa)As, [Al] = 0.4; (vii) 2.5 nm undoped GaAs; (viii) 50 nm, 2 x 10^{16} cm^{-3} GaAs; (ix) 0.5 µm, 2 x 10^{18} cm^{-3} n^+GaAs top contact layer.

Figure 3 shows only the low voltage section of the I(V) characteristics. Further peaks in the current are observed at 420 and 890 mV, corresponding to tunnelling into the second and third bound states of the well. The following features of Figure 3 are particularly noteworthy. Firstly, the magnetic field increases the amplitude of the main LO phonon related peak. Secondly, a weak secondary peak (E_1) emerges from the main resonant peak in the I(V) curve as the magnetic field is increased. Similar subsidiary peaks also evolve from the LO phonon feature with increasing B. Thirdly, the magnetic field increases the peak/valley ratio. The fan chart in Figure 4 shows the evolution of the magnetoquantum peaks in the I(V) characteristics. Transitions for which $k_\perp$ (or n) is not conserved are governed only by energy conservation. Therefore, for elastic and LO phonon emission we have

$$\varepsilon_0 = \varepsilon_1 + \frac{p\hbar eB}{m^*} (+\hbar\omega_L) .$$

To a good approximation $\varepsilon_0 - \varepsilon_1$ increases linearly with total applied voltage from its initial (negative) value at zero bias. Hence the chart shows two clearly identifiable groups of lines. In the limit $B \rightarrow 0$, peak E_1 extrapolates back to the main resonant peak at 70 mV. The peaks LO_p extrapolate back to the LO phonon satellite at 160 meV. The peak marked E_1 corresponds to non-resonant tunnelling involving an elastic (or quasi-elastic) scattering-induced transition from the n = 0 Landau level in the emitter contact to the n = 1 Landau level in the well. Such a transition could be induced by ionised impurities or interface roughness (elastic) or by acoustic phonon emission (quasi-elastic). The main resonant peak, which shifts only slightly with B, corresponds to tunnelling processes for which n is conserved. The peaks marked LO_p correspond to transitions of electrons from the n = 0 Landau level in the emitter to the pth Landau level in the well, together with the emission of an LO phonon.

It is interesting to note that the effect of the quantising magnetic field is to suppress the (quasi-) elastic scattering induced transitions at certain voltages (below and above E_1) and enhance it at other voltages (at

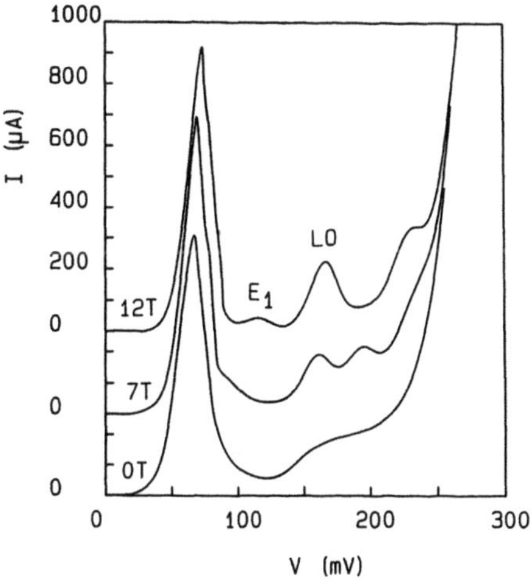

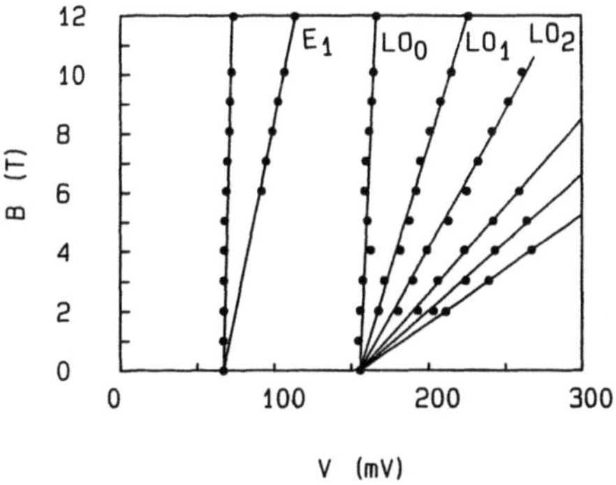

Fig. 3 (left)

The current-voltage characteristics at 4 K and at various magnetic fields B||J of a 100 μm diameter mesa fabricated from structure C (11.7 nm well) described in the text. The curves show only the region of the first resonance. The oscillations and current bistability are suppressed by a small resistor (25 Ω) in parallel with the device.

Fig. 4 (right)

Fan chart showing the magnetic field dependence (B||J) of the peaks in I(V) shown in Figure 3. The elastic and inelastic (LO phonon) scattering processes giving rise to the oscillations are discussed in the text.

E_1). In zero magnetic field such processes, which change $k_\perp$, are energetically allowed for all voltages beyond the main resonant peak in the tunnel current. At these voltages the lowest energy bound state in the well is below the energy of the electrons in the emitter. However, a large magnetic field quantises the electronic motion in the plane of the barriers, thus giving rise to sharp peaks in the densities of states. At certain voltages beyond the main resonant peak, and in the presence of a quantising magnetic field, energy conservation inhibits (quasi-)elastic scattering into the well. This explains the enhancement of the peak/valley ratio with increasing B which is evident in Figure 3.

The difference in voltage between the main resonant peak and the LO phonon satellite and between the various peaks LO in the presence of a magnetic field, are considerably larger than the LO phonon energy, $\hbar\omega_L/e$ (= 36 meV) and the cyclotron energy, $\hbar\omega_c$, respectively. This is fully consistent with the expected distribution of potential across the device: in the voltage range between ~100 and 300 mV only about 30% of the total applied voltage is dropped across the accumulation layer of the emitter, the emitter barrier and the half-well width. The remainder is dropped across the other half of the well, the collector barrier and the depletion layer of the collector contact.

We conclude this section by noting that the magnetoquantum peaks in the region of the valley current that were recently reported by Goldman et al. [7] and tentatively attributed to elastic ionised impurity scattering transitions with non-conservation of $k_{\perp}$, are more likely to be due to LO-phonon emission (i.e. the LO_n series of magnetoquantum peaks reported here).

4. Electrically and Magnetically Quantised States in Wide Wells

In this section we investigate resonant tunnelling in a double barrier structure whose well width, 60 nm, is considerably larger than those described in the previous sections. This structure (D) comprises the following layers in order of growth from the n^+GaAs substrate:
(i) 2 μm, 2 x 10^{18} cm^{-3} n^+GaAs buffer layer; (ii) 50 nm, 2 x 10^{16} cm^{-3} GaAs; (iii) 2.5 nm undoped GaAs; (iv) 5.6 nm undoped (AlGa)As, [Al] = 0.4; (v) 60 nm undoped GaAs; (vi) 5.6 nm undoped (AlGa)As, [Al] = 0.4; (vii) 2.5 nm undoped GaAs; (viii) 50 nm, 2 x 10^{16} cm^{-3} GaAs; (ix) 0.5 μm, 2 x 10^{18} cm^{-3} n^+GaAs top contact.

Figure 5 shows the I(V) characteristics of this structure obtained at 4 K. Note that the current is plotted logarithmically over 5 orders of magnitude. Sixteen distinct regions of negative differential conductivity are observed up to 1.0 V and several additional features are clearly observable at higher voltages. These resonances are associated with the large number of closely-spaced bound states that occur in a well of this width. The higher voltage peaks are probably associated with "over the barrier" resonances. The existence of well-defined "standing wave" resonances implies that some electrons traverse a distance of 120 nm (twice the well width) without appreciable scattering even when they are injected into the well with kinetic energies of several hundred meV.

Structures of this kind are ideal for studying the effect of a confining potential (quantum well) on the electron eigenstates in the crossed electric and magnetic field ($\underline{E}\perp\underline{B}$) configuration. The electric field in the well is perpendicular to the barriers so the magnetic field must be applied parallel to the barrier interfaces. The effect of a magnetic field on the current-voltage characteristics of this configuration ($\underline{B}\perp\underline{J}$) is shown in Figure 6. In order to enhance the resonant structure, the second derivative d^2I/dV^2 versus V is plotted for various values of magnetic field. In high fields it can be seen that the resonances form two groups, which are shown separated by an arrow in the plot for B = 7 T in Figure 6.

At low magnetic field the electron orbits impinge on the barrier interfaces on both sides of the well. But at sufficiently high fields, skipping states develop in which the electron is repeatedly reflected from only one interface. However, we note that the previously observed skipping states in single-barrier heterostructures [9] were in a heavily-doped n^+ region in which the electric field was essentially zero. In the present case there is a large electric field in the quantum well. The skipping motion is then cycloidal with a uniform drift velocity $v_d = E/B$ parallel to the interface, superposed on the circular motion of angular velocity $\omega_c = eB/m^*$ in the plane perpendicular to $\underline{B}$.

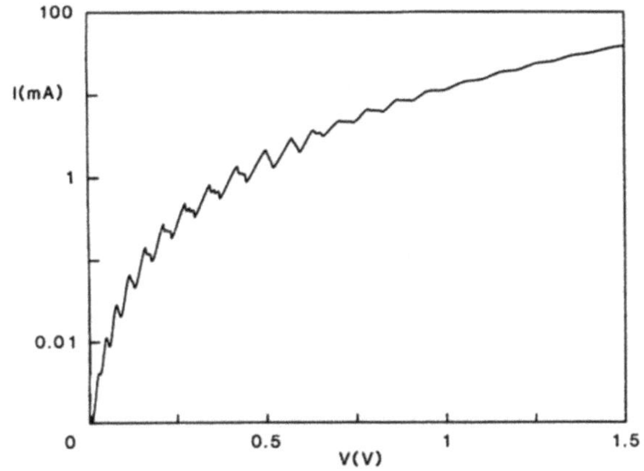

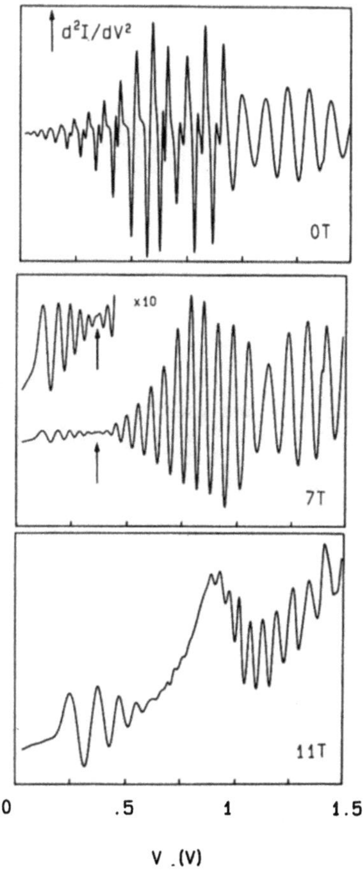

Fig. 5 (above)

The current-voltage characteristic at 4 K (I is plotted logarithmically) of a 100 μm diameter mesa fabricated from the double barrier structure D (60 nm well) described in the text.

Fig. 6 (right)

Plot of d^2I/dV^2 versus V for structure D (60 nm well) at 4 K for various magnetic fields, B⊥J. For B $\gtrsim$ 2 T, a series of oscillations due to tunnelling into magnetically quantised interface states can be observed at low voltages. For the trace at 7 T, the critical voltage V_c for the disappearance of the skipping orbits is indicated by an arrow.

We assume that the electrons tunnelling through the emitter barrier emerge into the well with a kinetic energy which is small compared with the kinetic energy gained during the subsequent cycloidal motion. The circular component of the cycloidal motion then has a radius $R = v_d/\omega_c$ and the orbit centre is located the same distance R from the emitter-barrier interface. The width of such an orbit along the direction of the electric field is $2R = 2v_d/\omega_c = 2m^*E/eB^2$. The critical case, when the cycloidal skipping orbit adjacent to the emitter-barrier interface just touches the collector barrier, is given by $2R = w$, where w is the well width. We may put $E = V/d$, where $d = w + 2b + \lambda_L + \lambda_R$ and λ_L, λ_R are screening layer widths in the emitter and collector electrodes. Thus tunnelling into cycloidal skipping states only occurs for $V < V_c$, where the critical voltage $V_c = eB^2wd/2m^*$. This B^2 dependence of V_c is observed experimentally. For B = 7 T and taking $\lambda_L = \lambda_R \simeq 10$ nm (d = 91 nm, $m^* = 0.067m_e$) we find $V_c \simeq 350$ mV. This value is shown by an arrow in Figure 6 and clearly lies at the division between the group of resonances due to single-interface skipping states ($V < V_c$) and the group due to states in which the electron motion is restricted by reflection at the interfaces at both sides of the well ($V > V_c$).

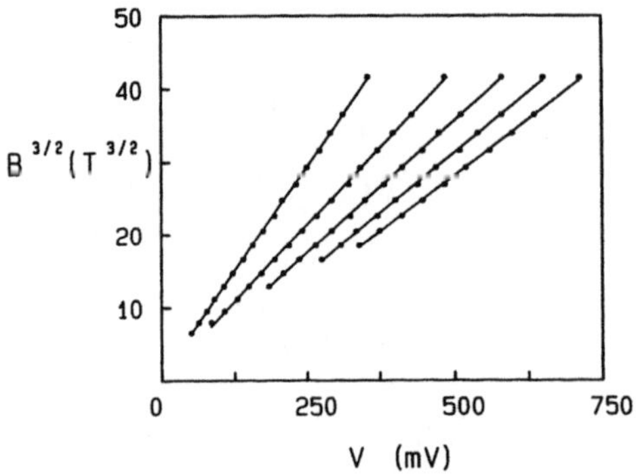

Fig. 7

Plot of $B^{3/2}$ versus V for the cycloidal skipping state resonances corresponding to n = 0,1,2,3,4 (left to right).

We have analysed the positions of the resonances (minima in d^2I/dV^2 versus V) for the cycloidal skipping states over a wide range of magnetic fields and voltages. When plotted as $B^{3/2}$ versus V, we obtain the linear dependences shown in the fan chart of Figure 7. These results may be understood in terms of the quantisation of the cycloidal orbits discussed above. When quantised, the energy of circular motion $\frac{1}{2}m^*\omega_c^2R^2$ takes allowed values $(n + \varphi)\hbar\omega_c$, where the phase factor $\varphi = 3/4$ in the WKB approximation. Putting $R = v_d/\omega_c$ gives $\frac{1}{2}m^*v_d^2 = (n + \varphi)\hbar\omega_c$, which can be written

$$\frac{m^{*2}E^2}{2e\hbar B^3} = n + \varphi. \quad (1)$$

This shows that for a given applied voltage (fixed E) the resonances are periodic in $1/B^3$. This is in contrast to the results obtained for skipping states in n^+ material in single-barrier structures [9] where the resonances were periodic in 1/B. (1) is consistent with the observed linear variation of $B^{3/2}$ with V given in the fan chart of Figure 7. Also, using the WKB phase factor, the slopes of the fan-chart lines calculated from (1) agree with experiment to within ~15%. This is satisfactory bearing in mind the uncertainty in the distance d over which the applied voltage is dropped.

Acknowledgements

This work is supported by SERC and CNRS. One of us (ESA) is supported by CNPq, Brazil. We acknowledge useful discussions with T. C. L. G. Sollner, J. J. Harris and D. Lacklison.

References

1. V. J. Goldman, D. C. Tsui and J. E. Cunningham, Phys. Rev. Lett. 58, 1256 (1987); ibid. 59, 1623 (1987).
2. T. C. L. G. Sollner, Phys. Rev. Lett. 59, 1622 (1987).
3. C. A. Payling, E. S. Alves, L. Eaves, T. J. Foster, M. Henini, O. H. Hughes, P. E. Simmonds, F. W. Sheard and G. A. Toombs, Proc. 7th Int. Conf. on the Electronic Properties of Two Dimensional Systems, Santa Fe, 1987. To be published in Surf. Sci. (1988). See also C. A. Payling, E. Alves, L. Eaves, T. J. Foster, M. Henini, O. H. Hughes, P. E. Simmonds, J. C. Portal, G. Hill and M. A. Pate, Proc. 3rd Int. Conf. on Modulated Semiconductor Structures, Montpellier, France. J. Physique C5, 289 (1987).

4. G. A. Toombs, E. S. Alves, L. Eaves, T. J. Foster, M. Henini, O. H. Hughes, M. L. Leadbeater, C. A. Payling, F. W. Sheard, P. A. Claxton, G. Hill, M. A. Pate and J. C. Portal, 14th Int. Symposium on Gallium Arsenide and Related Compounds, Crete, 1987. To be published by Institute of Physics.
5. V. J. Goldman, D. C. Tsui and J. E. Cunningham, Phys. Rev. B35, 9387 (1987).
6. F. W. Sheard and G. A. Toombs, Appl. Phys. Lett., to be published, April 1988.
7. V. J. Goldman, D. C. Tsui and J. E. Cunningham, Phys. Rev. B36, 7635 (1987). See also Proc. 3rd Int. Conf. on Modulated Semiconductor Structures, Montpellier, France. J. Physique C5, 467 (1987).
8. L. Eaves, G. A. Toombs, F. W. Sheard, C. A. Payling, M. L. Leadbeater, E. S. Alves, T. J. Foster, P. E. Simmonds, M. Henini and O. H. Hughes, Appl. Phys. Lett. 52, 212 (1988).
9. B. R. Snell, K. S. Chan, F. W. Sheard, L. Eaves, G. A. Toombs, D. K. Maude, J. C. Portal, S. J. Bass, P. Claxton, G. Hill and M. A. Pate, Phys. Rev. Lett. 59, 2806 (1987).

Perpendicular Transport in GaAs-GaAlAs High Electron Mobility Transistors

J. Smoliner[1], *E. Gornik*[1], *G. Weimann*[2], *and K. Ploog*[3]

[1]Institut für Experimentalphysik, Universität Innsbruck, Technikerstr. 25, A-6020 Innsbruck, Austria

[2]Forschungsinstitut der Deutschen Bundespost, Am Kavalleriesand 3, D-6000 Darmstadt, Fed. Rep. of Germany

[3] Max-Planck-Institut für Festkörperforschung, Heisenbergstr. 1, D-7000 Stuttgart 80, Fed. Rep. of Germany

Abstract :

The thickness of conventional GaAs–GaAlAs High Electron Mobility Transistors (HEMTs) is in the order of 1000 Å or smaller. Therefore, the perpendicular transport in such structures will strongly be influenced by tunneling effects. Investigating the tunneling current perpendicular to the GaAlAs barrier, which had a typical thickness of 500 Å on our samples, oscillatory behavior was observed in dI/dV on HEMT structures having a shallow alloyed gate contact. Using a Fowler–Nordheim tunneling theory, we were able to determine the conduction band discontinuity from the observed oscillations. The fit of the data gave a value of $\Delta E_c/\Delta E_g=0.61\pm0.04$ for aluminum concentrations of 30%, 36%, and 40%.

On samples having a semitransparent Au Schottky–gate contact the bandstructure was varied by illumination. Sharp peaks were observed in the derivative of the tunneling current after illumination at liquid helium temperature. Using a self consistent model, these peaks could be explained by resonant tunneling via subband states in the GaAlAs.

The subband energies in the two–dimensional electron gas (2DEG) were measured by tunneling spectroscopy on samples, where the tunneling process starts from an accumulation layer, and conventional structures, where the electrons tunnel from a metal electrode into the 2DEG. Self–consistent calculations were performed to determine the depletion charge from the measured subband energies. Furthermore the influence of a back–gate voltage was investigated both experimentally and theoretically.

Introduction :

Recently n–GaAs – GaAlAs – n–GaAs heterostructures have been studied extensively in order to understand barrier properties. Watanabe *et al.*[1] have used a capacitance – voltage (CV) profiling technique, which was developed by Krömer[2], to measure the band–gap discontinuity. They found $\Delta E_c/\Delta E_g$ to be 0.62, taking a correction for the interface charge into account. Arnold and co–workers[3] used thermionic emission data to determine the barrier height. A conduction–band discontinuity of 65% of the band–gap discontinuity for aluminum concentrations of 54% 70%, and 100% could be deduced from his data. Batey *et al.*[4] determined the valence– and conduction–band discontinuity independently, studying electron and hole transport properties in n–GaAs – GaAlAs – n–GaAs and p–GaAs – GaAlAs – p–GaAs capacitors. They have measured $\Delta E_c/\Delta E_v$=60:40 using thermionic emission data. Hickmott *et al.*[5] have combined current–voltage (IV) and CV measurements on n^-–GaAs –GaAlAs – n^+–GaAs capacitors with different GaAlAs layer thicknesses to determine the conduction–band discontinuity. Their result for $\Delta E_c/\Delta E_g$ was 0.63 ± 0.03.

Springer Series in Solid-State Sciences Vol. 83: **Physics and Technology of Submicron Structures**
Editors: H. Heinrich • G. Bauer • F. Kuchar

Not only the barrier properties, but also the properties of the two–dimensional electron Gas (2DEG) at the GaAs–GaAlAs interface have been studied extensively both theoretically and experimentally. Stern and Sarma[6] used self–consistent calculations to investigate the influence of the conduction–band discontinuity, acceptor doping in the GaAs , and several other parameters on the subband structure in the 2DEG. Khondker and Anwar[7] presented a simplified theory, assuming the potential in the channel to be piecewise linear. A comparison between the simplified theory and Stern's results showed that both theories agree very well. Earlier, Delagebeaudeuf[8] calculated the electron concentration in the 2DEG for GaAs–GaAlAs field–effect transistors with Schottky–gates. Vinter[9,10] improved this phenomenological model and took the subbands in the GaAs and GaAlAs as well as the partial neutralisation of donors in the n–doped GaAlAs into account.

Experimental investigations of surface bound states by tunneling spectroscopy were performed earlier on several kinds of junctions, but not yet on GaAs–GaAlAs field–effect transistors. BenDaniel and Duke[11] have shown theoretically that the tunnel conductance contains a contribution proportional to the density of states of the two–dimensional energy bands, which was demonstrated on Al–oxide–Bi junctions. Using tunneling spectroscopy Tsui[12,13] has investigated surface bound states in a narrow accumulation layer at the InAs InAs–oxide interface. The structure, which he had observed in dI/dV was due to phonons in the InAs and the existence of a subband at the InAs – InAs–oxide interface. Applying a magnetic field, he was able to observe Landau levels in the electric subbands by peaks in d^2I/dV^2. These measurements allowed a direct determination of the effective mass of the surface electrons. Later similar experiments were carried out on PbTe oxide–Pb, InGaAs–oxide–Pb and $Si–SiO_2$ junctions[14–16].

Determination of the conduction–band discontinuity :

Resonant Fowler–Nordheim tunneling through a n^-–GaAs – GaAlAs – n^+–GaAs single barrier heterostructure was observed first by Hickmott *et al.*[17]. In his structure electrons tunnel from an accumulation layer through a trapezoidal barrier. Oscillations were observed in the IV characteristics and could be explained quantitatively by Fowler–Nordheim tunneling[18,19]. We have investigated conventional GaAs – GaAlAs field effect transistors, where electrons tunnel from the 2DEG through the GaAlAs barrier into a shallow alloyed n^+–GaAlAs contact. Periodic structure, which is also due to Fowler–Nordheim tunneling, occurs in dI/dV on these samples. For the first time these data were used to calculate the conduction–band discontinuity ΔE_c from the positions of the observed oscillations.

Three types of GaAs–GaAlAs–GaAs samples were used in our experiments (see table I). All samples had a p–doped GaAs layer, with an acceptor concentration lower than $1 \cdot 10^{15}$ cm^{-3}, followed by a GaAlAs spacer, then n–doped GaAlAs, and finally a cap layer of undoped GaAs. The contacts to the 2DEG were made using a AuGe alloy (Au:Ge=8:1). A gate contact was formed by shallow diffusion of AuGe into the GaAlAs layer. The barrier thickness d_b was determined by CV–measurements taking the displacement of the 2DEG from the GaAs–GaAlAs interface into account. All IV, dI/dV, and d^2I/dV^2 measurements were made at 4.2 K with a four–terminal ac conductance bridge using a modulation frequency of 140 Hz and a modulation voltage of 1 mV.

The bandstructure was calculated after Stern [20]. An electron in the 2DEG is described by a quantized wave vector k_z perpendicular to the barrier, and $k_{\|}$ parallel to the barrier. The $k_{\|}$ values vary between 0 and the Fermi wave vector, k_f. k_z is a function of the subband energy E_0, which depends upon the electron concentration n_s. Shubnikov de–Haas measurements have shown that n_s saturates with gate voltage in forward bias[21]. Since the change in n_s with positive gate voltage is small, n_s is approximately independent of the applied gate voltage. Therefore E_0 and k_z are assumed to be constant. Under these conditions an electron in the 2DEG can be replaced by a free electron having a constant k_z vector perpendicular to the barrier. If the barrier is assumed to be piecewise linear, the wave function of the incident electron is $A_0 \exp[ik_0(x-x_0)] + B_0 \exp[-ik_0(x-x_0)]$. Inside the barrier the wave function is a

Table I :
Sample parameters : d_{sp} is the spacer layer thickness, d_b is the barrier thickness, n_s is the electron concentration, N_d is the doping in the GaAlAs, μ is the 4.2 K mobility and x is the aluminum concentration in the GaAlAs in %.

Sample	d_{sp}(Å)	d_b(Å)	n_s(cm^{-2})	N_d(cm^{-2})	μ(cm^{-2}/Vs)	x(%)
1503/1	210	593	$3.0 \cdot 10^{11}$	$3.5 \cdot 10^{18}$	730000	30
4991/7	213	542	$2.8 \cdot 10^{11}$	$1.0 \cdot 10^{18}$	470000	36
1360/4	220	460	$2.6 \cdot 10^{11}$	$1.2 \cdot 10^{18}$	610000	40
1360/8	220	527	$2.8 \cdot 10^{11}$	$1.2 \cdot 10^{18}$	610000	40

linear combination of Airy functions, and beyond the barrier the wave function can be described as $A_n \exp[ik_n(x-x_n)]$. Matching the wave functions and solving the system of linear equations, the transmission coefficient is evaluated as $T(V,k)=k_n/k_0(|A_n|^2/|A_0|^2)$ and depends on the k–vector of the incident electron and the applied voltage V. This method is described by Liu *et al.*[22]. The tunneling current density for one electron is

$$j = \int \frac{e\hbar}{(2\pi)^3 m^*} k_z \delta(k_z - k_0)\, T(V,K)\, d^3k \ . \qquad (1)$$

The integration has to be carried out over all allowed k vectors of the incident electron. The delta function is introduced because k_z was assumed to be constant in our system (bound state in z–direction). Then the tunneling current density is evaluated as

$$j = \frac{e\hbar\pi\, k_0}{(2\pi)^3 m^*} k_f^2\, T(V,k) \ . \qquad (2)$$

In this approach, the tunneling current is directly proportional to the transmission coefficient of the barrier.

If the concentration of electrons n_s and the depletion charge n_s is known, one can calculate V_{sp}, V_1, E_0, and E_f (fig.1) after Stern[20]. If the barrier height V_b is known, the tunneling current can also be evaluated. We have varied the conduction band discontinuity V_b for measured values of n_s, n_d, and d_b in order to fit the position of the calculated minima in dI/dV to the measured minima positions. Figure 2 shows a comparison between the calculated minima of d(logI)/dV and our experimental data for sample 1360–4. All of the measured minima agree well with theory. The fit was carried out on samples with aluminum concentrations of 30%, 36%, and 40%. The results are shown in fig.3. The measured barrier heights for the three aluminum concentrations were 226 meV, 272 meV, and 304 meV respectively. It can be seen that the ratio of $\Delta E_c/\Delta E_g$ is constant for all samples.

n_d can be calculated, if the acceptor concentration N_a in the GaAs is known. N_a was measured to be in the order of $1 \cdot 10^{15} cm^{-3}$ by Hall measurements on special samples grown under the same conditions as the heterostructure samples. So, one gets $n_d = 1 \cdot 10^{11} cm^{-2}$ and a mean value of 0.61 ± 0.04 for $\Delta E_c/\Delta E_g$, which is in good agreement with the recently published results[1–5]. The value of N_a may cause errors in n_d, because the acceptor concentration may vary slightly from sample to sample. Curves 2 and 3 in fig.3 show the influence of a n_d variation on the fitted value of ΔE_c. Using extreme values of $n_d = 0.4 \cdot 10^{11} cm^{-2}$ and $n_d = 1.6 \cdot 10^{11} cm^{-2}$ we get $\Delta E_c/\Delta E_g = 0.55$ and $\Delta E_c/\Delta E_g = 0.67$, respectively. However, we want to point out that this method can be used to determine $\Delta E_c/\Delta E_g$ more precisely, if the shallow alloyed gate contact is replaced by a thick n^+–GaAs layer, and if the depletion charge in the GaAs is known

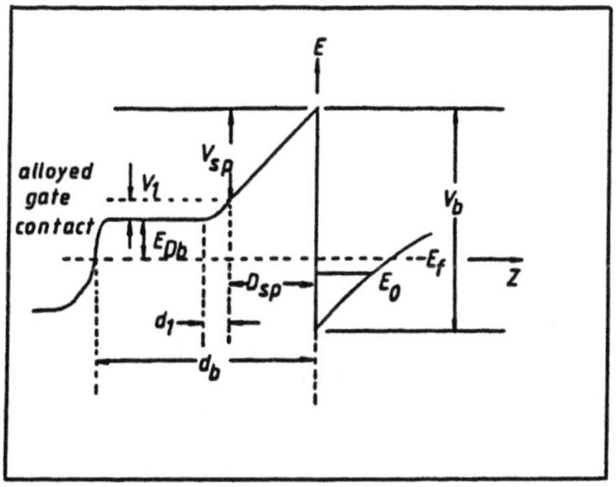

Fig.1 : Bandstructure for a typical sample. The shallow alloyed gate contact reaches into the GaAlAs, so that the resulting barrier thickness is d_b. E_0 is the subband energy, E_f the Fermi energy, and V_b is the barrier height. D_{sp} is the spacer layer thickness and V_{sp}/e is the voltage drop over the spacer. V_1/e is the voltage which drops over the area of ionized donors, which has the thickness d_1. E_{db} is the deep donor level.

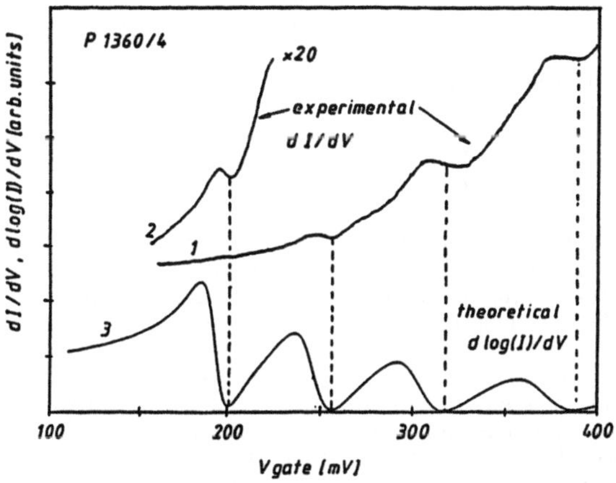

Fig.2 : Comparison of the fitted theoretical d(logI)/dV (3) and the experimental dI/dV curves (1,2). Curve 2 is the same as curve 1 but shifted and amplified by a factor of 20. The theoretical d(logI)/dV curve is plotted simply to allow the curve to fit on the same scale as the experimental one.

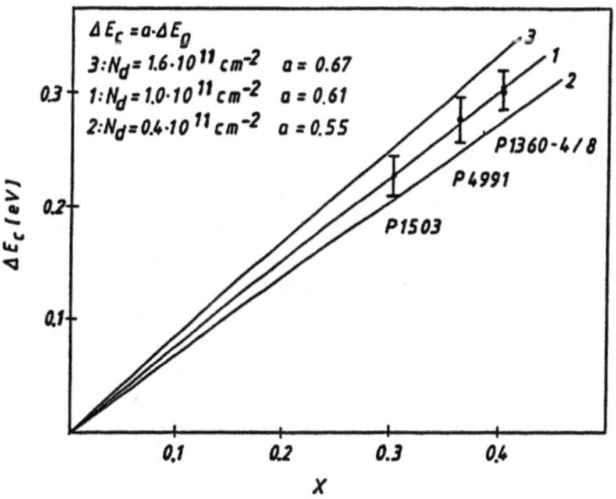

Fig.3 : Fitted conduction band discontinuity ΔE_c versus aluminum concentration x. The influence of n_d is shown in curves 2 and 3.

exactly. The main advantage of this method is that it needs no absolute values of the tunneling current, since the barrier height can be determined from the positions of the oscillations alone. This means that leakage currents and series resistances have no influence on the results.

Subbands in the GaAlAs :

We have investigated GaAs–GaAlAs field–effect transistor structures, where electrons tunnel from the 2DEG through the GaAlAs into a semitransparent Schottky–gate contact before and after illumination. Self–consistent calculations show that these peaks are due to resonant tunneling via subband states in the GaAlAs.

Both samples, which were used for this experiments, were conventional HEMT structures as described earlier, but the cap layer was n–doped ($1 \cdot 10^{18} cm^{-3}$) on one of the samples (1539). The sample parameters are given in table II. The semitransparent Au–gate was formed by evaporation in a cryo–pumped vacuum system. Before illumination the IV–characteristics and its derivative showed only exponential behavior. After the samples were illuminated with a red light–emitting diode for several seconds, peaks could be observed in dI/dV (fig.4). The first peak is resolved best, followed by a weak peak or shoulders in forward and reverse bias. Furthermore the linewidth is different in forward and reverse bias. On sample 1360 the peaks are better resolved than on sample 1539.

Vinter's calculations [9,10] show that the occupation of subbands in the GaAlAs depends on the conduction–band discontinuity, if n_s and the depletion charge is given. As our samples had an aluminum concentration of 35% and 40%, which is much higher than Vinter's value of 22%, the subbands in the GaAlAs will not be occupied, because the bottom of the conduction–band in the GaAlAs is above the Fermi energy. This is illustrated in fig.5.

If a positive voltage is applied, the subband levels in the GaAlAs decrease and one should expect structure in the IV–characteristics due to subband resonances. In the present experiment,however, no structure could be observed. After illumination n_s is strongly increased (see table II), while the depletion charge is decreased drastically. Further, we assume that the ionized donors in the GaAlAs are partially neutralized by the incident light. In this case the impurity scattering rate is reduced. This may explain why the peaks in dI/dV can only be seen after illumination. It is shown in fig.5 that now the subband energies in the GaAlAs are considerably lower due to the increased value of n_s, but it can be seen from Shubnikov de–Haas measurements that the subbands are still empty, because there is no evidence of parallel conduction.

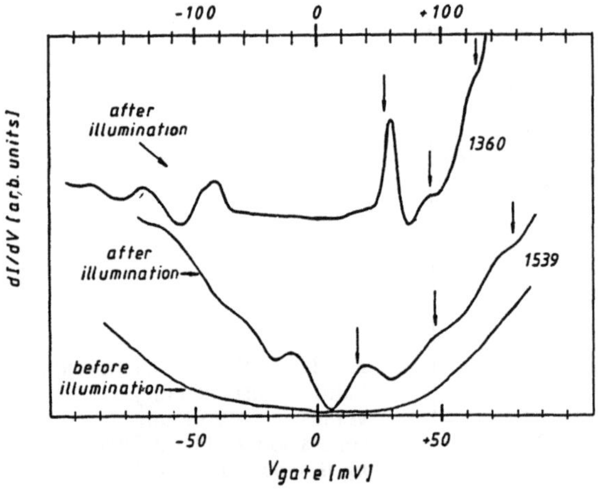

Fig.4 : Experimental dI/dV curves for sample 1360 and 1539. The 1360 curve belongs to the upper voltage scale and the 1539 curve to the lower voltage scale. For sample 1539 dI/dV before and after illumination is shown. The arrows indicate the theoretically calculated peak positions.

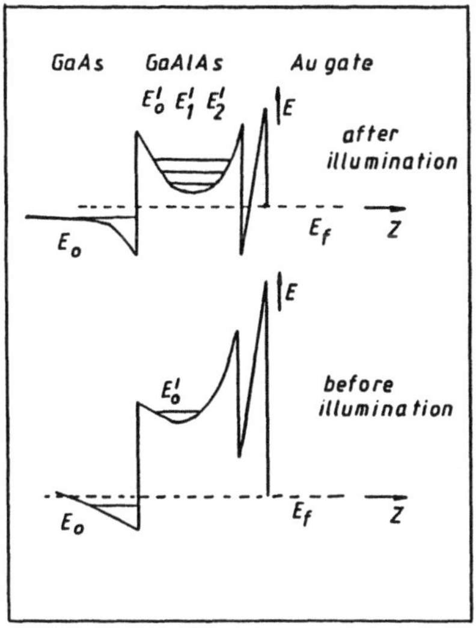

Fig.5 : Schematic bandstructure before and after illumination for a typical sample. E_f is the Fermi energy, E_0 the lowest subband in the GaAs, and E_0', E_1', and E_2' are subband levels in the GaAlAs.

Table II :
Sample parameters : x is the aluminum concentration, d_{sp} is the spacer layer thickness, d_d is the thickness of the doped GaAlAs, which had a Si–concentration N_d, n_s and n_{si} are the electron concentrations before and after illumination.

Sample	d_{sp}(Å)	d_d(Å)	n_s(cm^{-2})	$n_{si}$$cm^{-2}$	N_d(cm^{-3})	x(%)
1360	220	375	$3.0 \cdot 10^{11}$	$5.7 \cdot 10^{11}$	$1.2 \cdot 10^{18}$	40
1539	280	400	$2.6 \cdot 10^{11}$	$4.5 \cdot 10^{11}$	$1.0 \cdot 10^{18}$	35

If the sample is negatively biased, electrons are injected from the Fermi energy in the metal through the GaAlAs barrier. Each time the Fermi level crosses a subband in the GaAlAs, some structure in the IV–characteristics is expected. Looking at the data of sample 1360, one can see that the linewidth of the first peak in forward bias is considerably smaller than in reverse bias. We think that this is due to the fact that the tunneling process starts from quasi–bound state and therefore all tunneling electrons have the same k_z vector perpendicular to the barrier. In reverse bias the incident electrons do not have the same wave vector k_z and therefore the lines are broadened. For sample 1539 the linewidth differences are not that significant. This may be due to the fact that sample 1539 had a n–doped cap layer where the impurity scattering rate is higher than in the undoped cap layer of sample 1360.

With the following simplifications we have used a self–consistent model to calculate the positions of the peaks in dI/dV: All electrons in the 2DEG are assumed to be in the lowest subband and the subbands in the GaAlAs are assumed to be empty. As it is known that illumination drastically decreases the depletion charge, we have set the depletion charge in the GaAs equal to zero. The dielectric constants are assumed to be the same in the GaAs and the GaAlAs. All effective mass effects were neglected. In the doped GaAlAs we have assumed a uniform charge density N_d^*. As N_d^* can hardly be

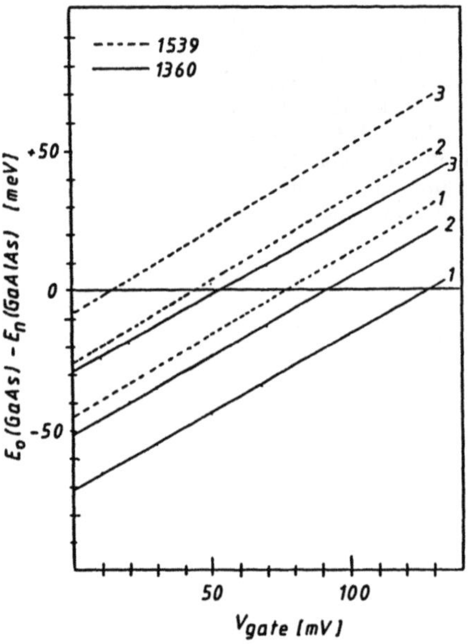

Fig.6 : Calculated energy differences between the lowest subband in the GaAs and the n–th subband in the GaAlAs. The numbers indicate the subband index. The dashed lines belong to sample 1539 and the full lines to sample 1360. Each time the energy difference is zero, resonant tunneling occurs and a peak is expected in dI/dV.

deduced from the doping in the GaAlAs, especially when the samples were illuminated, we have used it as a fitting parameter. The presence of deep and shallow donor levels was not taken into account. For the conduction–band discontinuity we have used a value of 0.61 [23]. Solving Poisson's equation and taking the potential to be piecewise linear, we have made an Airy–function approach similar to Ohnishi's work [24]. While Ohnishi has calculated the tunneling current through double barriers directly, we have only calculated the energy levels both in the GaAs and the GaAlAs self–consistently in forward bias. N_d^* is adjusted so that the theoretically calculated peak positions fit best to the experiment.

Fig.6 shows the calculated subband differences between the lowest subband in the GaAs and the n–th subband in the GaAlAs versus gate–voltage for both samples. Each time the energy difference is zero, subband resonances occur and structure is expected in dI/dV. It can be seen that the energy differences show a linear behavior with gate–voltage. For sample 1539 the fit gave $N_d^*=2.2\cdot10^{17}cm^{-3}$. For sample 1360 N_d^* was determined to be $2.9\cdot10^{17}cm^{-3}$, which is considerably lower than the doping in the GaAlAs. This is due to the existence of neutral deep donor levels before illumination and the additional neutralisation of donors by the illumination process.

For these N_d^* values the theoretically calculated subband resonances occur at 16 mV, 47 mV, and 79 mV for sample 1539 and at 56 mV, 93 mV, and 127 mV for sample 1360. This fits well the values of 20 mV, 50 mV, 73 mV, and 60 mV, 90 mV, 120 mV for sample 1539 and sample 1360, respectively. We think that the different N_d^* values are due to the different aluminum concentrations and due to the somewhat different doping in the GaAlAs. Furthermore the barrier height of a Schottky contact, which is different for Au on undoped and n–doped GaAs, may have an influence on this value.

Determination of the depletion charge :

We have measured the subband energies in GaAs–GaAlAs field effect transistors, where the electrons tunnel from an accumulation layer into the 2DEG and a conventional structure, where tunneling takes place from a metal electrode. Using self–consistent calculations the depletion charge in the GaAs could be determined from the measured subband energies. Furthermore, the influence of a back–gate voltage was investigated.

For this experiments we have used the sample type 1503 (see table I). An Au back–gate was evaporated after polishing the backside of the samples, and the contacts to the 2DEG were made using a AuGe alloy. For the gate contacts AuGe was used also, but on some samples we have only evaporated the contacts to achieve a Schottky contact with low barrier height (fig.7a). Other samples were heat treated to diffuse the AuGe into the GaAs so that a n–doped GaAs cap layer is formed. If the GaAs cap layer is n–doped, an accumulation layer exists at the GaAs–GaAlAs interface, where the lowest subband energy is E_0' (fig.7b). From capacitance measurements we conclude that on the samples having a shallow diffused gate contact only the GaAs layer became conductive. Although Ge may have penetrated into the GaAlAs, the GaAlAs is still frozen out at 4.2 K.

The IV–curves of the tunneling current and its derivatives are shown in fig.8. The measurements were made for negative back–gate voltages of 0 V, –100 V, –200 V, and –300 V. Fig. 8a shows the IV characteristic for a sample having a shallow alloyed gate contact. Steps can be seen in the IV–curve and the corresponding peaks in dI/dV are plotted in fig.8b. If a back–gate voltage is applied the steps in the IV–characteristics and the peaks in dI/dV are shifted and better resolved. The dI/dV curves of a sample having a Schottky gate are shown in fig.8c. On these samples steps only occur in dI/dV, and the steps were also better resolved with rising back–gate voltage. Fig.8d shows the same as fig.8c, but here the curves are extended and amplified to show the first step more clearly. We want to note that the total thickness of the GaAlAs (505 Å) on the used samples is relatively thin for conventional field effect transistors. On other samples the GaAlAs barrier was thicker, and the tunneling current was too small to be measured with our conductance bridge.

In the case of the Schottky gate electrons tunnel from the metal through the barrier into the n–th subband of an inversion layer. After BenDaniel and Duke[11], dI_n/dV is calculated as

$$dI_n/dV = 2Ae\, \rho_n(E-E_n)\, 1/\tau(E)\ , \qquad (3)$$

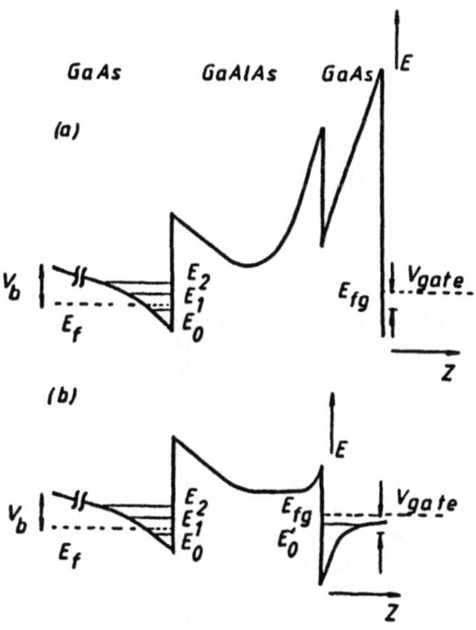

Fig.7 : The bandstructure for a typical sample with Schottky contact is shown in the upper part. E_0, E_1, and E_2 indicate the subbands, V_b is the back–gate voltage, E_f the Fermi energy, V_{gate} the gate voltage, and E_{fg} is the Fermi energy in the gate contact. The bandstructure for a sample with n–doped GaAs surface is shown below. E_0' is the subband energy in the accumulation layer.

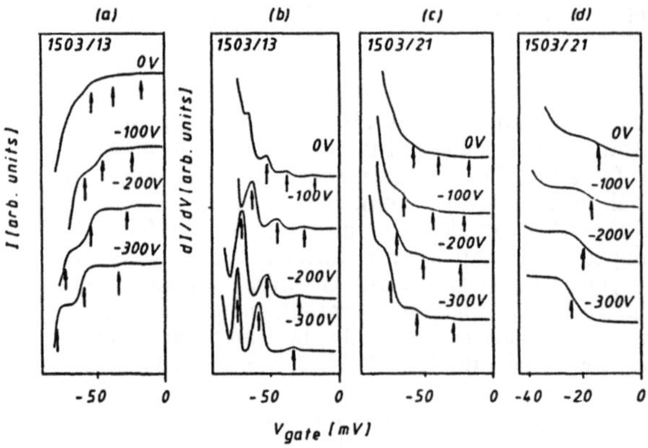

Fig.8 : The IV and dI/dV curves for sample 1503/13 and 1503/21 are shown for back–gate voltages of 0 V, –100 V, –200 V, and –300 V. All curves are scaled to fit on the same plot. Sample 1503/13 (a,b) has a n–doped gate contact, and sample 1503/21 has a Schottky gate. In fig.8d the dI/dV curve of sample 1503/21 is amplified to show the first peak more clearly. The arrows indicate the theoretically calculated step positions.

where A is the area of the junction, e the electron charge, $1/\tau$ is the tunneling probability and ρ_n is the two–dimensional density of states in the n–th subband. One can see that dI/dV is directly proportional to the density of states, and therefore dI/dV will have steps, each time the Fermi energy in the gate electrode E_{fg} crosses the next subband energy. If a negative back–gate voltage V_b is applied, the electrons will be better confined. The external field causes an increased energy of the lowest subband and the energy differences between the subbands will become larger. Therefore the steps in dI/dV are better resolved with rising back–gate voltage.

A second type of samples was prepared by a shallow diffusion process. The GaAs top layer was doped with Ge, which is n–type doping for GaAs. Then we can assume that an accumulation layer is created at the interface, and the electrons in the n–doped GaAs will have an energy between the subband energy E_0' and E_{fg}. As the Fermi level is not much above the subband energy, we make the rough approximation that all tunneling electrons have the same energy E_{fg}. This can be argued because the observed linewidth in dI/dV is small (fig.8b), which reflects the energy distribution of the incident electrons. Under this assumptions the tunneling current into the n–th subband is :

$$I_n = 2Ae \int \rho_n(E-E_n)\, 1/\tau(E)\, \delta(E-E_{fg})\, dE\,, \tag{4}$$

$$I_n = 2Ae\, \rho_n(E_{fg}-E_n)\, 1/\tau(E_{fg})\,. \tag{5}$$

Since in this approximation the incident electrons have the same energy, which is described by the δ–function, the tunneling current and not dI/dV is directly proportional to the density of states. This means that steps will not occur in dI/dV but in the IV–characteristics for a n–doped gate contact.

The energy levels in the 2DEG mainly depend on the the electron concentration n_s and the depletion charge n_d. As the wave functions penetrate into the GaAlAs barrier, the conduction band discontinuity also influences the energy levels, but this is only a second order effect compared to the influence of n_s and n_d. Furthermore our calculations have shown that all other sample parameters, especially the barrier–thickness, have almost no influence on the results. In principle the depletion charge can be calculated if the acceptor concentration in the GaAs is known[26]. In GaAs–GaAlAs field effect transistors the doping of the GaAs buffer layer depends mainly on the impurities in the semiinsulating substrate, which diffuse into the GaAs buffer layer. Therefore, a direct determination of the doping level is not possible. In addition the doping may vary over the buffer layer thickness. If a negative back–gate voltage is applied, the external field has the same effect as an additional depletion

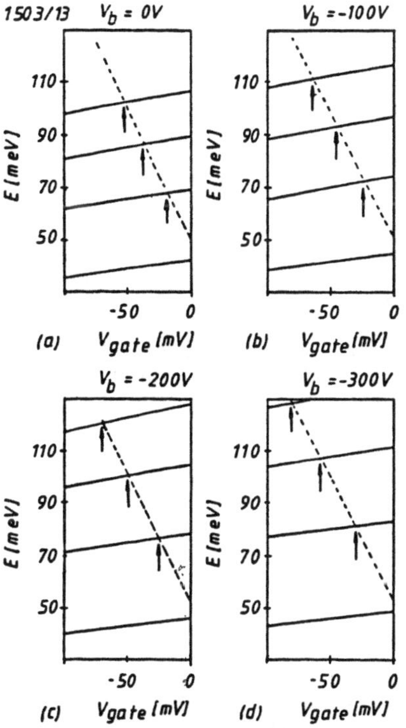

Fig.9 : Theoretically calculated subband energies versus gate voltage for sample 1503/13 (solid lines) for $n_d=1.5\cdot10^{11}cm^{-11}$ and a sample thickness of 220 μm. The electron concentration for zero applied gate voltage and back–gate voltages of V, –100 V, –200 V, and –300 V was 2.09, 1.93, 1.82, and $1.73\cdot10^{11}cm^{-2}$ respectively. Each time the Fermi energy in the gate contact E_{fg} (dashed line) crosses a subband, which is indicated by arrows, structure is expected in dI/dV.

charge n_d with $n_d=\epsilon\epsilon_0 V_b/d_s$, where d_s is the sample thickness, and $\epsilon\epsilon_0$ is the dielectric constant.

Using an Airy function approach [22,24,25], we have calculated the subband energies self–consistently, in order to determine where steps in IV or dI/dV should be observed. Starting with the assumption that with zero applied voltage the Fermi energy is constant over the sample, structure in dI/dV is expected each time the Fermi energy in the gate contact E_{fg} crosses a subband in the 2DEG. Note, that the first step in IV or dI/dV is due to tunneling of electrons into the second subband. Fig.9 shows the calculated subband energies for $n_d=1.5\cdot10^{11}cm^{-2}$ and back–gate voltages of 0 V, –100 V, –200 V, and –300 V. Several features are apparent: the subband energies decrease almost linearly if the gate is negatively biased, and an applied back–gate voltage results in an increased subband energy. This means that the first step in IV and the corresponding peak in dI/dV is shifted to higher gate voltages. The Fermi energy in the gate E_{fg} is also plotted in fig.9a–d. Each time E_{fg} crosses a subband, which is indicated by arrows, structure is expected in dI/dV. For the determination of the depletion charge n_d has to be varied until the theoretically calculated steps in IV or dI/dV fit best to the experiment. We have calculated n_d for several back–gate voltages taking the decreased electron concentration into account. As a result we find that on the samples taken from the center of the wafer the theoretically obtained step positions fit perfectly to the experiment for $n_d=1.5\cdot10^{-11}\pm0.1\cdot10^{-11}cm^{-2}$ for all applied back–gate voltages. On samples taken from the rim of the wafer our calculation gave $n_d=1.7\cdot10^{11}\pm0.1^{11}cm^{-2}$. A comparison between the experimental data in fig.8b and the theoretically calculated peak positions in fig.9a–d show that the experiment agrees well with theory.

As this method makes a quite precise measurement of the depletion charge possible, it could be used to study the influence of n_d on the electron mobility. Furthermore, as the purity conditions during the sample growth process are reflected in n_d, the device quality can be directly controlled.

Summary :

In summary we have observed oscillations of the transmission coefficient (Fowler–Nordheim tunneling) of the GaAlAs barrier on conventional GaAs–GaAlAs field effect transistors having a shallow alloyed gate–contact. Using a simple one–electron picture, we were able to determine the conduction band discontinuity ΔE_c from the positions of the observed oscillations. The ratio of $\Delta E_c/\Delta E_g$ was found to be 0.61 ± 0.04, which agrees well with the recently published results [1–5].

Investigating the tunneling current through the GaAlAs barrier on samples having a semitransparent Schottky–gate, sharp peaks were observed in dI/dV after illumination. This effect can be explained by resonant tunneling via subband states in the GaAlAs. Self–consistent calculations were performed to determine the peak positions theoretically. Using the concentration of ionized donors in the GaAlAs as fitting parameter, the theoretically calculated peak positions were found to be in good agreement with the experiment for $N_d^* = 2.9 \cdot 10^{17} cm^{-3}$ and $2.2 \cdot 10^{17} cm^{-3}$ on sample 1360 and 1539, respectively.

On samples having a relatively thin GaAlAs barrier, the subband energies in the two–dimensional electron gas were measured by tunneling spectroscopy. From the experimental data the depletion charge could be determined using self consistent calculations. We found n_d to be $1.5 \cdot 10^{11} cm^{-2}$ in the center of the wafer and $n_d = 1.7 \cdot 10^{11} cm^{-2}$ on the rim of the wafer. Furthermore the influence of a back–gate voltage on the subband energies was investigated both experimentally and theoretically.

Acknowledgements :

This work was sponsored by Stiftung Volkswagenwerk, project No. I–61840. The authors are grateful to M.Helm, R.A.Höpfel, K.Berthold, R.Lassnig, and S.Lyon for valuable discussions. Further we whish to thank H.Kuen for his technical assistance and C.Obmascher for preparing some of the samples.

References :

[1] M.O.Watanabe, Y.Yoshida, M.Mashita, T.Nakanisi, A.Hojo, J.Appl. Phys. 57, 5340, (1985)

[2] H.Krömer, W.Y.Chien, J.S.Harris Jr, D.D.Edwall, Appl. Phys. Lett. 36, 295, (1980)

[3] D.Arnold, A.Ketterson, T.Henderson, J.Klem, H.Morkoc, J. Appl. Phys. 57, 2880, (1985)

[4] J.Batey, S.L.Wright, D.J.DiMaria, J. Appl. Phys. 57, 484, (1985)

[5] T.W.Hickmott, P.M.Solomon, R.Fischer, H.Morkoc, J. Appl. Phys. 57, 2844, (1985)

[6] F.Stern, S.D.Sarma, Phys. Rev. B 30, 840, (1984)

[7] A.N.Khondker, A.F.M.Anwar, Solid State Electron. 38, 847, (1987)

[8] D.Delagebeaudeuf, L.T.Linh, IEEE Trans. Electron Dev. ED–29, 955 (1982)

[9] B.Vinter, Appl. Phys. Lett. 44, 307, (1984)

[10] B.Vinter, Solid State Comm. 48, 151, (1983)

[11] D.J.BenDaniel, C.B.Duke, Phys. Rev. 160, 679, (1967)

[12] D.C.Tsui, Phy. Rev. Lett. 24, 303, (1970)

[13] D.C.Tsui, Phys. Rev. B 4, 4438, (1971)

[14] D.C.Tsui, G.Kaminsky. P.H.Schmidt, Phys. Rev. B 9, 3524, (1974)

[15] Pong–Fei Lu, D.C.Tsui, H.M.Cox, Appl. Phys. Lett. 45, 772, (1984)

[16] U.Kunze, J.Phys. C, 17 5677, (1984)

[17] T.W.Hickmott, P.M.Solomon, R.Fischer, H.Morkoc, Appl. Phys. Lett. 44, 90, (1984)

[18] J.Maserjian, J.Vac. Sci. Technol. 11, 996, (1974)

[19] K.H.Gundlach, Solid State Electron. 9, 949, (1966)
[20] F.Stern, Appl. Phys. Lett. 43, 949, (1983)
[21] K.Hirakawa, H.Sakaki, J.Yoshino, Appl. Phys. Lett. 45, 253, (1984)
[22] W.W.Liu, L.Fukuma, M.Fukuma, J. Appl. Phys. 60, 1555, (1986)
[23] J.Smoliner, R.Christanell, M.Hauser, E.Gornik, G.Weimann, K.Ploog, Appl. Phys. Lett. 50, 1727, (1987)
[24] H.Ohnishi, T.Inata, S.Muto, N.Yokoyama, A.Shibatomi, Appl. Phys. Lett. 49, 1248, (1986)
[25] Y.Ando, T.Itoh, J. Appl. Phys. 61, 1497, (1987)
[26] T.Ando, A.B.Fowler, F.Stern, Rev. Mod. Phys. 54, 437, (1982)

Part III

Quantum Interference Effects, Mesoscopic Systems

Quantum Interference Effects in Disordered Sub-micron Wires and Rings

R.A. Webb, S. Washburn, H.J. Haucke, A.D. Benoit, C.P. Umbach, and F.P. Milliken

IBM Research Division, T.J. Watson Research Center,
P.O. Box 218, Yorktown Heights, NY 10598, USA

The low temperature transport properties of small disordered normal conductors exhibit a wealth of quantum interference phenomena. When the sample size becomes comparable to the distance electrons can propagate without losing phase coherence, these quantum interference effects manifest themselves as random conductance fluctuations in a wire and periodic oscillations in a loop as the magnetic field applied to the sample is changed. We will briefly review the current status of this field by discussing the origin of these quantum interference effects and how the measurement technique affects the details of the data analysis. A discussion of the non-local properties of small structures is presented as well as some of the problems associated with determining the phase coherence length.

1. CONDUCTANCE FLUCTUATIONS

The model for resistance in a metal is that the sea of carriers flows through the lattice, and its progress is slowed or resisted by imperfections (referred to collectively as impurities) in the lattice. A simple classical model might assume that the carriers behave as billiard balls rolling in uneven terrain – the more uneven the impurity potential the more resistive the metal. The quantum-mechanical superposition of carrier wave-functions contributes to the effective resistance of a metal sample just as the classical billiard ball effects contribute. The classical resistance of the metal can be obtained from the quantum-mechanical model if the superpositions are ignored [1]. In large samples the superpositions have random sign and amplitude from point to point in the metal, and so, averaging to zero over the whole sample, they contribute nothing to the resistance beyond the billiard ball terms. At low enough temperatures where the phase coherence length L_ϕ significantly exceeds the mean free path length ℓ, the interference plays a noticeable role in determining the resistance. For instance the weak localization effects seen in all disordered metals at low temperatures result from such superpositions which on average always increase the resistance of the sample [2].

It has been noticed rather recently that when the sample size is comparable to L_ϕ, the random contributions of the superpositions appear in the measured resistance of the sample [3-33]. In fact the random fluctuations in the conductance have a certain universal character [20-33]. All phase coherent samples ($L = L_\phi$) will exhibit fluctuations in conductance as the chemical potential is varied [4] or the magnetic field is varied [3]. The fluctuations will have amplitude $\Delta G \simeq e^2/h$. An ensemble of phase coherent samples having the same shape will have conductances which vary randomly, and the distribution of conductances will have width $\Delta G = Ze^2/h$ where Z is a shape dependent constant (of order 1) which can be calculated to reasonable accuracy.

The source of the conductance fluctuations in a magnetic field is the Aharonov-Bohm effect [35]. The various current paths through the metal enclose random quantities of magnetic flux

Springer Series in Solid-State Sciences Vol. 83: **Physics and Technology of Submicron Structures**
Editors: H. Heinrich • G. Bauer • F. Kuchar

which in turn yield random phase-shifts in the carrier wave-functions. The superpositions which contribute to the conductance grow and shrink randomly as the field changes. If one forms a loop sample, then the area enclosed by the loop contains flux which causes a net phase shift between those carriers that pass along one arm of the loop relative to those passing though the other arm. The net shift is $\Delta\phi = B\bullet(\text{area})/(h/e) = \Phi/(h/e)$. The 2π symmetry of the wave-function means that, as the field increases linearly, the current through the loop oscillates with period $\Phi = h/e$ [17]. In general then the conductance of the loop is composed of the classical term plus quantum mechanical corrections $\Delta g(B)$ of the form [18, 25]

$$\Delta g(B) = \sum_{n=0}^{\infty} g_n(B) \cos\left\{ \frac{2\pi n\Phi}{h/e} + \alpha_n(B) \right\} . \tag{1}$$

The amplitudes $g_n(B)$ fluctuate randomly in B on various field scales, and similarly the phase factors $\alpha_n(B)$ change randomly on about the same field scales. The field scales are determined by the criterion that scrambling the phases requires that a net flux h/e be enclosed by the paths through the wire. For g_0, this is equivalent to a flux h/e through the metal. Since the electrons which contribute to h/e oscillations travel twice as far in the metal, the flux required to change g_1 is about half a great as for g_0, and so on for g_2, ... The cascade of harmonics in h/e can be extracted from the resistance of a sample in a straightforward manner. The magnetoresistance of a loop of gold is displayed (top trace) in Fig. 1(a), and the Fourier transform of these data is in Fig. 1(b). Using a digital filter, we can select the term in the sum we wish to study, and the first three terms are displayed individually in the same figure. Qualitatively, the above argument about the field scales is borne out. The scale to change g_0 (minimum to maximum) is $\Delta B = 0.1$T, and the scales to change g_1 and g_2 are reduced by factors of about 2 and 4 respectively. These terms appear in the Fourier transform (see Fig. 1(b)) as a set of peaks which decay and broaden as n increases.

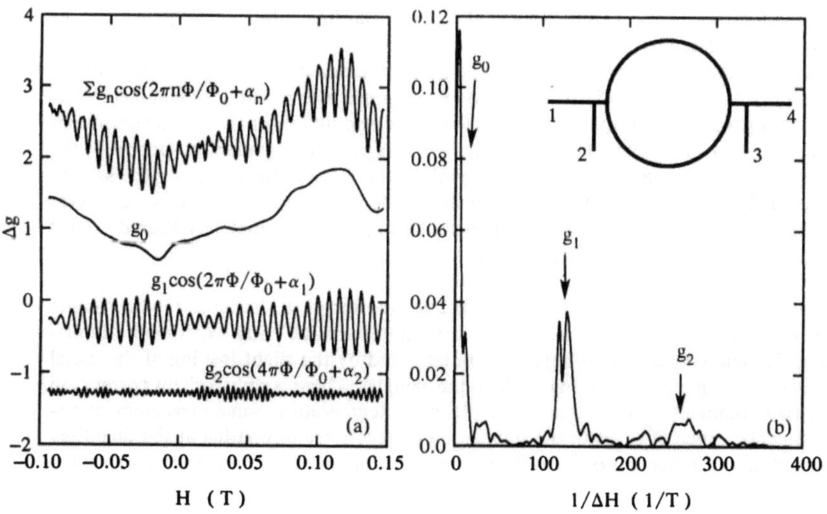

Fig. 1. (a) The conductance fluctuations of a gold loop (diameter 820 nm and wire width 50 nm): (top) the total conductance fluctuations, (2nd) the term g_0 obtained with a digital filter, and (2) and (3) the second and third terms extracted similarly from the top trace. The data were obtained at T = 0.032 K , and the average resistance was 29 Ω . (b) The Fourier transform of the top trace in fig.1. The terms in Eq. 1 are manifested as peaks as marked

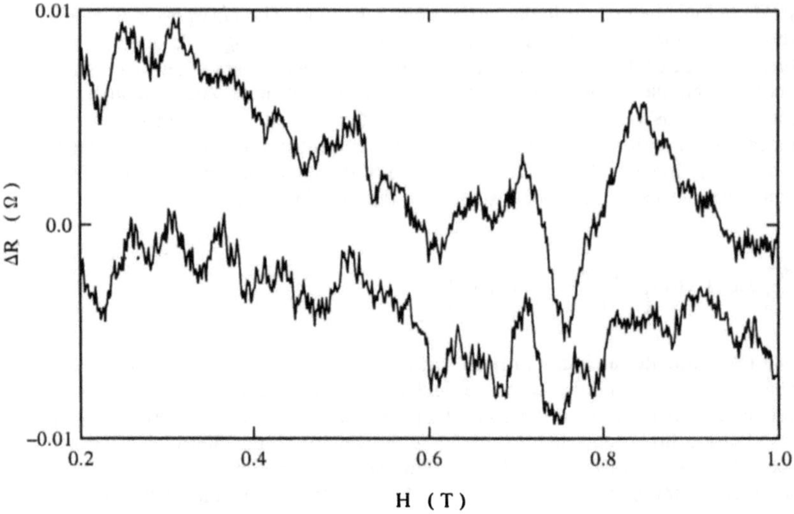

Fig. 2. Two traces of the magnetoresistance of a gold sample illustrating the sensitivity of the fluctuation pattern to changes in the impurity potential field. The sample was a loop of gold (diameter 250 nm and wire width ranging from 30 to 120 nm); the average resistance was 7.7 Ω and the temperature was 0.3 K

The pattern of fluctuations $\Delta g(B)$ reflects, in a convoluted way, the positions and sizes of the impurities in the metal [18, 24]. In the classical conductance, moving one impurity can have no effect on the total g. In the quantum mechanical case, the fluctuation pattern results from the net phases accumulated by the electrons after collisions with impurities so that moving the impurities changes the net phases and scrambles up the fluctuation pattern. The result from the theory is that in fact (in 1 and 2 dimensions) moving so much as one impurity completely rewrites the pattern [24]. At first one might guess that moving only one of the myriad scatterers in the sample can cause changes no larger than $\Delta g^{(1)} \sim 1/N_{scat}$. The argument in support of the theory is that in 1 or 2 dimensions, the diffusive path of the carrier tends to cover the sample so that all of the carriers visit all of the impurities – the effect of each scatter is, in effect, amplified. The fluctuation pattern in a phase coherent wire, then, is completely altered – $\Delta g^{(1)} \sim 1$ – by slight changes in the impurity potential.

One annoying feature of this amplification is that any motion of the impurities in the sample cause the measurements not to be reproducible. Examples from an early experiment are displayed in Fig. 2. The two traces are separated in time and by the slight jostling of the metal caused by disconnecting and connecting leads to the sample. There is only a slight correlation between the two nominally identical (same sample, same temperature, same measurement circuit) measurements. Such changes in the fluctuation pattern were infuriating at the time they were discovered since the fluctuations and their sensitivity to impurity motion were unexpected.

2. CONDUCTANCE MEASUREMENTS

With few exceptions the theory has been constructed without much consideration for the measurement methods. Until recently, all of the calculations have assumed a two-probe measurement whereas the experiments have usually been four-probe measurements. Upon doing the calculations appropriate to model the four-probe experiments, the theoreticians have managed to explain many of the results which had been puzzling when interpreted in light of the two-probe theory [27 - 33]. The two-probe model of current I flowing between reservoirs having chemical potentials μ_1 and μ_2 yields a conductance $G = I/(\mu_1-\mu_2)$. The measurements, by and large, were performed by measuring the voltages along a sample in which current flowed. The voltage and current were then divided, as if the sample were a classical resistor, to obtain the conductance $G = I/(V_1-V_2)$. This dichotomy between voltage measurements and the two-probe theory is at the heart of some confusion about the fluctuation measurements.

3. MAGNETIC FIELD ASYMMETRIES

One of the most misunderstood aspects of the transport properties of small, multi-terminal wires and rings is that the measured electrical resistance is in general *not* symmetric about zero magnetic field if more than two-probes are attached to the sample [12]. The confusion results from lack of understanding of what the voltage probes actually measure, and how they change the sample. If the quantum-mechanical wave properties of the electron play an important role in the measured transport properties of a multi-terminal sample, the relationship between measured voltages and internal electric fields in the sample has a non-local contribution [8, 29-33]. The electron waves are sensitive to interference effects that occur within a phase coherent section of the sample and any leads that are connected to the sample are obviously part of the sample. Thus interference effects that come from electron excursions into the voltage probes contribute to the measured properties of the sample. Classically, these effects are small. In phase coherent samples, however, they lead to large contribution which on first observation seem to violate time-reversal symmetry and the concomitant Onsager-Casimir symmetry relations [36].

An example of the importance of measurement probes in contributing to the asymmetry in the magnetoconductance of a 0.8 μm average diameter gold ring is shown in Fig. 3(a). Four magnetoconductance traces on the *same* sample with three different ways of measuring voltage and applying current are displayed. For clarity of display, the h/e oscillations have been digitally filtered out, and only the aperiodic conductance fluctuations are shown. The same qualitative features, however, apply to the h/e oscillations as well. The top trace was measured by applying current to lead 1 and removing it via lead 4. The voltage difference was measured between lead 2 and lead 3. The conductance $G_{14,23}$ is clearly *not* symmetric about zero field. The other two traces result from permutations of current and voltage leads as indicated below each curve. Each trace exhibits a different pattern of fluctuations, and all of them are asymmetric about zero field. The dashed curve on the top trace illustrates that after four days of lead switching experiments the data is extremely reproducible. These asymmetries have been observed in every experiment on conductance fluctuations in rings and lines in our laboratory over the last four years.

A particularly clear explanation for these asymmetries has been given by BÜTTIKER [27]. He has pointed out that Onsager-Casimir symmetry relations of the local conductivity tensor do not take into account what is measured; they only relate the properties of the local conductivity tensor, $\sigma_{\alpha\beta}(H)=\sigma_{\beta\alpha}(-H)$, where α and β index the spatial coordinates. The proper way to state the symmetry properties of electrical conductors is through the reciprocity theorem, $R_{12,34}(H)=R_{34,12}(-H)$, where now the indices refer to current and voltage leads. This symmetry property is present in Fig. 3(a): compare the positive field data in the top trace with the negative field data in the bottom trace. In fact it is clear from lead reversal measurements that the symmetric and anti-symmetric parts of the conductance fluctuation are determined by the following formulas [12]

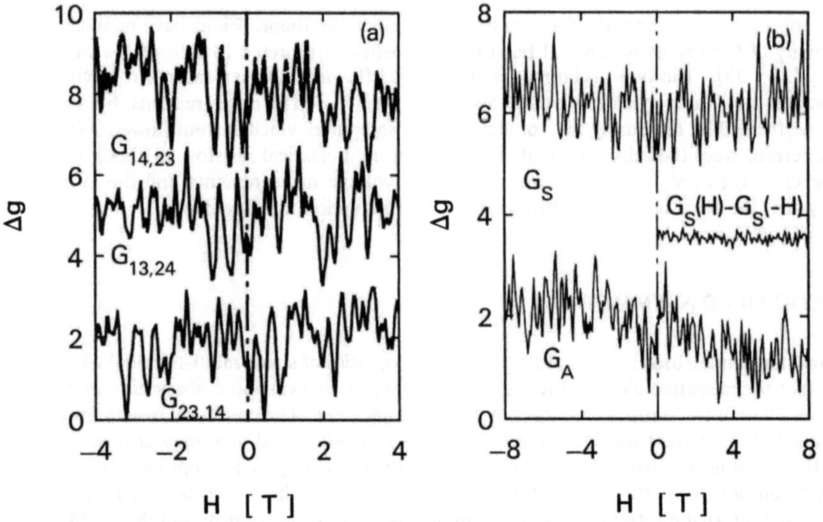

Fig. 3. (a) Magnetoconductance data at 40 mK from a four-probe gold ring sample over ± 4 T magnetic field range for three different permutations of current and voltage probes. Dashed curve on top trace demonstrates the long term reproducibility of the measurements. (b) The symmetric and anti-symmetric contributions to the fluctuation data obtained by adding (G_S) and subtracting (G_A) the top and bottom traces shown in (a). Also shown is the residual asymmetry in G_S

$$G_S = 2/[R_{14,23}(H) + R_{23,14}(H)]\,, \tag{2}$$

$$G_A = 2/[R_{14,23}(H) - R_{23,14}(H)]\,. \tag{3}$$

Although it is tempting to think of G_S as equivalent to a "longitudinal conductance" G_{xx} and of G_A as a "transverse conductance" G_{xy}, this identification is not correct when there is a non-local contribution to the measured resistance. In a four-probe measurement, there are six different combinations of current and voltage measurements, and each of the three pairs of lead reversal measurements will, in general, give a different G_A [27].

Figure 3(b) displays the results obtained by adding and subtracting the top and bottom traces of Fig. 3(a) over a ±8 T field range. The residual asymmetry in G_S, $G_S(H)$-$G_S(-H)$, is also shown; its amplitude is very nearly equal to the noise in the measurement. The data show that the anti-symmetric part of the conductance fluctuation is also a random function of magnetic field. The rms value of the symmetric part of the fluctuations is 0.56 e^2/h and the anti-symmetric part is 0.44 e^2/h. Both have the same characteristic magnetic field scale, H_C, for the fluctuations [12]. Thus the symmetries predicted by BÜTTIKER [27] are indeed the ones found in the experiments and the asymmetries in a magnetoresistance measurement are no longer mysterious.

4. NON-LOCAL EFFECTS

Although the predictions from the two-probe conductance fluctuation theory [17-26] seem to describe much of the physics observed in small structures, several recent experiments have clearly shown that apparent conductance fluctuations can far exceed e^2/h in four-probe measurements [8, 9]. Careful measurements of the dependence of the measured conductance fluctuations on the ratio L/L_ϕ, where L is the distance between voltage probes in a four-probe measurement, have demonstrated that the sample under investigation is *not* determined by the length L but rather by the length L_ϕ when $L<L_\phi$. This means that when the spacing between voltage probes becomes less than the phase coherence length, the measured voltage fluctuations become independent of the distance between probes while the rms value of the conductance fluctuations diverges as $L \to 0$. This effect is another example of non-local contributions to the resistance of metal wires and loops.

Figure 4 displays the length dependence of the voltage fluctuations measured in two samples: a long Sb wire 0.08 nm thick, and 0.1 nm wide, and a long gold wire 0.03 nm thick and 0.3 nm wide [8]. The samples have eight leads attached with average spacings that vary from 0.2 μm to 4.0 μm. The symmetric and anti-symmetric parts of the voltage fluctuations were determined from ± 3 T field sweeps for each of the 15 different combinations of voltage probe spacings. The phase coherence length L_ϕ was estimated from the weak localization behavior observed in one of the segments of the Sb line. We point out that not all segments actually exhibit a clear weak localization magnetoresistance because the conductance fluctuations frequently distort the shape of the curve. Therefore our estimate of L_ϕ is not reliable to better than 20%-50%. Using the measured values of 1.10 μm for L_ϕ in the Sb line and 1.6 μm for L_ϕ in the Au line, and assuming that the conductance fluctuation for voltage probes spaced L_ϕ apart is e^2/h, we compute the rms value of the voltage fluctuation for this phase coherent segment, ΔV_ϕ, and use it to normalize all the voltage fluctuation data. The open circles are the symmetric part of the voltage fluctuation, which for $L/L_\phi>1$ grows in magnitude as $(L/L_\phi)^{1/2}$. Translated into conductance fluctuations, this Δg decays as $(L/L_\phi)^{-3/2}$ in agreement with the predictions of the two-probe conductance fluctuation theory [21, 23]. For $L/L_\phi<1$, however, the rms value of the voltage fluctuations become independent of the distance L between voltage probes. This behavior is immediately understood when one remembers that, owing to the wave-like properties of the electron, the interference phenomena we are measuring come from a phase coherent section of the sample. This means that when $L<<L_\phi$ the effective extent of the sample is no longer determined by the placement of the voltage probes but by L_ϕ itself. The anti-symmetric part of the voltage fluctuations is found to be independent of length (within experimental error) [8] for all segments tested in agreement with theory [30-33].

The non-local contribution to the measured voltage fluctuations is best understood by the following experiment. Take a long line with many voltage probes connected to it. Put current in the first probe and remove the current through the next closest probe. Measure the voltage with any two-probes away from the classical current path (this might be $G_{12,34}$ in the inset of Fig.1(b)). The average voltage will be nearly zero. When a magnetic field is applied to the sample the voltage will fluctuate randomly due to phase coherent carriers propagating into the classically forbidden regions of the sample [8, 15]. The results of such an experiment performed on the above sample are also shown in Fig. 4 for both the symmetric and anti-symmetric voltage fluctuations. The distance L is now the distance between the end of the classical current path and the closest voltage probe. Both the symmetric and anti-symmetric contributions to the voltage fluctuation decay exponentially with increasing length with an approximate dependence on length given by $\Delta V/\Delta V_\phi \simeq \exp(-aL/L_\phi)$ with $a = 1.2 \pm 0.2$. This exponential decay in not surprising, but that scale for the decay is L_ϕ is surprising (in classical physics, the width sets the scale). This also demonstrates that carriers propagate into all leads of the sample and that adding probes to a phase coherent sample actually changes the sample [15].

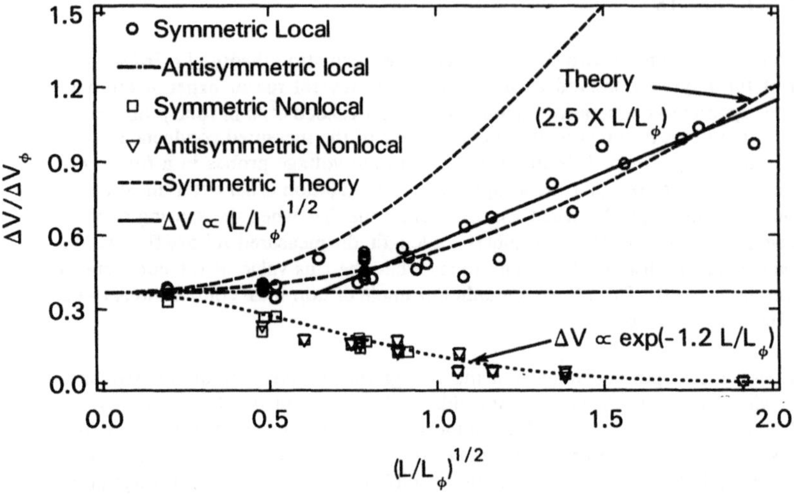

Fig. 4. The length dependence, normalized to L_ϕ, of the symmetric and anti-symmetric parts of the local and non-local voltage fluctuations (normalized to ΔV_ϕ) found in both Au and Sb lines. The recent theoretical predictions are also shown (dashed line) along with a comparison to the ensemble averaging prediction, $\Delta V/\Delta V_\phi \propto (L/L_\phi)^{1/2}$ (solid line)

There have been several recent theories [30-33] that have attempted to obtain the form of the curve that describes the length dependence of the symmetric part of the local conductance fluctuations. All three theories are four-probe theories and yield similar curves. All are based upon the same formalism that led to the original two-probe conductance fluctuation theory. These theories contain the contributions from both the particle-particle and particle-hole channels at H=0. Our experiments are performed at large magnetic fields; the particle-particle contribution to the fluctuations is absent, and thus all the theoretical $\Delta V/\Delta V_\phi$ should be reduced by $1/\sqrt{2}$. In addition, experimentally we find that the prefactor in front of the total conductance fluctuations for wires is generally on the order of 0.6 to 0.8 [3-16]. The dashed curve in Fig. 4 is representative of all the predictions from the recent theories. We have adjusted the theoretical prefactor in front of $\Delta V/\Delta V_\phi$ to make the curve fit our data at L=0. While qualitatively the theoretical curve has the same functional form as our experimental data, the dependence upon L/L_ϕ is different by a factor of $\simeq 2.5$. We do not believe that such an error is possible from the experimental determination of L_ϕ. In fact, in recent work, DIVINCENZO [33] has calculated the decay of the non-local part of the voltage fluctuation. He finds that the voltage should decay as $\exp(-aL/L_\phi)$ with $a = 1.0 \pm 0.1$ in good agreement with our measurements. Thus it appears that our experimental phase coherence lengths are not in error by a factor of 2.5. It is also interesting to note that, if the phase coherence length is determined from the decay of the magnetic field auto-correlation function, the value of L_ϕ is found to be about twice as large. For reasons discussed below, we believe that there is a real problem with the theory for the auto-correlation function, and that a weak localization determination of L_ϕ gives much better agreement with the magnitude of the observed h/e and random conductance fluctuations [15].

5. L_ϕ - AUTO-CORRELATION FUNCTION

We have now several methods by which to estimate L_ϕ: weak localization, the fluctuation amplitudes g_n, and the decay of the non-local terms. From the theory [23], we have that the decay of the auto-correlation function $\langle \Delta g(H) \Delta g(H+\Delta H) \rangle$ is related to the phase coherence length. The characteristic field for the decay of the auto-correlation is $H_C = zh/ewL_\phi$ where w is the width of the wire being studied. This is a fourth method by which we can infer the value of L_ϕ in a particular wire. The amplitude of the fluctuations, the width of the auto-correlation and the decay of the non-local fluctuations are all applicable in the same magnetic field ranges. The weak-localization effects are useful only near H=0, and moreover, these effects are much more sensitive to time reversal symmetry. Ferromagnetic spins, for instance, will destroy weak localization, but not quench the fluctuations.

We may then use the various methods to attack the discrepancy between theory and experiment which was mentioned above. Measurements [10] made in the regime $L > L_\phi$ indicate that $L_\phi(WL) \simeq L_\phi(g_n)$. Figure 5 contains the results of measurements on an Sb loop. The triangles illustrate the temperature dependence of L_ϕ obtained from the auto-correlation assuming z = 1.0. Also displayed are L_ϕ from weak localization and from the amplitude g_1. In order to extract L_ϕ from the fluctuation amplitude, we have employed the standard formula $g_1 = b\exp(-L/L_\phi)$ with b = 0.2 as found in numerous experiments on metal loops in our laboratory. This leads to measured L_ϕ which are within 15% of those determined from studies of the magnetoresistance associated with weak localization in the wires (after account is taken of the four-probe measurement method and the finite length of the wires). More to the point, the agreement between $L_\phi(g_1)$ and $L_\phi(WL)$ obtains over more than a decade in temperature and a factor of 2 in L_ϕ. The auto-correlation function, on the other hand, yields results which are larger by a factor of about 2 than those from the previous two measurements.

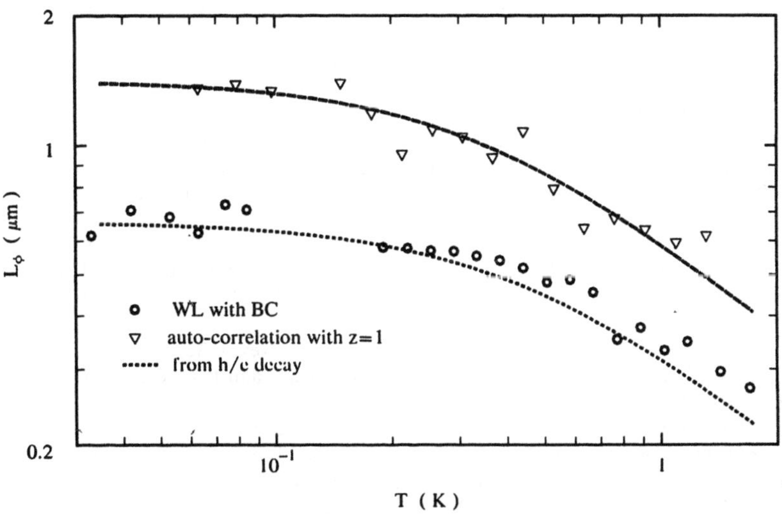

Fig. 5. Comparison of three methods of measuring L_ϕ for the same Sb loop. The diameter and wire width were 850 nm and 38 nm , and R = 876 Ω . The circles are L_ϕ inferred from weak localization (after accounting properly for the boundary corrections (BC) for the measurement). The dotted line is $L_\phi(T)$ obtained from the exponential decay of the oscillation amplitude (g_1). The triangles are L_ϕ determined from the width of the auto-correlation function with z = 1.0, and the dashed line guides the eye.

From various theoretical calculations, one finds values for factor z which range from 1 to 2.4 [19, 23, 34]. The estimates tend to the high end of the range as L_ϕ/w increases to be much greater than unity – as the sample becomes more one-dimensional. In attempt to shade the results as favorably as possible toward the theoretical prediction we will take the factor z to be 1.0 even though the ratio L_ϕ/w for our wire is between 2 and 10. Using this hedged bet for z, we still obtain values for $L_\phi(H_C)$ which are a factor of 2 greater than the values obtained from the other measurements. We re-emphasize that all of the measurements are made on the same samples [10].

The anomaly in the auto-correlation function is not restricted to our laboratory, but neither is it universal. Some experiments on MOSFETs and heterostructures are in agreement with the value z=2.4 [9, 37], while other experiments [38] on heterostructures find $z \simeq 0.4$ as we do. The theoretical prediction for z depends somewhat on the relationship between the thermal smearing length $L_T = \sqrt{(\hbar D/k_B T)}$ and the phase coherence length [23, 37]. For $L_T >> L_\phi$ $z \simeq 0.4$ which would nicely reconcile the $L_\phi(H_C)$ with the other measurements. Unfortunately, our experiments do not meet this criterion; in fact, $L_T < L_\phi$ for all $T > 0.05$ [K].

It would appear from this analysis that there is a problem of a peculiar nature with the theory which permits it to achieve estimates which agree with certain measurements but not with others. All of the calculations are founded in the same perturbation theory for the conductance, and at least one of the three "right" answers is obtained in the four-probe version of the calculation. The exact source of the discrepancy wants further investigation from both experiment and theoretical analysis.

REFERENCES

1. N.D. Mermin and N.W. Ashcroft, *Solid State Physics* (Holt, Rinehart, and Winston, New York, 1976), p. 214, *et seq.*
2. For a discussion of weak localization, see B.L. Al'tshuler, A.G. Aronov, D.E. Khmel'nitskii, and A.I. Larkin, in *Quantum Theory of Solids* ed. I.M. Lifshits (MIR, Moscow, 1982); or D.E. Khmel'nitskii, Physica **126B**, 235 (1984).
3. R.A. Webb, S. Washburn, C.P. Umbach, and R.B. Laibowitz, in *Localization, Interaction, and Transport Phenomena in Impure Metals*, G. Bergmann, Y. Bruynseraede, and B. Kramer Eds. (Springer, Heidelberg, 1985); C.P. Umbach, S. Washburn, R.B. Laibowitz, and R.A. Webb, Phys. Rev. B **30**, 4048 (1984).
4. J.C. Licini, D.J. Bishop, M.A. Kastner, and J. Melngailis, Phys. Rev. Lett. **55**, 2987 (1985).
5. R.A. Webb, S. Washburn, C.P. Umbach, and R.B. Laibowitz, Phys. Rev. Lett. **54**, 2696 (1985). V. Chandrasekhar, M.J. Rooks, S. Wind, and D.E. Prober, Phys. Rev. Lett. **55**, 1610 (1985); S. Datta, *et al.* Phys. Rev. Lett. **55**, 2344 (1985).
6. S. Washburn, C.P. Umbach, R.B. Laibowitz, and R.A. Webb, Phys. Rev. B **32**, 4789 (1985).
7. C.P. Umbach, C. van Haesendonck, R.B. Laibowitz, S. Washburn, and R.A. Webb, Phys. Rev. Lett. **56**, 386 (1986).
8. A.D. Benoit, C.P. Umbach, R.B. Laibowitz, and R.A. Webb, **58**, 2343 (1987).
9. W.J. Skocpol, P.M. Mankiewich, R.E. Howard, L.D. Jackel, D.M. Tennant, and A.D. Stone, Phys. Rev. Lett. **56**, 2865 (1987); Phys. Rev. Lett. **58**, 2347 (1987).
10. F.P. Milliken, S. Washburn, C.P. Umbach, R.B. Laibowitz, and R.A. Webb, Phys. Rev. B **36**, 4465 (1987), and unpublished.
11. R.A. Webb, S. Washburn, and C.P. Umbach, in press Phys. Rev. B.
12. A.D. Benoit, S. Washburn, C.P. Umbach, R.B. Laibowitz, and R.A. Webb, Phys. Rev. Lett. **57**, 1765 (1986).
13. C.P. Umbach, P. Santhanam, C. van Haesendonck, R.A. Webb, Appl. Phys. Lett. **50**, 1289 (1987).

14. S. Washburn, H. Schmid, D. Kern, and R.A. Webb, Phys. Rev. Lett. **59**, 1791 (1987).
15. H. Haucke, A.D. Benoit, C.P. Umbach, S. Washburn, H. Schmid, D. Kern, and R.A. Webb, preprint
16. S. Washburn and R.A. Webb, Adv. Phys. **35**, 375 (1986).
17. Y. Gefen, Y. Imry, and M. Ya. Azbel, Phys. Rev. Lett. **52**, 129 (1984).
18. M. Büttiker, Y. Imry, R. Landauer, and S. Pinhas, Phys. Rev. B **31**, 6207 (1985).
19. A.D. Stone, Phys. Rev. Lett. **54**, 2692 (1985).
20. B.L. Al'tshuler, JETP Lett. **41**, 648 (1985).
21. B.L. Al'tshuler and D.E Khmel'nitskii, JETP Lett. **42**, 359 (1985).
22. B.L. Al'tshuler and B.I. Shklovskii, JETP **64**, 127 (1986).
23. P.A. Lee and A.D. Stone, Phys. Rev. Lett. **55**, 1622 (1985); P.A. Lee, A.D. Stone, and H. Fukuyama, Phys. Rev. B **35**, 1039 (1987).
24. B.L. Al'tshuler and B.Z. Spivak, JETP Lett. **42**, 447 (1985); S. Feng, P.A. Lee, and A.D. Stone, Phys. Rev. Lett. **56**, 1960 (1986).
25. A.D. Stone and Y. Imry, Phys. Rev. Lett. **56**, 189 (1986).
26. Y. Imry, Europhys. Lett. **1**, 249 (1986); B.L. Al'tshuler, V.E. Kravtsov, and I.V. Lerner, JETP Lett **43**, 441 (1986); JETP **64**, 1352 (1986); K.A. Muttalib, J.L. Pichard, and A.D. Stone, Phys. Rev. Lett. **59**, 2475 (1987).
27. M. Büttiker, Phys. Rev. Lett. **57**, 1761 (1986).
28. M. Ma and P.A. Lee, Phys. Rev. B **35**, 1448 (1987).
29. M. Büttiker, Phys. Rev. B **35**, 4123 (1987).
30. Y. Isawa, H. Ebisawa, and S. Maekawa, preprint; J. Phys. Soc. Japan, **56**, 25 (1987).
31. S. Hershfield and V. Ambegaokar, preprint.
32. C.L. Kane, P.A. Lee, and D. DiVincenzo Phys. Rev. in press.
33. D. DiVincenzo and C.L. Kane, submitted to The Physical Review, and D. DiVincenzo, unpublished.
34. R. Landauer and M. Büttiker, Phys. Rev. **36**, 6255 (1987).
35. Y. Aharonov and D. Bohm, Phys. Rev. **115**, 485 (1959).
36. H.G.B. Casimir, Rev. Mod. Phys. **17**, 3 (1945).
37 C.W.J. Beenakker and H. van Houton, Phys. Rev. B, in press.
38 G.L. Timp, unpublished.

Universal Conductance Fluctuations and Quantum Interference Effects in Microstructures

A.D. Stone

Applied Physics, Yale University, P.O. Box 2157,
New Haven, CT 06520, USA

Abstract The low-temperature conducting properties of microstructures are observed to vary in a random manner as a function of magnetic field or carrier density. These fluctuations are sample-specific, but time-independent and reproducible within a given sample. It is now understood that these fluctuations result from random quantum interference of diffusing electrons. We review the recently-developed theory of these novel fluctuation phenomena, which predicts that the order of magnitude of the absolute fluctuations in the conductance of any metallic system is e^2/h, as long as the conductor is probed on length scales of the order of the inelastic diffusion length. Differences between the behavior of the fluctuations measured in a two-probe as compared to a multi-probe geometry are discussed, and some recent results relevant to the multi-probe case are presented. A recent hypothesis relating these time-independent fluctuation effects to low-frequency noise in dirty metals is reviewed, and a decisive experiment for testing the theory is proposed.

1. Introduction

In this article I shall review the current state of the theory of the novel fluctuation phenomena which have been discovered recently in well-conducting metal and semiconductor microstructures. The basic motivation for the theoretical investigations to be described below was to understand the origin of the "reproducible noise" which was observed in the low-temperature conductance of microstructures both as function of magnetic field and carrier density[1,2]. One of the earliest and most striking observations was that of Umbach et al.[3], who studied the magnetoresistance of sub-micron Au and AuPd rings in an attempt to see a normal-metal Aharonov-Bohm effect[1,2]. The initial experiments observed a complicated non-monotonic behavior of the resistance as a function of magnetic field, which was reproducible upon cycling the field in a given sample, but varied in shape from sample to sample, and which increased in size as the temperature was lowered. This reproducible pattern did not oscillate with field with a single dominant periodicity, and even more strikingly, experiments on small (singly-connected) wires revealed apparently quite similar structure, indicating that the phenomenon was not simply-related to the familiar Aharonov-Bohm Effect.

At roughly the same time a variety of experiments on quasi-one-dimensional semiconducting microstructures were finding reproducible noise in the resistance both as a function of magnetic field and of gate voltage, or equivalently, carrier density [4,5], in the highly-conducting (metallic) limit. Because the carrier densities are much lower

Springer Series in Solid-State Sciences Vol. 83: **Physics and Technology of Submicron Structures**
Editors: H. Heinrich · G. Bauer · F. Kuchar

in these devices than in true metals, they are much closer to the fully one-dimensional limit, in which carriers at the fermi energy have only one or a small number of transverse states to occupy; therefore this behavior was interpreted as arising essentially from one-dimensional density of states effects, and was not seen as related to the effects observed in metals. However these effects were also sample-specific and increased with decreasing temperature or decreasing size of the system.

Because these effects were sample-specific, they naturally raised a basic theoretical question: What is the magnitude of the fluctuations induced in the conductance of a normal metal due to changes in its microscopic impurity configuration, or due to the imposition of an external probe such as a magnetic field, or gate voltage? In order to answer this question a basic change was needed in the theorist's *weltanschaung*, and this ocurred in two stages. First, the theorist had to abandon the standard notion that the only quantity of interest when calculating a linear response coefficient was its average over an ensemble of macroscopically similar, but microscopically different specimens. Second, the full theory of these fluctuations had to take into account the actual measuring configuration and device geometry of a given experiment; in particular, it was necessary to develop somewhat different approaches to describe "two-probe" and "multi-probe" measurements.

2. Review of Drude Theory of the Conductivity of Metals

In order to explain the physical origin of these novel fluctuations, it is useful to review briefly the elementary Drude theory of the conductivity of metals. For an infinite perfect crystal it is only the disorder in the lattice which gives rise to resistance. Roughly speaking this disorder comes from two sources, vibrations of the lattice atoms away from their equilibrium positions in the perfect crystal, and impurities. Both of these cause deviations of the crystal from perfect periodicity, giving rise to scattering of the conduction electrons with a total scattering rate $1/\tau_s = 1/\tau + 1/\tau_{in}$, where τ is the elastic (impurity) scattering time, and τ_{in} is the inelastic (phonon) scattering time. Assuming that this scattering exerts a damping force on the electrons proportional to their velocity, one can calculate the steady-state current in response to an electric field, and one finds $J = \sigma E$ with $\sigma = ne^2\tau_s/m$, where n is the electron carrier density, e is its charge, and m is its mass. The conductance is $G = \sigma A/L$, where A is the cross-sectional area of the sample and L is the length. Since $\tau_{in} \to \infty$ as $T \to 0$, but τ does not, in the standard picture of a metal the conductance increases with decreasing T until it saturates at a finite residual conductance, $G = \sigma_0 A/L$, where $\sigma_0 = ne^2\tau/m$.

This picture has a questionable feature when applied to very small conductors at low temperature, since it treats elastic and inelastic scattering as having equivalent effects on the conductivity and yet the two processes are physically very different. Elastic scattering from impurities is a fundamentally *static* phenomenon, the impurities represent a fixed random potential through which the electron waves propagate. When a wave is incident on an array of fixed but randomly placed scattering centers, in general multiple-scattering is very important; and interference of multiply-scattered waves makes a large contribution to the total transmitted intensity. Since the array is random, the interference will lead to a complicated, rapidly-varying spatial distribution of the transmitted intensity; such a distribution is familiar in laser optics, where it is known as a *speckle pattern.* On the other hand, scattering from phonons is a *dynamic* phenomenon; successive electrons will scatter from different distorted configurations of the lattice and any interference effects will be lost since an experiment essentially time-averages over many uncorrelated configurations. In other words, inelastic scattering introduces irreversibility and true phase incoherence into the conduction process, whereas elastic scattering allows interference effects to occur, albeit interference effects with a complicated, apparently random character. This suggests that in small samples the Drude picture should break down at sufficiently low temperatures, because a regime is entered in which the probability that the electron will undergo even a single phase-randomizing inelastic collision in traversing the sample is negligible, and then the random interference due to elastic scattering might be manifested in the behavior of the conductance.

3. Definition of the Metallic Regime

This point of view naturally leads one to consider the problem of the elastic diffusion of waves. One can think of the wave motion through the impurity potential as a random walk where the step length is the elastic mean free path, $l = v_f\tau$, (v_f is the fermi velocity). We shall define a metal for our purposes as any conductor in which the sample length, L, is much longer than elastic mean free path, and much shorter than the localization length, ξ. The sample-specific effects will occur in microstructures when the distance the electron can diffuse elastically without undergoing an inelastic collisions, $L_{in} = \sqrt{D\tau_{in}}$, is much longer than l, and will be largest when $L \leq L_{in}$. In this regime there are two types of interference effects which influence the transport properties of microstructures, the fluctuation effects discussed above, often referred to as "universal conductance fluctuations"(UCF); these will be the main subject of this article. However there also the "weak localization" effects, which have been widely-studied since they were predicted and observed about nine years ago[6,7].

4. Contrast between Weak Localization and UCF

Weak localization effects are generic, not sample-specific, and relate to the behavior of the ensemble-averaged conductance. The essential physical idea[6] is that for a wave performing a random walk the probability of return to the origin is larger than for a classical random walk, because one has to add first the amplitudes to return along all paths and then square that sum. In that sum, most of the terms have no definite phase relationship, however there are pairs of terms which always interfere constructively, since they arise from time-reversed pairs of classical paths, which always have the same phase in the sum over paths. Therefore, when the diffusion is phase-coherent, the conductance is somewhat smaller, when averaged over impurity configurations, than it would be in Drude theory, which neglects this constructive backscattering. The theory of weak localization predicts many important and interesting effects, such as a *decrease* in the average conductivity as $T \rightarrow 0$, a negative average magnetoresistance on very small field scales, and the normal-metal Aharonov-Bohm Effect with fundamental period h/2e[1,2,7], however these effects are not our central focus at present. We emphasize that although both UCF and weak localization are interference effects, they differ fundamentally, because the latter arise from a special subset of the interference terms which have the same sign in all samples, and which therefore contribute to the *average* conductance. Whereas, as we shall see below, the UCF arise from interference terms from all diffusive trajectories, which have no fixed sign and thus are zero on average. Furthermore, since weak localization constructive backscattering arises from time-reversal symmetry, these effects only occur at relatively low fields, whereas the UCF effects persist to extremely high fields[1]. Finally, UCF are not in any sense localization effects, they occur in all metals, even in three dimensional systems which would remain metallic in the infinite volume limit.

5. Landauer Approach to Quantum Transport

In order to understand why conductance fluctuations as a function of magnetic field or fermi energy arise from the random interference effect described above, it is useful to adopt the Landauer approach to the resistance of a finite, *open* disordered conductor. We begin by considering the simplest model of a two-probe conductor, having warned the reader that the theory will be somewhat different for the multi-probe case. Landauer[8] proposed many years ago that one should think of a quantum resistor, in the limit where it only produces elastic scattering, as one complex scattering center regulating the current flow between two macroscopic reservoirs (where all the energy dissipation occurs). When a voltage is applied the left-hand reservoir will have a higher chemical potential, $\mu_1 = \mu_2 + eV$ than that of the the right-hand reservoir, μ_2, and a current will flow from 1 to 2. Landauer argued that one should think of each reservoir as emitting an incoherent stream of electrons into the left and right-going momentum states (respectively) of the leads connecting the sample and the reservoirs (the leads are taken to be perfectly ordered for convenience). Assume that the reservoirs are at low temperature so that their fermi distributions are well-approximated by step functions at μ_1 and μ_2 respectively. Then the current carried by the left-going states of reservoir 2 and right-going states of reservoir 1 will cancel for all states below μ_2. In the energy interval eV between μ_2 and μ_1 there are right-going filled states from reservoir 1, but no left-going filled states from reservoir 2. Thus there is a net right-going current given by the product of the charge of the electron, the number of states in that interval, the velocity of each momentum channel, v_i, and the transmission probability into all channels from that state:

$$I = e\sum_i^M v_i \frac{dn_i}{dE} eV \sum_j^M T_{ij} \,. \tag{1}$$

Using the fact that for a quasi-1d density of states, $dn_i/dE = 1/hv_i$, (1) yields an expression for the *dimensionless conductance*, g

$$g = G/(e^2/h) = \sum_{i,j}^M T_{ij} = Tr(\mathbf{t}\mathbf{t}^\dagger) \,, \tag{2}$$

where $\mathbf{t}$ is the transmission *matrix* describing the scattering between the various "channels", i.e. states of different transverse momentum at the fermi surface. If the system is exactly one-dimensional then there are only two ingoing and outgoing channels at each energy, so the scattering matrix of the resistor just consists of two reflection and two transmission amplitudes (at the fermi energy). If the system has finite transverse area, A, then in general there will be a certain number, M, of propagating channels, and the transmission and reflection amplitudes become M x M matrices. It is important to remember that M is of order Ak_f^2, where k_f is the fermi wavevector (typically $k_f \sim \AA^{-1}$ in a metal); thus even in ultra-small metal wires a few hundred Å in cross-section, $M \sim 10^4 - 10^6$, (2) contains $\sim 10^{10}$ terms, and may behave quite differently than an ideal 1d system. Although the above argument arriving at (2) is only heuristic, it is possible to derive the same result rigorously from linear response theory, given certain boundary conditions (infinite perfectly-ordered leads at fixed voltage), which are chosen to simulate the effects of the massive electrodes (or reservoirs)[9,10].

6. Can a Perfect Conductor Have a Resistance?

Equation (2) was first derived by Fisher and Lee [9] (and by Economou and Soukoulis in the 1d case[11]), and its correctness was questioned because it has the feature that in the limit when the elastic scattering in the sample goes to zero, the resistance in (2) does not go to zero, but goes to a constant, $R_0 = [Me^2/h]^{-1}$; i.e. according to (2) the resistance of a "perfect" conductor is not zero. Landauer's original argument was only for the 1d case, and instead of arriving at (2), he found $g = T/(1-T)$[8], which gives the intuitively expected behavior, zero resistance when T=1. However, his argument relied essentially on making a distinction between the voltage drop across the resistor(if it were somehow measured in the ideal leads,without affecting the conductance), and the voltage difference between the reservoirs serving as current source and sink. For a two-probe measurement this distinction is not meaningful. In fact Imry has argued that the resistance,R_0, should be viewed as a true contact resistance, which is unavoidable in a two-probe measurement[12]. Conversely, in the case of a typical four-probe measurement it is clear from the results to be discussed below, that it is unreasonable to suppose that the voltage drop across some region of a small sample can be measured without the presence of the voltage probes strongly affecting the results of the measurement. It is an open question whether there exists an experimental technique for measuring the conductance as defined in Landauer's original derivation; whereas the existence of the contact resistance,R_0, in two-probe measurements, has recently been convincingly demonstrated by the observation in ballistic point contacts of quantized conductance steps corresponding precisely to the values of R_0 predicted by (2) (as the channel number is varied)[13]. Therefore the use of (2) in the two-probe theory is justified; the many issues relating to the question of the correct Landauer formula, have recently been reviewed elsewhere [10].

7. Qualitative Origin of the Reproducible Noise

Having established that the conductance is given by the many-channel generalization of the transmission coefficient through the sample, it is natural then to hypothesize that the origin of the reproducible noise is a random interference effect of the type discussed in general terms above. The conductance is just the sum of all the transmission coefficients, and each transmission coefficient is simply-related to the Fourier transform of the Green function evaluated at the fermi energy, with spatial arguments chosen at points $\underline{r}_0, \underline{r}_L$ on opposite sides of the sample, $T_{ij} \sim |G(\underline{r}_0, \underline{r}_L, E_f)|^2$[9,10]. We can imagine evaluating this Green function as a path integral in the semi-classical (WKB) approximation in which only the classical paths (and Gaussian fluctuations) will contribute, so that very schematically we have

$$g \sim T_{ij} \sim \sum_{\alpha,\beta} exp[i(S_\alpha - S_\beta)] \; , \tag{3}$$

where the cross-terms in the sum over paths represent interference of paths α and β between the same points on opposite sides of the system, and these terms have a fixed but arbitrary phase.

These cross-terms are then expected to be zero when averaged over impurity configurations, but in a given sample, with a fixed impurity potential, they make some contribution, the typical magnitude of which remains to be determined. However, if this contribution is not negligible, then it apparently will lead to qualitatively the correct experimental behavior. In particular, any perturbation which affects the phase of the electron along paths traversing the sample will cause these interference terms to vary in a complicated manner which should induce fluctuations in g. A magnetic field is such a perturbation; in the approximation in which the magnetic bending radius is much less than the elastic scattering length, it just changes the phase S_α by an additional increment $(e/\hbar)\int \underline{A} \cdot d\underline{S}$ along the original trajectory. Changing the field a sufficient amount will rearrange the phases in (3) completely causing g to "fluctuate", however returning the field to its original value will also return g to its original value just as is observed in the experiments. Similarly, changing the carrier density, i.e. the fermi energy, will also rearrange the phases in (3) in a complex manner leading to a chaotic variation in g.

In both cases there must be a natural scale B_c and E_c determining how much one needs to change the field or the fermi energy in order to alter substantially the phase of electrons traversing the sample, thus leading to a substantial variation in g. For the case of the magnetic field this scale is determined by demanding that the relative phase shift due to the field acquired along two different electronic trajectories traversing the sample between the same two points become of order unity. This is equivalent to demanding that the flux enclosed by two typical paths be of order the quantum of flux,$\phi_0 = h/e$. Under the reasonable assumption (which can be rigorously confirmed[14]) that the area enclosed by two random walks across the sample scales with the area, A, of the sample, this simple argument predicts that $B_c \approx \phi_0/A$. A similar simple argument yields the estimate $E_c \approx hD/L^2$[15]; E_c is then the inverse time to diffuse across the sample.

8. Numerical Calculation of $g(B)$

In order to test whether the complicated behavior suggested by the previous argument, and observed experimentally, would naturally follow from the solution of a microscopic model, I performed numerical calculations of the conductance of small samples, using exactly the model for which (2) was derived, a finite disordered region attached to infinite perfectly-ordered leads[16]. The disordered region was described by an Anderson random tight-binding model, of the type originally studied by Lee and Fisher[17]; and the transmission coefficients needed for evaluating $g(B)$ using (2) were obtained using the same recursive Green function technique. The new feature of these simulations was that $g(B)$ was studied for a *fixed impurity configuration*, instead of averaging the results over many configurations. The results showed that a complex, noise-like dependence of the transmission on magnetic field, qualitatively similar to that observed experimentally, was indeed a property of the solutions of the one-particle Schrödinger equation in a random potential(see Fig. 1). Moreover, by studying the scaling of the spacing of the peaks and valleys as a function of the area of the sample, the simulations showed that their separation varied inversely with the sample area, with a typical spacing corresponding to a few flux quanta through that area, just as was predicted by the simple heuristic argument given above.

However there was one basic question left open by the qualitative agreement of the simulations with experiment. How was one to compare the magnitude of the fluctuations in the simulations with those observed experimentally? When stated as fractional fluctuations in the resistance, the fluctuations found in the simulations[16] were two orders of magnitude *larger* than those observed experimentally. Conversely, the simulations were limited by computational constraints to treating samples with at most 50 transverse channels, whereas, as noted above, the experimental samples corresponded to about 10^5 channels, so if the relative fluctuations decreased rapidly with the number of channels (as suggested by some simple statistical arguments to be discussed below), then the result obtained from the simulations was too *small* to account for the magnitude of the experimental effect.

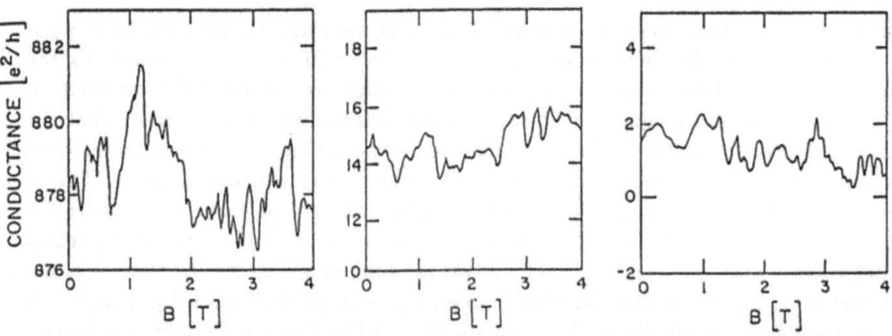

Fig. 1: Comparison of aperiodic magnetoconductance fluctuations in three different systems. (Left) $g(B)$ in 0.8 micron diameter gold ring, reprinted with permission of Washburn and Webb[1] (the rapid Aharonov-Bohm oscillations have been filtered out). (Right) numerical calculation of $g(B)$ for an Anderson model using the technique of [16]. (Center) $g(B)$ for a quasi-1d Silicon MOSFET, data from [24],reprinted with the permission of Skocpol et al. Conductance is measured in units of e^2/h, magnetic field in Tesla. Note the three orders of magnitude variation in the background conductance while the fluctuations remain order unity.

9. The Ergodic Hypothesis

How to calculate analytically, or even estimate the size of the magnetoresistance fluctuations in a given sample was not immediately clear, for the following reason: as the magnetoresistance pattern was completely different for each impurity configuration, $g(B)$ must depend very sensitively on the coordinates, $\{R_i\}$, describing all the impurities, hence finding the function $g(\{R_i\}, B)$ analytically, seemed like an impossible task. On the other hand, the qualitative explanation of the fluctuations given above, that they occur because the field rearranges the phases in (3) in a complicated manner, suggested that a statistically similar rearrangement might occur if instead of changing the field for fixed impurity configuration, one changed the impurity configuration for a fixed value of the field. Therefore the root-mean-squared (*rms*) fluctuations in g, induced by changing field (or fermi energy), might be expected to be of the same order as the statistical fluctuations, $rms(g)$, induced by changing the impurity configuration. This postulate was called an *ergodic hypothesis*[18], because it stated that an average of any statistical property of the conductance over a large enough field range (many times B_c), would be equal to an ensemble average. The approximate validity of this ergodic hypothesis was easily established by comparing the two averages in numerical simulations[18,19].

10. Classical Estimate of $rms(g)$

Using the ergodic hypothesis, it was possible to estimate the size of the $g(B)$ fluctuations, by estimating the size of the statistical fluctuations of the conductance to be expected as a function of impurity configuration. The simplest approach, which will be referred to as the *classical estimate*, is to start from the zero temperature Drude formula, $\sigma_0 = ne^2\tau/m$ and note that σ_0 depends on the impurity configuration only through the scattering time τ, which is inversely proportional to the impurity density, n_i. Therefore the fractional conductance fluctuation $<\delta g^2> / <g>^2 \sim <\delta\tau^2> / <\tau>^2 \sim <\delta n_i^2> / <n_i>^2$, where the last proportionality only holds for small relative fluctuations. It is reasonable to assume that the impurities are typically distributed in an uncorrelated manner with uniform probability throughout the sample, thus their relative density fluctuations are like those of an ideal gas, and should be of order the ratio of the volume, V_1, containing on average one impurity, to the total volume L^d of the system. Since $V_1 \leq l^d$, we obtain the classical estimate,

$$<\delta g^2> / <g>^2 \propto (l/L)^d \quad \Rightarrow \quad <\delta g^2> \propto L^{(d-4)} . \tag{4}$$

According to this classical estimate the relative conductance fluctuations self-average inversely with the volume in all dimensions, and even the absolute fluctuation goes to zero as $L \to \infty$ for $d < 4$, (where the latter result was obtained by using Ohm's Law for the average metallic conductance). This estimate, if used to predict the order of magnitude of the fluctuations observed experimentally, predicts much too small a value.

11. Numerical Calculation of the Size-dependence of $rms(g)$

Although it was not possible to simulate numerically conductors with as many scattering channels as the experimental systems, it was straightforward to study the variation

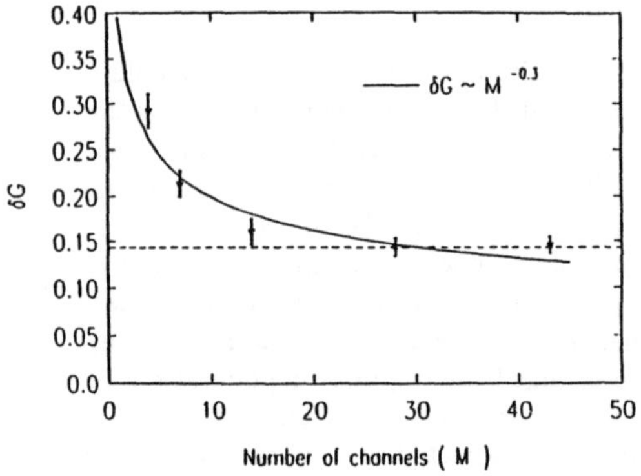

Fig. 2: Size-dependence of the fluctuation amplitude, $\delta G = rms(G)$ (in units of e^2/h), measured by scaling at fixed shape in 2d. The solid line is a fit to the weak decay $\delta G \sim M^{-0.3}$ whereas the dashed line indicates a saturation to a constant value of order e^2/h which is the result obtained from the analytic analysis.

of $rms(g)$ as function of the sample size (number of channels) for samples with 5 to 50 channels. The calculation was done for 2d samples and the shape of the sample was kept fixed as the size was varied (so that $< g >$ remained constant), and at each size the fluctuations of $g(B)$ around its average were calculated over a range of $10\phi_0$ for many different impurity configurations[16]. The results are plotted in Fig. 2 below. We see that although there is some initial variation of $\delta G = rms(G)$, rapidly its value appears to become very weakly size-dependent. This behavior strongly differed from that expected from the classical estimate above. In fact the simulations appeared most consistent with a saturation of $rms(g)$ to a size-independent value, but it was not possible to rule out some weak size-dependence.

12. Analytic Calculation of $rms(g)$ and Correlation Functions

If one accepts the ergodic hypothesis, it then becomes possible to calculate the relevant statistical quantities analytically, using impurity-averaged perturbation theory in the small parameter $(k_f l)^{-1}$. One need only calculate the *statistical* variance, $var(g) =< \delta g^2 >$ in order to calculate the typical size of the $g(B)$ fluctuations in a given sample ($\delta g = g - < g >, rms(g) = \sqrt{var(g)}$). This approach was taken by Lee and Stone[18], and independently by Altshuler [20], and generalized to non-zero temperature by Lee et al.[19] and Altshuler and Khmelnitskii[21]. In addition to calculating the magnitude of the fluctuations, it was possible to obtain their typical spacing in field or fermi energy, by defining the *conductance correlation functions*:

$$F(\Delta B) =< \delta g(B)\delta g(B + \Delta B) > \tag{5}$$

$$F(\Delta E) =< \delta g(E_f)\delta g(E_f + \Delta E) > \tag{6}$$

where the angle brackets indicate an ensemble average. When $\Delta B = 0, \Delta E = 0, F = var(g)$ for that value of the field and fermi energy, however as the value of $\Delta B, \Delta E$ increases in a given sample, the fluctuation from the average will tend to vary randomly in sign compared to its original value, and on average the product will tend to zero over a scale B_c, E_c determined by the typical scale of variation of $g(B)$ or $g(E_f)$. Thus the half-width of these correlation functions will give the typical spacing of the peaks and valleys in $g(B)$ or $g(E_f)$. In analyzing experimental data these correlation functions can also be calculated using the ergodic hypothesis to replace the average over samples by an average over field points spaced much more widely than B_c[19]. These correlation functions were calculated using diagrammatic techniques; a detailed discussion of the calculation is given in ref. [19].

13. Fluctuations in the Two-Probe Theory at T=0

The basic results of the analytic calculations for the two-probe geometry can be summarized as follows:

$$var(g) \sim 1 \qquad\qquad L \leq L_{in} \,. \tag{7}$$

For samples smaller than the inelastic diffusion length the absolute conductance fluctuations are constant and $rms(G) \approx e^2/h$, independent of the average conductance or the size of the system. I emphasize that the size-independence of $rms(g)$ for $L < L_{in}$ only holds in a two-probe geometry. Although of order unity, the exact value of $rms(g)$ is not a universal number, it depends on the shape of the conductor, the presence of spin-orbit scattering, and of a magnetic field [18-20]. For example, the value of the statistical variance of g measured for an ensemble in a magnetic field is exactly half of its value in the absence of a field. It is the fact that $rms(g)$ is independent of the *average* value of g (and therefore to a large extent of material parameters) that is the most surprising aspect of the result, and in this important sense the fluctuations are "universal". The excellent agreement of this prediction of the theory with experiments on different systems with average conductance varying over two orders of magnitude is illustrated in Fig. 1 above. The analytic prediction for $rms(g)$ is also in precise agreement with the numerical simulations discussed above; this is illustrated in Fig. 1 and detailed comparison is given in [18]. A fundamental explanation for the universality of $rms(g)$ in the two-probe theory has been given recently by appealing to the general properties of random matrices[22,23]. This analysis shows an exact analogy between the fluctuation properties of small conductors and the statistical properties of the spectra of complex nuclei; a topic of great interest in nuclear physics for more than 30 years[23]. Unfortunately space does not allow an adequate discussion of this approach in the present article.

The typical spacing of the fluctuations in field and fermi energy, obtained from the half-width of the correlation functions, are

$$B_c = c_1(\phi_0/A) \tag{8}$$
$$E_c = c_2(hD/L^2) \tag{9}$$

where c_1, c_2 are constants of order unity; these results confirm the intuitive arguments given earlier. The constants c_1, c_2 depend on shape and can be calculated (as can the value of $rms(g)$) by numerically performing some simple summations given in [19], however the correct numerical value is expected to depend on the measuring configuration for sample sizes of order L_{in}, so precise comparison of correlation functions for the two-probe theory with experimental results for samples measured in a four-probe configuration is unreliable. However for samples much larger than L_{in} a precise comparison is possible, and for smaller samples the multi-probe generalization of the theory will also allow such a comparison; both of these points will be discussed below. Plots of the theoretical correlation functions and comparisons with numerical results showing very good agreement are given in [18,19].

14. Fluctuations at Finite Temperature

For samples larger than L_{in} the classical self-averaging behavior is restored, the analytic result in this limit is

$$var(g) \sim (L_{in}/L)^{4-d} \qquad L \gg L_{in} \tag{10}$$

(where I have omitted discussion of some subtleties relating to the 1d case [19,20]). Thus the relative conductance fluctuations decrease inversely with the volume, leading to self-averaging as $L \to \infty$(for any non-zero temperature). Note however the crucial point that the volume scale which enters in the correct theory is not the volume containing one impurity, $V_1 \sim l^d$, but instead is the phase-coherent volume, $V_{in} \sim (L_{in})^d$,(which goes to infinity as $T \to 0$). This provides another way of stating the fundamental result. If one argues that the total conductance of the sample is obtained by addition (in parallel and series) of the conductance of smaller regions of the sample each of which fluctuates in an uncorrelated manner, then the classical estimate corresponds to the hypothesis that the conductance of each volume of size l^d fluctuates in an uncorrelated manner, whereas the correct statement is that due to the long-range phase-coherence of the wave-functions (even in the presence of elastic scattering), it is only the much larger volumes of size V_{in} that actually have uncorrelated fluctuations. Hence the fluctuations are much larger than predicted by the classical estimate of (4). Given this point of view it is not surprising that the result for $var(g)$ in (10), in the limit $L \gg L_{in}$, can be obtained simply by assuming each phase-coherent volume has an intrinsic fluctuation $\delta g \sim 1$ and then using parallel and series addition of resistances to estimate the total fluctuation for the sample[22,24].

For example, in the simplest case of a wire which is narrower but much longer than L_{in}, with $< g >\gg 1$, we have $\delta g/g \approx \delta R/R$, where R is the total resistance, which we assume to be the sum of the resistances, R_{in}, of each segment of length L_{in}. Hence $\delta R = \sum_i \delta R_i \approx \sqrt{L/L_{in}}(\delta g_{in} R_{in}^2)$. Since the conductance fluctuation of each region, $\delta g_{in} \approx 1$, and $R_{in} \sim L_{in}^2$, we obtain $\delta g \sim (L_{in}/L)^{3/2}$, consistent with the rigorous 1d result given above. This behavior of the conductance fluctuations as a function of temperature has been well-confirmed by a number of experiments[1,24,25].

The correlation ranges B_c and E_c, for samples much larger than L_{in}, are just as given above for the T=0 case, except that either L_{in} or $L_T = \sqrt{hD/kT}$ replaces all sample dimensions longer than L_{in}[19]. Recently Beenakker and Van Houten [26] have calculated the coefficient c_1 needed to obtain the exact value of B_c in this case, as well as the coefficient of $var(g(T))$. In this limit these values should be independent of measuring configuration to a good approximation, and should provide a means of obtaining from experimental measurements of the correlation function accurate estimates of L_{in}, and in the quasi-1d case, the sample width. Therefore UCF measurements should eventually provide a useful tool in characterizing microstructures. At present there seems to be a larger discrepancy (approximately a factor of two) between theoretical predictions and experimental measurements than is expected on statistical grounds[27], so further work may still be required before such applications are completely reliable.

15. UCF and the h/e Aharonov-Bohm Effect

The theory of UCF also plays an important role in explaining the observability and nature of the normal-metal Aharonov-Bohm Effect with fundamental period h/e which has recently been observed in small metal and semiconducting doubly-connected microstructures. This topic has received extensive discussion in two recent reviews[1,2], so I will only treat it briefly, focusing on its connection to the theory of UCF. As noted above, periodic magnetoconductance oscillations with fundamental frequency h/2e had been predicted based on weak localization theory, and observed in metal cylinders and arrays of rings some time ago[1,2]. Conversely, arguments based on the 1d Landauer formula suggested that the fundamental period in a single ring (without ensemble- averaging) should be h/e [28]. However the initial experiments of Umbach et al.[3] on small single rings described above failed to detect convincingly *any* dominant periodicity to the magnetoresistance oscillations.

The explanation for these inconclusive results is now clear. In the experiments the magnetic field penetrates both the aperture and the annulus of the ring. From UCF theory we know that the flux in the annulus will give rise to magnetoresistance oscillations with no dominant period; a Fourier power spectrum will be peaked at zero frequency with half-width of order $B_c \approx (h/e)/A$, where A is the area of the annulus. In a ring, on top of this "low-frequency noise", there will be the periodic oscillation, due to the flux in the aperture, with fundamental period $B_0 = (h/e)/a$, where a is the area of the aperture. The low-frequency noise due to the aperiodic $g(B)$ fluctuations will be convolved with the periodic component causing a complex splitting of the periodic component. If the two scales B_c and B_0 are comparable in size, i.e. if $A \approx a$, then the dominant frequency of the periodic component will be unrecognizable. This was the case in the earlier experiments. Later experiments, motivated partially by the numerical simulations described above[16], employed larger aspect ratios and clearly demonstrated the presence of h/e A-B oscillations in single rings[1,2]. Their statistical behavior is very similar to the aperiodic fluctuations; they have rms magnitude of order e^2/h, and can be calculated from a conductance correlation function of the type defined above. The earlier experiments which observed only the h/2e weak localization oscillations were in samples much larger than L_{in}, and thus the h/e oscillations, which like the aperiodic fluctuations are zero on average, had self-averaged to an unobservable level. Recently the detailed theory of the h/e A-B Effect in normal metals, including the generalization to four-probe configurations has been worked out by several groups[15,29,30].

16. Limitations of the Two-probe Theory

The distinction between two-probe and multi-probe measurements is not a firm one experimentally; all experiments attach voltage probes somewhere on the apparatus.

Roughly speaking a two-probe measurement refers to an experiment in which the voltage probes are attached between regions of the sample which one expects to be approximately equipotentials, and which are relatively far from the region which is expected to dominate the total resistance. The theory described above however makes a firm distinction by only imposing boundary conditions corresponding to attaching two perfect leads. The question then is under what circumstances are real experiments likely to be described by the two-probe theory? The full answer is not clear at this point, but it is reasonable to assume that if the voltage probes are attached at points separated by a distance much greater than L_{in}, then they don't affect the measured conductance fluctuations much, because we have seen that in this regime we can treat each phase-coherent region as if it fluctuates independently, and there are many such regions which are not within L_{in} of the voltage probes. On the other hand, once the spacing of the probes,L, approaches L_{in}(which is often the case in microstructures at low temperature), there should be some effect of electrons diffusing into the probes. That is why the exact value of $rms(g)$ calculated for $L \approx L_{in}$ in the two-probe theory cannot be compared to four-probe experiments, although it correctly predicts the right order of magnitude in *all* experiments. Moreover, in the limit where $L \ll L_{in}$, the "sample" between the points where the probes are attached is only a small fraction of the phase-coherent region which can contribute to interference of electrons, and any correspondence between the two-probe theory (which does not allow phase-coherent excursions into the perfect leads and back) and the experimental measurements is likely to have broken down.

The seriousness of this problem is indicated by two experimental observations. First, in all samples in which probes are attached with spacing of order or less than L_{in} the magnetoconductance, $g(B)$, defined simply as the ratio of the imposed current to the induced voltage drop, is found to be asymmetric around B=0. However it is easy to show that the conductance given by (2) must be symmetric in field, due to an exact symmetry of the S-matrix. Second, it is observed that voltage probes placed directly across from one another perpendicular to the current flow exhibit voltage fluctuations around zero average voltage drop (it is necessary then to consider only the field-symmetric part to eliminate the Hall effect). If regarded as conductance fluctuations of the portion of the sample between the probes this obviously leads to values of $rms(g)$ much greater than unity[24,25]. Of course, it is now more sensible to treat the voltage fluctuations themselves as a property of the full phase-coherent region, including portions of the voltage probes, and regions of the sample not between the probes up to a distance of order L_{in}. Therefore it is necessary to generalize the theory to a multi-probe geometry in which the spacing of the "voltage probes" is not automatically set equal to the phase-coherent region contributing to the fluctuations, as it is in the two-probe theory.

17. Multi-probe Landauer Formula

Several groups independently undertook to generalize the conductance fluctuations theory to a multi-probe geometry, either by incorporating boundary conditions representing many leads into the perturbative calculations based on the Kubo formula[32-34], or by generalizing the Landauer formula of (2) to include many leads[35], which then could be used to perform microscopic calculations numerically[36]. Because of its simplicity, we shall focus on the generalized Landauer formula, first proposed by Büttiker[35] on the basis of phenomenological arguments.

Büttiker considered a disordered conductor of arbitrary shape to which are attached four(possibly disordered) probes which ultimately are connected by perfect leads to four reservoirs at different potentials, $\{V_n\}$. He assumed that all the leads were macroscopically the same, i.e. their was no qualitative distinction between current and voltage leads. Then he evaluated the current flowing into or out of each reservoir in response to the potential differences by an argument identical to that given above to justify (2): the current flowing between two reservoirs with a potential difference $\Delta V = V_1 - V_2$ was taken to be $(e^2/h)T_{12}\Delta V$, where T_{12} is the transmission coefficient from one to two in the one-channel case, and the trace of the transmission matrix times its hermitian conjugate in the many-channel case. This immediately yields a straightforward generalization of (2) to a system with N leads, for which (in linear response) the currents are related to the voltages by a conductance matrix, g:

$$I_m = \sum_{n=1}^{N} g_{mn} V_n \qquad \begin{cases} g_{mn} = (e^2/h)\, Tr\{t_{mn} t_{mn}^{\dagger}\} & m \neq n \\ g_{nn} = (e^2/h)\, Tr\{r_{nn} r_{nn}^{\dagger} - 1\} & \end{cases} \tag{11}$$

where the conductance coefficients $\{g_{mn}\}$ are given in terms of the transmission and reflection matrices between leads m and n (note that m,n are lead indices, channel indices have been suppressed for simplicity). Recently Stone and Szafer[10] have rigorously derived this formula from linear response theory in the absence of a magnetic field, and have shown that this approach is equivalent to the approaches based on the Kubo formula (except for the manner in which the two approaches introduce inelastic scattering). Büttiker has assumed that this formula is valid even in the presence of an arbitrary magnetic field; a substantial assumption whose range of validity is unknown[10]. However, the diagrammatic calculations indicate that it is correct in good metals for weak fields[32], and so, following Büttiker[35], we shall employ this formula to discuss the symmetry of voltage fluctuations in well-conducting microstructures with $\omega_c \tau \ll 1$.

18. Explanation of the Asymmetric Voltage Fluctuations

A merit of the multi-probe Landauer formula, (11), is that it can easily be used to resolve the apparent paradox presented by the asymmetry of the $g(B)$ fluctuations observed in the multi-probe measurements. The main objection to observing an asymmetric $g(B)$ was that in the case where the voltage probes are separated some distance along the sample, it is natural to assume that an experiment at fixed current must be measuring a diagonal element of the resistivity tensor, which must be field-symmetric due to time-reversal symmetry as embodied in the Onsager relations. However, as we shall see from (11), time-reversal symmetry does *not* imply that the voltage measured at fixed current between two widely-separated probes must be symmetric in magnetic field.

To see this we consider (11) for a multi-probe system with the typical experimental boundary conditions, a current I_1 is fed into lead 1 and taken out of lead 2, and the voltage difference $V_3 - V_4$ is measured between two different leads. The quantity of interest then is $R_{12,34}$, where we can define all possible resistances by

$$R_{m'n',mn} \equiv \frac{V_m - V_n}{I} \qquad \begin{cases} I = I_{n'} = -I_{m'} \\ I_l = 0 \quad \text{for} \quad l \neq n', m' \ . \end{cases} \tag{12}$$

The currents $\{I_m\}$ in (11) can be considered as forming an N-component vector; since the matrix $\mathbf{g}$ is fixed by the scattering matrix of the disordered region, if this is known, the N components of the voltage "vector" can be obtained in terms of the $\{g_{mn}\}$ by inversion. Care must be taken in performing the inversion because g is not invertible in the entire N-dimensional vector space. It is easy to see that the condition that only the voltage differences determine the current flow requires that g have an eigenvector with zero eigenvalue, and it is necessary to subtract from g the projection operator onto that subspace to give an invertible matrix, $\overline{\mathbf{g}}$[10]. This can be done straightforwardly, and the $\{R_{m'n',mn}\}$ can be obtained in terms of the elements of $\overline{\mathbf{g}}^{-1}$. The result is

$$R_{m'n',mn} = (\overline{\mathbf{g}}^{-1})_{mn'} + (\overline{\mathbf{g}}^{-1})_{nm'} - (\overline{\mathbf{g}}^{-1})_{mm'} - (\overline{\mathbf{g}}^{-1})_{nn'} \ . \tag{13}$$

The question then is what does time-reversal symmetry imply about the symmetry of these resistances? Time-reversal symmetry applied to the scattering problem states that given one solution of the scattering problem, another solution is obtained by interchanging incoming and outgoing flux amplitudes, complex conjugating, and reversing the magnetic field(ignoring spin). This immediately leads to the condition on the scattering matrix $\mathbf{S}(B) = \mathbf{S}^T(-B)$. Since from (11) the conductance coefficients $\{g_{mn}\}$ are simply related to the $\{t_{mn}\}$, which in turn are simply submatrices of the S-matrix, one sees that time-reversal symmetry is fully satisfied by the condition $g_{mn}(B) = g_{nm}(-B)$. It is easily shown that the coefficients $\overline{\mathbf{g}}^{-1}$ also have this symmetry[10]:

$$\mathbf{S}(B) = \mathbf{S}^T(-B) \quad \Rightarrow g_{mn}(B) = g_{nm}(-B) \quad \Rightarrow \quad \overline{\mathbf{g}}^{-1}_{mn}(B) = \overline{\mathbf{g}}^{-1}_{nm}(-B) \ . \tag{14}$$

Analyzing (13) using this relation, one sees that if voltages are measured on the *same* leads that carry the current, $(m = m', n = n')$, then time-reversal symmetry implies that the resistance is symmetric in field (recovering the result quoted for the two-probe theory in greater generality). However, in a four-probe measurement, in which the voltage and current leads are distinct, $(m \neq m' \neq n \neq n')$, it is evident that *time-reversal symmetry does not require the resistance to have any definite symmetry under field reversal.* In typical circumstances in homogeneous macroscopic samples the conductance coefficients will have additional approximate symmetries that combined with time-reversal symmetry imply that in a standard six-lead Hall bar geometry, the longitudinal resistance is symmetric in field, and the Hall resistance is anti-symmetric. The notion that a four-probe longitudinal resistance must be symmetric in field arises from mistakenly assuming that this behavior of macroscopic, homogeneous samples follows from time-reversal symmetry alone.

Time-reversal symmetry does imply certain exact symmetries of any *set* of four-probe measurements; and these have been known for some time to experimentalists studying inhomogeneous conductors[37]. In such studies one defines a *symmetric* resistance, $R_S = R_{mn,m'n'} + R_{m'n',mn}$ and an *antisymmetric* resistance, $R_A = R_{mn,m'n'} - R_{m'n',mn}$, which are then obtained from two voltage measurements in which current and voltage leads are interchanged. It is easy to see from (13) and (14) that time-reversal symmetry applied to the multi-probe Landauer formula does imply that these quantities are symmetric and antisymmetric respectively under field reversal. Thus the correct test of whether the magnetoresistance fluctuations in a four-probe experiment violate time-reversal symmetry (and therefore presumably involve some eliminable experimental artifact) is to measure R_S, R_A by appropriate lead-interchanging. This was done at the same time as the theoretical treatment given above was being developed, by Benoit et al.[31], and they obtained remarkable agreement between the symmetries demanded by time-reversal invariance, and the experimentally observed resistance fluctuations. The voltage asymmetry shows clearly that a microstructure in the phase-coherent limit must always be treated like an inhomogeneous conductor in which any voltage measurement typically involves an admixture of diagonal and off-diagonal transport coefficients.

19. Voltage Fluctuations in Multi-Probe Devices

A number of experiments were done to measure voltage fluctuations on multi-probe microstructures with many leads spaced at distances,L, both greater than and less than L_{in}[1,24,25]. The essential observation was that for $L \gg L_{in}$ the self-averaging behavior of (10) was observed, which corresponds (by the argument give above) to $rms(R) \propto \sqrt{L}$; and for $L \approx L_{in}$ the resistance fluctuations were such that the $rms(g) \approx 1$ as predicted in the two-probe theory, and shown in Fig.1. But for $L \ll L_{in}$ the voltage (or resistance) fluctuations were roughly constant. When symmetric and anti-symmetric resistances were measured, it was found that only the symmetric part had this behavior, whereas the antisymmetric part was always roughly constant (even for $L \gg L_{in}$), and therefore was much smaller than the symmetric part in samples much larger than L_{in}[25]. That is why in longer samples often the fluctuations can still be observed, but they appear symmetric in field. This behavior of the voltage fluctuations is shown in Fig. 3 below, where data are given for the symmetric and antisymmetric voltage fluctuations (normalized to the voltage fluctuations for probes spaced by $L_{in} \equiv L_\phi$, which correspond to $\delta g \sim 1$), as a function of L/L_{in}, for a multilead Sb microstructure (pictured in the inset)[25].

It is possible to use the formulation described above to perform microscopic calculations of the resistance fluctuations in such a system. These calculations are quite involved, and unfortunately the results do not suggest any simple physical principles,

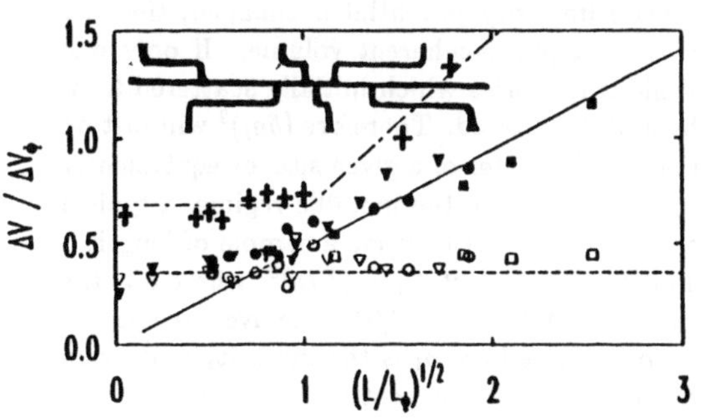

Fig. 3: Measured rms voltage fluctuations as a function of length, normalized as discussed above, reprinted from [25]. The symmetric contributions are represented by solid symbols; the antisymmetric by open symbols. Crosses are the results of the numerical simulations of [36]. The solid line indicates the $\sqrt{L}$ scaling discussed above.

similar to (7) and (10) above for the two-probe theory, which completely explain the qualitative behavior. Nonetheless the theoretical calculations do reproduce the qualitative and quantitative features of the data (up to factors of two), including the difference in behavior between the symmetric and antisymmetric resistance fluctuations[32-34]. In Fig. 3 we have compared the results of numerical simulations of the multi-probe Landauer formula by Baranger et al.[36], with those of [25], indicating the qualitative and approximate quantitative agreement.

20. Sensitivity of g to Small Changes in the Impurity Potential

The original theory of UCF considered the question of the statistical fluctuations of the conductance of a sample induced by a total rearrangement of the impurity configuration of the sample, and found that they were relatively large and had the universal properties described by (7) and (10). A natural subsequent question is how large are the statistical fluctuations in the conductance induced by a small change in the impurity potential? This question was answered independently by Altshuler and Spivak[38], and by Feng et al.[39]. The latter chose to state the problem in terms of an impurity potential consisting of randomly-placed point scatterers at positions $\{R_i\}$, with delta-function scattering potentials, in which one scatterer is moved some distance, d. They then defined the quantity $\delta g_1 =< [g(R_1, R_2, \ldots, R_n) - g(R_1 + d, R_2, \ldots, R_n)]^2 >^{1/2}$, which has the interpretation of the typical change in the conductance induced by moving a single impurity a distance d. Again, such a quantity can be calculated either numerically, or using impurity-averaged perturbation theory, and the result[39] for $d \gg k_f^{-1}$ at T=0 is

$$(\delta g_1)^2 \approx \frac{1}{(k_f l)^{d-1}} (l/L)^{d-2}. \tag{15}$$

This result implies that in two dimensions the effect of moving a *single* impurity is independent of the size of the system, and consequently of the total number of impurities, and for $k_f l \approx 1$, the effect is of the same order as if all the impurities had been rearranged.

Although initially surprising, this result can be understood by the following simple physical argument. The fluctuations arise from rearrangements in the phase of the electronic trajectories which interfere after traversing the sample. If all the phases are rearranged, as they are when the entire impurity potential is changed, then we know from UCF theory that $var(g) \approx 1$ in a phase-coherent volume. If only one impurity is moved, then only the electronic trajectories which initially scattered from that impurity will have their phases substantially altered. Therefore $(\delta g_1)^2$ will just be proportional to the fraction of trajectories which scatter of a given site, or equivalently to the fraction of sites visited by a single trajectory. In the metallic regime, a typical trajectory is a random walk of step length,l, which will traverse a sample of length L in $N_l = (L/l)^2$ steps, the number of impurity sites is just $N_i = (L/s)^d$, where s is the average spacing of the impurities; thus $f \sim (N_l/N_i) \approx L^{2-d} s^d / l^2$. The average spacing is related to the mean free path by $s^d = \sigma_i l$, where here σ_i is the impurity scattering cross-section. In the limit where each impurity is a strong scatterer, the cross-section reaches the unitarity limit, $\sigma_i \sim \lambda^{d-1} \sim k_f^{1-d}$. Substituting this into the equation for f yields (15) above.

21. UCF and Low-frequency Noise in Metals

Feng et al.[39] applied (15) to predict the magnitude of the low-frequency noise expected from slow impurity motion in dirty metals. Their detailed predictions will not be discussed here. The basic approach was to adopt the framework of the standard model of 1/f noise[40], and then calculate the amplitude of the noise arising from the UCF mechanism, given these standard assumptions. The application to large conductors was accomplished by using the result for δg_1 given above to calculate the resistance change due to a given number of moving scatterers per phase-coherent volume, and then use the semi-classical averaging behavior described by (10) to calculate the total resistance fluctuation, which is then equal to the total noise power, integrated over frequencies. The theory when combined with standard assumptions does predict a noise power consistent with that observed in dirty metals at low temperatures, and recent experiments on appropriate systems have revealed regularities in the behavior of the low-frequency noise which arise naturally in the UCF theory, and not in other models[41]. However, these comparisons of theory and experiment are hindered by the necessity of introducing many auxiliary assumptions about the number of moving impurities, their frequency distribution, temperature-dependence, etc., in order to arrive at a prediction. An alternative approach is to focus on microstructures in which telegraphic "switching noise", presumably due to single-impurity motion, has been observed; and try to study the systematics of this noise[42]. However this approach suffers from the intrinsic lack of statistical power in such a method, since it is extremely difficult to measure enough different "switchers" to get reliable statistics as a function of controllable parameters such as temperature or field. Therefore I would like to propose a clean and definitive experimental test of the UCF-based noise theory.

22. Definitive Experimental Test of UCF Noise Theory

As noted above, the value of $var(g)$, defined as the *statistical* fluctuations from sample to sample, is reduced to exactly half its value by the imposition of a sufficiently large magnetic field. Technically this is due to the fact that for every diffuson-like diagram in the perturbation theory there is a cooperon-like diagram with exactly the same value in zero magnetic field[18-20]. However, just like the weak-localization corrections to the average conductance, these cooperon-like contributions are sensitive to a weak magnetic field, and go to zero for a sufficiently large magnetic field, reducing $var(g)$ by exactly a factor of two. An identical structure occurs in the calculation of $(\delta g_1)^2$. Since the UCF noise magnitude is proportional to $(\delta g_1)^2$, this means that the magnitude of the UCF-based noise power will be reduced by exactly a factor of two by the imposition of a magnetic field, as long as the field is weak enough to leave the impurity motion giving rise to the noise undisturbed. It is thus necessary to calculate the field scale, B_c^*, over which the noise will be reduced to 3/4 its value (the relative noise only approaches 1/2 asymptotically as $B \to \infty$). I have recently done such a calculation[43], based on the formalism of Feng et al.[39], and find for the relevant case ($L \gg L_{in}$)

$$B_c^* \approx c^*(h/e)/(L_{in})^2 \tag{16}$$

where c* is a constant which depends weakly on the relative size of L_T and L_{in}, typically $c^* \approx 0.1 - 0.2$. Note that this relation then predicts a substantial reduction in the noise at low temperatures for weak magnetic fields, making the assumption that the imposition of a field has not affected the impurity motion a plausible one. Because the noise is predicted to be sensitive to such a weak field, the observation of such an effect very strongly supports the hypothesis that quantum interference over a phase-coherent volume is the source of the noise (just as the same sensitivity to weak fields indicated that weak localization and UCF were fundamentally interference effects). And finally, because the experiment only tests the behavior of the relative noise power, all the unknown parameters in the Feng et al. theory drop out of the calculation.

Acknowledgements

I would particularly like to thank P.A. Lee and Y. Imry for their valuable insights and collaboration on much of the work presented here. I also thank my other collaborators H. Baranger, D. Divincenzo, S. Feng, H. Fukuyama, K. Muttalib, J.L Pichard and A. Szafer. I am also very grateful to the experimentalists for keeping me in touch with reality, particularly R. Webb, S. Washburn and W. Skocpol, as well as D. Prober and M. Weissman. I also acknowledge helpful discussions with M. Büttiker and R. Landauer. This work was partially supported by NSF Grant DMR-8658135.

References

1. S. Washburn and R.A. Webb, Adv. Phys. 35, 375 (1986).
2. A.G. Aronov and Yu.V. Sharvin, Rev. Mod. Phys. 59, 755 (1987).
3. C.P. Umbach, S. Washburn, R.B. Laibowitz and R.A. Webb, Phys. Rev.B30, 4048 (1984).
4. W.J. Skocpol, L.D. Jackel, R.E. Howard, H.G. Craighead, L.A. Fetter, P.M. Mankiewich, P. Grabbe, and D.M. Tennant, Surf. Sci 142, 14 (1984).
5. J.C. Licini, D.J. Bishop, M.A. Kastner and J. Melngailis, Phys. Rev. Lett.55, 2987 (1985).
6. G. Bergmann, Phys. Rep.107,11, (1984).
7. P.A. Lee and T.V. Ramakrishnan, Rev. Mod. Phys.,57, 287 (1985).
8. R. Landauer, Phil. Mag. 21, 863 (1970).
9. D.S. Fisher and P.A. Lee, Phys. Rev. B23, 6851 (1981).
10. A.D. Stone and A. Szafer, IBM Journal of Res. Dev., May, (1988), in press.
11. E.N. Economou and C.M. Soukoulis, Phys. Rev. Lett. 46, 618 (1981).
12. Y. Imry: In Directions in Condensed Matter Physics, ed. by G.Grinstein and G.Mazenko, World Scientific, Singapore, 1986.
13. B. Van Wees, H. Van Houten, C. Beenakker, J. Williamson, L Kouwenhoven, D. Van Der Marel, and C. Foxon, Phys. Rev. Lett., 60, 848 (1988).
14. R. Landauer and M. Büttiker, Phys. Rev. B36, 6255 (1987).
15. A.D. Stone and Y. Imry, Phys. Rev. Lett. 56, 189 (1986).
16. A.D. Stone, Phys. Rev. Lett. 54, 2692 (1985).
17. P.A. Lee and D.S. Fisher, Phys. Rev. Lett. 47, 882 (1981).

18. P.A. Lee and A.D. Stone, Phys. Rev. Lett. 55, 1622 (1985).
19. P.A. Lee, A.D. Stone and H. Fukuyama, Phys. Rev. B35, 1039 (1987).
20. B.L. Al'tshuler, JETP Lett. 41, 648 (1985).
21. B.L. Al'tshuler and D.E.Khmel'nitskii, JETP Lett. 42, 359 (1986).
22. Y. Imry, Europhys. Lett.1, 249 (1986).
23. K.A. Muttalib,J.L. Pichard, and A.D.Stone, Phys. Rev. Lett. 59, 2475 (1987); and references therein.
24. W.J. Skocpol, P.M. Mankiewich, R.E. Howard, L.D. Jackel, D.M. Tennant and A.D. Stone, Phys. Rev. Lett. 56, 2865 (1986); Phys. Rev. Lett. 58, 2347 (1987).
25. A.D. Benoit, C.P. Umbach, R.B. Laibowitz, and R.A. Webb, Phys. Rev. Lett. 58, 2343 (1987).
26. C. Beenakker and H. Van Houten, Phys. Rev B37, 6544 (1988).
27. R.A. Webb, private communication and this volume.
28. Y. Gefen, Y. Imry and M.Ya. Azbel, Phys. Rev. Lett.52, 129 (1984).
29. Y. Isawa, H. Ebisawa and S. Maekawa J. Phys. Soc. Jpn 55,2523 (1986).
30. D.P. DiVincenzo and C.L. Kane, preprint.
31. A.D. Benoit, S. Washburn, C.P. Umbach, R.B. Laibowitz and R.A. Webb, Phys. Rev. Lett.57, 1765 (1986).
32. C.L. Kane, P.A. Lee and D.P. DiVincenzo, , Phys. Rev. B, in press.
33. S. Maekawa,Y. Isawa, H. Ebisawa, J. Phys. Soc. Jpn 56,25 (1987); Y. Isawa, H. Ebisawa, and S. Maekawa, Proceedings of University of Tokyo International Conference on Anderson Localization, Springer-Verlag, (1987).
34. S. Hershfield and V. Ambegaokar, preprint.
35. M. Büttiker, Phys. Rev. Lett. 57, 1761 (1986).
36. H.U. Baranger, A.D. Stone and D. DiVincenzo, Phys. Rev. B37 6521, (1988).
37. L.J. van der Pauw, Phillips Res. Repts. 13,1 (1958).
38. B.L. Al'tshuler and B.Z. Spivak, JETP Lett. 42, 447 (1985).
39. S. Feng, P.A. Lee, and A.D. Stone, Phys. Rev. Lett. 56, 1960 (1986).
40. P. Dutta and P. Horn, Rev. Mod. Phys. 53, 497 (1981).
41. G. Garfunkel, G. Alers and M. Weissman, Bull. Am. Phys. Soc., 33, 632 (1988); unpublished.
42. D. Beutler, T. Meisenheimer, and N. Giordano, Phys. Rev. Lett. 58, 1240 (1987).
43. A.D.Stone, Bull. Am. Phys. Soc., 33, 632 (1988); unpublished.

Four Terminal Resistance of a Multichannel Electron Waveguide

G. Timp, H.U. Baranger, P. deVegvar, R. Behringer, J.E. Cunningham, P.M. Mankiewich, and R.E. Howard

4G-18, AT&T Bell Laboratories, Holmdel, NJ 07733, USA

Abstract. We have measured the four terminal magnetoresistance of a multichannel electron waveguide, i.e. a submicron wire fabricated in extremely high mobility GaAs/AlGaAs heterostructure. Only a few transverse, one dimensional subbands or channels carry the current in this submicron wire because the width is comparable to an electron wavelength. We observe that the average resistance increases when the current path bends through a junction at or beyond the voltage terminals, and attribute the increase to a change (due to the bend) in the relative transmittance of the few transverse channels which carry the current. Our observations reveal that in these devices the average resistance is predominately associated with scattering from the leads used for the measurement.

1. Introduction

Electron coherence and confinement on a submicron scale give rise to gross deviations from Boltzmann transport which describes the resistance of large devices.[1] For example, large amplitude fluctuations (larger than e^2/h) in the magnetoresistance due to the Aharonov-Bohm effect, [2] [3] [4] gross asymmetries in the magnetoresistance about zero magnetic field, [5] [6] and negative resistance [6] have been observed in submicron structures at low temperature. In this regime, the resistance is nonlocal and a four terminal measurement is invasive. According to Buttiker, [7] the resistance is directly related to the quantum mechanical transmission and reflection of carriers. If a carrier propagates into a lead while maintaining phase coherence, then the lead is an integral part of the four terminal resistance and must be included in the computation of the resistance.

Here we review and augment an earlier report[6] on four terminal measurements of the magnetoresistance of an electron waveguide, i.e. a quasi-one dimensional (1D) wire which is only a few Fermi wavelengths wide and comparable in length to both the elastic mean free path and the phase coherence length. We observe that the *average* symmetrized magnetoresistance consistently increases (beyond the fluctuation amplitude) when the current path bends through a junction either at the voltage terminals or beyond the terminals in correspondence with our numerical simulation of the measurement. As the number of transverse subbands or channels decreases the effect is exaggerated. The increase in the *average* four terminal resistance is attributed to a change in the relative transmittance through the bend of the few current carrying channels, and to the lack of scattering by impurities. In contrast to previous work[3] [4] which focussed on the effect of impurity scattering in the leads on fluctuations in the magnetoresistance, our observations reveal that in these devices the *average* resistance can be predominately associated with scattering from the leads used for the measurement.

Springer Series in Solid-State Sciences Vol. 83: **Physics and Technology of Submicron Structures**
Editors: H. Heinrich · G. Bauer · F. Kuchar

2. Experimental Detail

The fabrication of the 1D wires has already been described elsewhere.[8] Devices have been fabricated in both δ-doped [9] and conventional modulation doped heterostructures prepared by gas source molecular beam epitaxy. Here we describe work on one δ-doped heterostructure in particular. This heterostructure comprises a 7nm GaAs cap layer, an undoped AlGaAs layer 49nm thick, a monolayer of dopant ($5\times10^{16}m^{-2}$), and a 34nm AlGaAs spacer layer grown on top of a GaAs buffer layer. The mobility and carrier density of the two dimensional (2D) electron gas at the GaAs/AlGaAs interface are determined at 280mK using a Hall bridge geometry 300μm wide and 2mm long. Figure 1a shows the magnetoresistances R_{xx} and R_{xy} used to deduce the mobility $\mu=180m^2/Vs$ and 2D carrier density $n_{2D}=3.2\times10^{15}m^{-2}$. The magnetic field associated with the depopulation of the second lowest spin-split Landau level (i=2 where $R_{xy}=h/e^2i$) is indicated in the figure. Electron beam lithography is used to pattern an etch mask onto the heterostructure. The etch mask protects the underlying doped AlGaAs from a subsequent partial etch which laterally confines the 2D electron gas to the region beneath the mask. Although the lithographic linewidth of the mask is typically 500nm, the conducting width of the wire beneath the mask is only about 100-200nm because of edge depletion. We have not observed any degradation of the mobility using this fabrication technique.

The four terminal resistance (of devices patterned as shown schematically in the inset to Fig. 2) is measured using an ac resistance bridge operating at 14Hz. If the same lead configuration is used, the measured magnetoresistance is correlated from day to day to better than 99.6%, even after the leads have been changed several times. The resistance measured at H=0 for T=4K does not generally scale with the separation between voltage terminals, e.g. it depends on bends in the current path (see below), but the high field quantized Hall effect measured at every junction in the pattern and the corresponding Shubnikov-de Haas oscillations observed in every segment (see Fig. 1) are characteristic of a homogeneous device.

In Fig. 1b and c we show the resistances $R_{39,68}$ and $R_{39,8\text{-}10}$. The convention used to index the leads is given in the inset to Fig. 2. For the measurement denoted by $R_{39,68}$, the voltage terminals (leads 6 and 8) on the device are separated by 700nm, and the current is injected (through leads 3 and 9) along a straight line through the junction between the voltage and current leads and exits straight through the second junction. The measurements depicted in Fig. 1b correspond to a device which is at least 130nm wide (device #1 hereafter), while the device of Fig. 1c (device #2) is at least 95nm wide. Reproducible fluctuations are observed in the magnetoresistance. As the width narrows, the amplitude and typical period associated with the fluctuations generally increase. Specifically, below 300mT the rms amplitude of fluctuations increases from approximately 70Ω to 400Ω, while the period changes from approximately 15mT to 25mT.

Because of the large fluctuations in the resistance, and because the resistance does not scale with length, we cannot characterize the mobility of the wire. (The value for mobility fluctuates between 50 and $2000m^2/Vs$.) Using the 2D mobility we estimate the elastic mean free path to be $L_e=\hbar\mu/e2\pi n_{i=2}\approx$5-10$\mu m$, where the carrier density, $n_{i=2}$, is inferred from the Hall effect at high field in the vicinity of i=2 (see Fig. 1b and c). This estimate effectively neglects small angle forward scattering, however (see below). The conducting width of the wire, W, and the phase coherence length, L_ϕ, are inferred from the magnetoresistance of nominally 2μm diameter annuli fabricated from the same material in the same way. Peaks in the Fourier spectra of the magnetoresistance of the annuli are observed at approximately $720T^{-1}$ and $1450T^{-1}$ corresponding to fluxes hc/e and hc/2e penetrating the area of the annulus defined by the average diameter.[10] The width of the

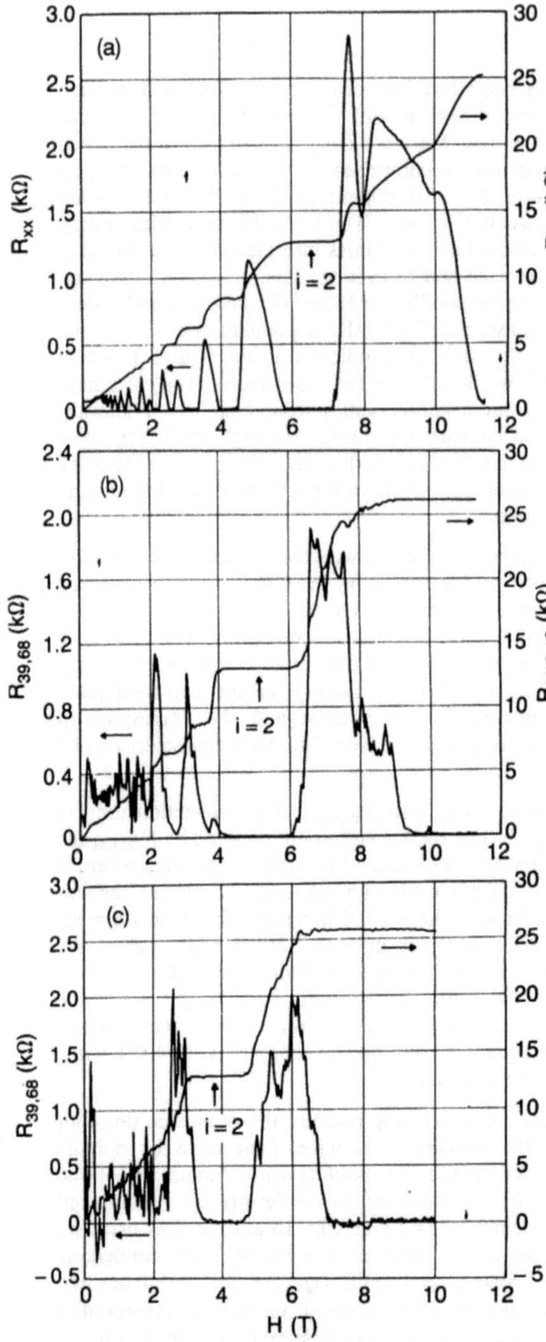

Fig. 1.
Figure 1a shows the magnetoresistances R_{xx} and R_{xy} of a large device approximately 300 μm wide. Figure 1b and c show resistance measurements $R_{39,68}$ and $R_{39,8\text{-}10}$ as a function of magnetic field at 280mK in 1D devices at least 130nm and 95nm wide respectively. The measurement $R_{39,68}$ classically corresponds to R_{xx}, while $R_{39,8\text{-}10}$ would correspond to the Hall resistance.

hc/e fundamental is assumed to be given approximately by $\Delta H^{-1} = 2\pi r W/(hc/e)$ where r is the radius, and W is (a lower bound on) the conducting width of the wire. Once W is estimated from ΔH^{-1}, a lower bound on the total number of transverse subbands, N_T, is calculated using a hard wall confining potential. For device #1 we find that $N_T \approx 4$-5 while for the device #2, $N_T \approx 2$-3. The number of obvious Hall plateaus found in a measurement such as $R_{39,8\text{-}10}$ is consistent with magnetic depopulation of N_T spin degenerate subbands. However, our estimate for N_T is always smaller (by about a factor of 2) than the number obtained if we identify the feature in $R_{39,8\text{-}10}$ at low field (near H = 400mT) with magnetic depopulation of the highest hybrid electic and magnetic one dimensional subband as Roukes proposes.[11] L_ϕ is estimated from the ratio of the peak Fourier amplitudes associated with hc/e and hc/2e obtained for $H > 20mT$ by assuming that the the ratio of the Fourier amplitudes is proportional to $exp(2\pi r/L_\phi)$ and ignoring coupling to the leads.[12] Our estimate for $L_\phi \gtrsim 3.2\mu m$ determined in this way is consistent with the observation that the fluctuations in the conductance are approximately e^2/h on a length scale where $L \approx L_\phi$.

3. Experimental Result

Figure 2 shows a typical example of the magnetoresistance observed near H=0 at 280mK in device #1. The transverse (to the magnetic field) magnetoresistance $R_{39,68}$ is grossly asymmetric, and $R_{39,8\text{-}10}$ is nonzero at H=0. The zero field position was determined from a separate Hall probe in close proximity with the device, but it should be apparent that no translation of the field axis could result in $R_{39,68}(H) = R_{39,68}(-H)$. Moreover, we observe that the resistance is negative near H=120mT.

The asymmetry in the magnetoresistance, the finite Hall effect at H=0, and the negative resistance observed in our four terminal measurements do not violate Onsagers symmetry relations because the resistance is actually anisotropic (it depends on the leads and impurity configuration) as well as nonlocal for $L < L_\phi$ where L is the distance between voltage probes, so that the usual four terminal resistance is not trivially related to the coefficients to which Onsager relations apply. That is, the measurements $R_{39,68}$ and $R_{39,8\text{-}10}$ do not generally correspond to R_{xx} and R_{xy} respectively. Because of the microscopic reciprocity of the transmission coefficients however, the resistance is invariant under a field (and magnetization) reversal when accompanied by an exchange of the current and voltage terminals.[7] It

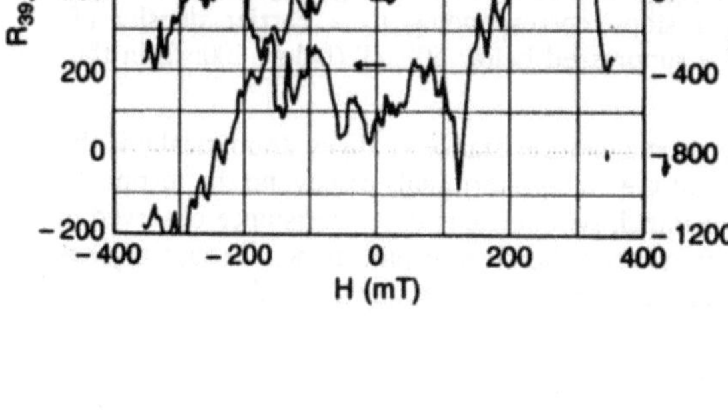

Fig. 2.
Figure 2 shows the low field magnetoresistance $R_{39,68}$ and $R_{39,8\text{-}10}$ through zero field obtained from device #1. The inset in the upper left corner shows the cross pattern of a typical device. The center to center separations between probes in the asymmetric cross pattern are (clockwise from 12) 2.2μm, 900nm, 2.5μm 700nm.

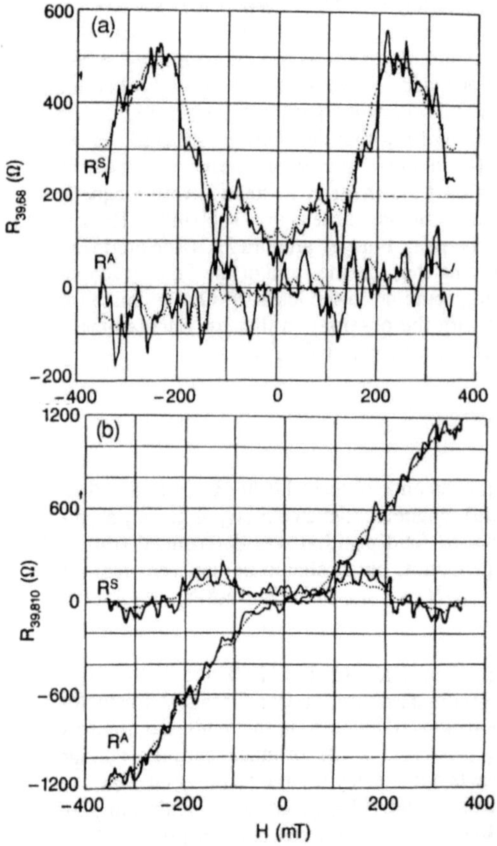

Fig. 3.
Figure 3a and b show the symmetric and antisymmetric decomposition of the $R_{39,68}$ and $R39,8\text{-}10$ respectively. The solid line corresponds to a temperature of 280mK, the dotted line to 1.7K

has been shown [7] that the decomposition $R^S = 1/2[R_{kl,mn}(H) + R_{mn,kl}(H)]$ and $R^A = 1/2[R_{kl,mn}(H)\text{-}R_{mn,kl}]$ is symmetric and antisymmetric respectively as a direct consequence of this symmetry.

When the current and voltage leads are interchanged, we find that $R_{kl,mn}(H) = R_{mn,kl}(\text{-}H)$ approximately. The results of the decomposition of $R_{39,68}$ and $R_{39,8\text{-}10}$ are shown in Fig. 3a and b for two temperatures, 280mK (solid) and 1.7K (dotted). R^S is approximately (within 5-10%) symmetric about H =0; R^A is approximately antisymmetric; there is no negative resistance in $R^S{}_{39,68}$; and the asymmetry is only weakly dependent on temperature. If we decompose $R_{39,8\text{-}10}$ we find that R^S can be negative, while R^A is linear in field beyond 500mT with a slope corresponding to a carrier density of $2.4\times10^{15}m^{-2}$. Generally, the Hall effect is suppressed below 500mT (below 100mT in the sample shown in Fig. 1).[11]

We observe that the average, symmetric magnetoresistance changes dramatically if the the current path bends through a junction at the voltage terminals or beyond the terminals. Figure 4a shows the changes in the four terminal, symmetric magnetoresistance observed in device #1 for voltage terminals 900nm apart when the current path bends through a junction 700nm from the voltage terminals ($R_{3\text{-}10,64}$), at one voltage terminal ($R_{3\text{-}12,64}$), and at

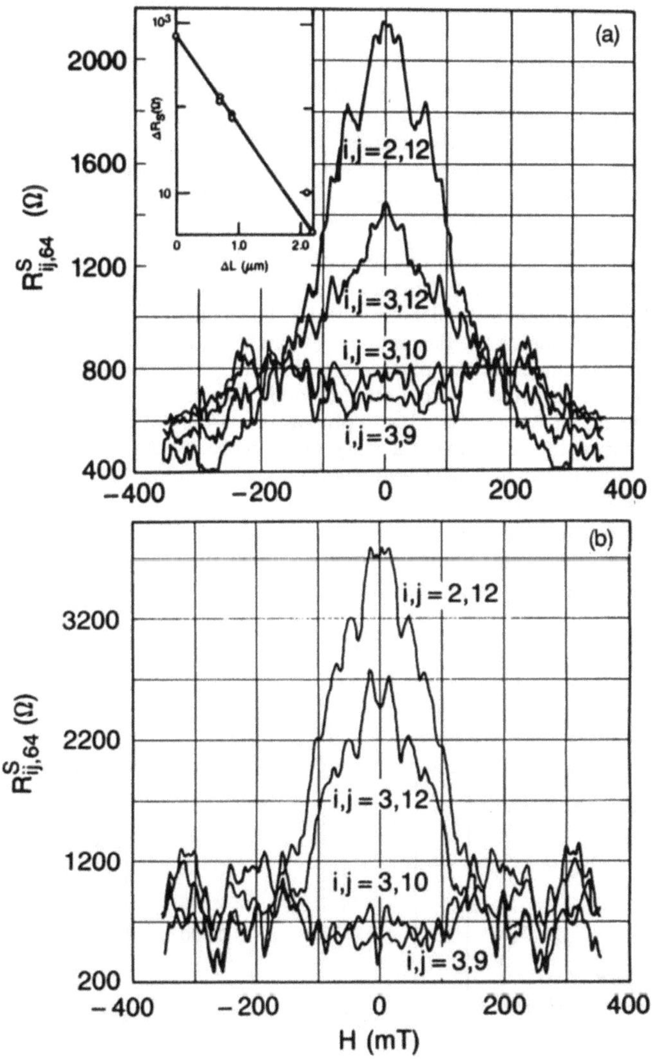

Fig. 4.
Figure 4a and b show R^S found at 280mK using different current terminals for the same voltage terminals separated by 900nm in device #1 and #2 respectively. The inset in Fig. 4a illustrates the decay of ΔR_S versus the distance between the bend in the current probe and the voltage terminal for device #1

both voltage terminals ($R_{2\text{-}12,64}$) relative to a geometry in which the current is injected through a junction along a straight line ($R_{39,64}$). The four terminal resistance is always observed to increase at zero field (in the 52 cases examined) whenever the current path bends at a voltage probe, if the mobility of the starting heterostructure exceeds $\mu=30m^2/Vs$ with $W\approx 130nm$ and $N_T\approx 5$, whereas no change is observed outside the amplitude of fluctuation when $\mu=4.7m^2/Vs$ with $W\approx 180nm$ corresponding to $N_T\approx$8-10. The data of Fig. 4b depicts the same measurements performed on device #2. We observe that for a narrower device with a fewer number of transverse channels, the change in resistance measured when the current bends through a junction at one voltage terminal, R_0, increases dramatically from about 700Ω to 2100Ω.

Furthermore, the average resistance consistently increases for $\mu \gtrsim 87 m^2/Vs$ and $N_T \approx 2$-5 when the current bends through a junction separated from the voltage terminals by at least 700nm, i.e. there is a nonlocal effect on the average resistance. The change in resistance due to a bend in the current path, ΔR_S, observed as a function of the distance between the nearest voltage terminal and the junction where the current path bends, ΔL, is plotted for device #1 in the left inset to Fig. 4a. The change is consistent with the law $\Delta R_S = R_0 exp(-\Delta L/L_s)$ where R_0 is the change in resistance measured when the current bends through a junction at one voltage terminal, and L_s is a characteristic length of $420 \pm 140 nm$.

We can discriminate scattering by a lead from that of an impurity because the position and geometry of the lead can be reproduced and manipulated between different devices. Generally, the same value of R_0 is obtained throughout a device independent of the leads used to measure it within the amplitude of the fluctuations. (In addition, the value of R_0 obtained in the annuli used to determine the two parameters, W and L_ϕ, are the same as the values obtained throughout the asymmetric cross pattern within the error.) Figure 5b summarizes the changes in the resistance associated with propagation around a bend as a function of the number of channels for devices with a range of 2D mobility. As shown in Fig. 5b, devices with different mobility can also show a similar R_0.

Finally, we notice that the change in resistance with magnetic field consistently decays within about 200mT for each of the devices examined. At these fields the Larmor radius $r_c = \sqrt{hc/eH} \approx W/2$, and the magnetic potential dominates over the electrostatic confinement.

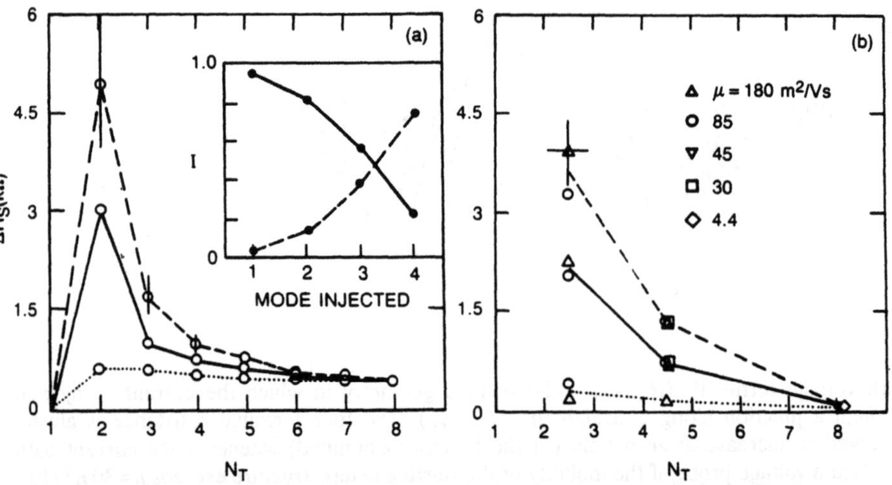

Fig. 5. Figure 5a and b depict respectively the calculated and measured change in the average resistance versus number of modes. The inset to Fig. 5a shows the normalized current transmitted out of each of the two exit ports of a "T" (straight through, solid; around a corner, dashed) when a specific mode is injected down the main channel. The error in the mode number shown in Fig. 5b represents the error in the determination of the lower bound on the mode number.

4. Numerical Simulation and Discussion

We attribute our observations to differences in the transmission coefficients for propagation straight through and around a bend in a junction. To justify this assertion we have calculated the transmission coeffcients at the Fermi energy and the resistance in a model multiprobe microstructure. The model microstructure, formed by strips with hard wall boundary conditions, is described by a nearest-neighbor tight-binding Hamiltonian[13] with two spin channels where each strip is 20 sites wide. The on-site energies can be randomly distributed to introduce disorder.

Our calculation of the resistance for these structures proceeds in three steps. First, we find the Green's function in the site representation via the recursion technique[14] used previously in diffusive multiprobe structures.[15] Second, we find the total transmitted intensity between any two leads, T_{kl}, from the Green's function in the channel representation. Corresponding to each channel in a lead there is a propagating eigenstate. Such states are plane waves labelled by a continuous momentum index k_x along the length of the lead and a discrete transverse momentum index k_y. The Fermi energy determines the total number of channels through the dispersion relation

$$E_f = E_j \equiv 2\cos(k_x a) + 2\cos[k_y(j)a], \tag{1}$$

where $k_y(j) = \pi j/(N+1)$ with N the channel index. In terms of the Green's function $G^+_{ij}(k,l)$ from channel i in lead k to channel j in lead l and the channel velocity, $v_j - \partial E_j/\partial k_x$, the transmitted intensities are[16]

$$T_{kl} = \sum_{i,j=1}^{N_T} \left| i(v_i v_j)^{1/2} G^+_{ij}(k,l) - \delta_{ij}\delta_{kl} \right|^2. \tag{2}$$

Finally, the current in lead k, I_k, is the sum of the pairwise currents between that lead and each of the others. The pairwise current is simply the probability, T_{kl}, times the chemical potential difference, V_l-V_k,

$$I_k = -(e^2/h) \sum_{l \neq k}^{n} T_{kl}(V_l - V_k), \tag{3}$$

as obtained by Buttiker through general arguments. [7] Equation (3) can be solved for the voltages subject to the constraint that the current is fed in one lead and out another, and vanishesin all other leads. (The solution requires some care, however, since the system of equations is singular – the vector $V_k = 1$ has eigenvalue zero.) The resistance, measured with voltage leads m and n and a current, I, fed through leads k and l, is then $R_{kl,mn} = (V_m - V_n)/I$.

To characterize the scattering caused by a junction, we determined the transmission coefficients for a three terminal device (a T structure) without disorder with four propagating modes. Suppose current is injected straight through the junction (down the top of the T) in a single mode. The inset to Fig. 5a shows the relative normalized intensity transmitted straight through and around the bend in a junction in all final modes as a function of the incident mode. The inset shows that the ground state propagates preferentially in the forward direction while the higher lying modes have a large probability to turn the corner (though they do not necessarily remain in the same mode). Heuristically, a mode turns the corner when its wavevector matches closely the wavevector of a mode in the side probe.

The discrepency between the transmission coefficients for propagation straight through and around a bend in a junction gives rise to changes in the average resistance. In Fig. 5a we show the calculated change in resistance in an eight probe structure as the current path is bent at a remote point (dotted line), once near a voltage terminal (solid line), and at both voltage terminals (dashed line). The model structure is comprised of a main channel with three cross strips without disorder, where the center to center separation of the leads is twice their width. The change in resistance as a function of number of propagating modes is obtained by averaging the resistance change at different energies which yield the same number of modes. The change in resistance is large for low channel number in all three cases but decays as the number of channels increases. Suprisingly, the magnitude of the change in resistance agrees with the experimental results shown in Fig. 5b, even though this naive calculation contains no material parameters. While the correspondence between the calculation and the experiment may be fortuitous (for example, we have not examined the effect of different confinement potentials on the resistance change), the agreement justifies our assertion that the changes (due to the bend) in the relative transmittance of the few channels which carry the current give rise to the observed changes in resistance.

The change in resistance for the remote bend is expected to decay as the bend gets further away from the voltage terminals since (elastic) impurity scattering should reduce any average non-local effect (though not a non-local fluctuation effect if phase coherence is maintained). Impurity scattering scrambles the mode structure of the current which results from scattering through a bend. Theoretically, the decay of ΔR_S is consistent with an exponential dependence on the elastic mean free path.[17] The more rapid decay seen in the experiment may arise from preferential forward scattering in the real wires.

In 2D GaAs/AlGaAs heterostructures, there is a large discrepancy between the elastic scattering time, τ_e, estimated from the mobility, and the lifetime determined from Shubnikov-deHaas measurements, τ_s. Typically, $\tau_e \approx 20\tau_s$.[18] [19] The difference arises because τ_e is essentially determined by large angle backscattering, while the scattering which determines τ_s is isotropic, originating from remote ionized impurities. Even though forward scattering is not measured by the mobility it will, nevertheless, scramble the effect of a remote bend on the modal structure of the current. Our observations suggest that in 1D the scattering is not isotropic but rather predominately forward scattering since $L_e/L_s \approx 10$.

5. Conclusion

We have shown that the average four terminal magnetoresistance of one dimensional wires fabricated in extremely high mobility GaAs/AlGaAs heterostructures changes dramatically when the current path bends through a junction at or beyond a voltage terminals. The change is only observed below approximately 200mT. We attribute the change in resistance to changes due to the bend in the relative transmittance of the few channels which carry the current because of the correspondence with our numerical simulation of the measurement. There are three important implications: firstly, it is apparent from the data of Fig. 4 that the resistance can be predominately determined by the leads used for the measurement; secondly, the magnetoresistance of two different devices can be highly correlated (see Fig. 5b) since the resistance can be due primarily to geometric features which may be reproduced from device to device; and thirdly, a realistic model for the magnetoresistance in such structures must include the probes used for the measurement.

References

1. Sean Washburn and Richard A. Webb, Advan. Phys. 35, 375 (1986).
2. G. Timp, A. M. Chang, P. Mankiewich, R. Behringer, J. E. Cunningham, T. Y. Chang and R. E. Howard, Phys. Rev. Lett. 59, 732 (1987).

3. W. J. Skocpol, P. M. Mankiewich, R. E. Howard, L. D. Jackel, D. M. Tennant and A. D. Stone, Phys. Rev. Lett. 58, 2347 (1987).

4. A. Benoit, C. P. Umbach, R. B. Laibowitz and R. A. Webb, Phys. Rev. Lett. 58, 2343 (1987).

5. A.D. Benoit, S. Washburn, C.P. Umbach, R.B. Laibowitz, and R.A. Webb, Phys. Rev. Lett. 57, 1765 (1986).

6. G. Timp, H. U. Baranger, P. deVegvar, J. E. Cunningham, R. E. Howard, R. Behringer, and P. M. Mankiewich, to appear in Phys. Rev. Lett.

7. M. Buttiker, Phys. Rev. Lett. 57, 1761 (1986), M. Buttiker, to appear in IBM J. Res. Develop., and R. Landauer, _Localization, Interaction, and Transport Phenomena,_ edited by G. Bergmann and Y. Bruynseraede, Springer-Verlag, Heidelberg, New York (1985) p. 38-50.

8. P. M. Mankiewich, R. E. Behringer, R. E. Howard, A. M. Chang, T. Y. Chang, B. Chelluri, J. E. Cunningham, G. Timp, J. Vac. Sci. Tech. B 6, 131 (1988).

9. J.E. Cunningham, W.T. Tsang, G. Timp, E.F. Schubert, A.M. Chang and K. Owusu-Sekyere, Phys. Rev. BI 37, 4317 (1988).

10. G. Timp, A.M. Chang, J.E. Cunningham, P.M. Mankiewich, R. Behringer, T.Y. Chang, R.E. Howard, Phys. Rev. Lett. 58, 2814 (1987).

11. M.L. Roukes, A. Scherer, S.J. Allen Jr., H.G. Craighead, R.M. Ruthen, E.D. Beebe, and J.P. Harbison, Phys. Rev Lett. 59, 3011 (1987).

12. G. Timp, A. M. Chang, P. deVegvar, R. E. Howard, R. E. Behringer, J. E. Cunningham and P. M. Mankiewich, Surf. Sci. 196, 68 (1988).

13. A.D. Stone, Phys. Rev. Lett. 54, (1985) 2692.

14. P.A. Lee and D.S. Fisher, Phys. Rev. Lett. 47, 882 (1981).

15. H.U. Baranger, A.D. Stone, and D.P. DiVincenzo, to be appear in Phys. Rev. B.

16. A. Szafer and A. D. Stone, to appear in IBM J. Res. Devlop.

17. H.U. Baranger, unpublished.

18. S. DasSarma and F. Stern, Phys. Rev. B 32, 8442 (1985).

19. F. F. Fang, T. P. Smith and S. L. Wright, Surf. Sci. 196, 1988 (1988).

Localisation and Transport in One-Dimensional Disordered Systems

B. Kramer

Physikalisch-Technische Bundesanstalt Braunschweig,
Bundesallee 100, D-3300 Braunschweig, Fed. Rep. of Germany

The theory of localisation and transport in one-dimensional disordered systems is reviewed. The localisation length is calculated. It is shown that the dc-conductivity vanishes at the absolute zero of the temperature, if the Fermi energy lies in the region of localised states. The Landauer formula for the conductance of a system of finite length is derived. It is shown that the averages of the resistance and the conductance increase, and decrease exponentially with the length of the system, respectively. The distributions of the localisation length, the resistance, and the conductance are discussed. It is shown that the localisation length is the quantity which obeys the central limit theorem, whereas the width of the distributions of the resistance and the conductance increase with increasing size of the system faster than the averages of these quantities in the thermodynamic limit.

1. INTRODUCTION

One-dimensional disordered systems play a key role in the understanding of the properties of solids since many features of the electronic states, and of the related transport properties, can be discussed rigorously. The main theoretical interest concerns the localisation of the states, i. e. their asymptotic behavior in space, the behavior of the conductivity of the infinite system, the conductance of the finite system near the absolute zero of the temperature, and the statistical properties of the transport quantities. In this article I discuss all of these problems at an introductory level.

There are a number of review articles on the subject of one-dimensional localisation which may be used for more detailed studies. I refer only to four of them. Firstly, there is the comprehensive review by ISHII [1] in which an extensive discussion of the exponential growth of particular solutions of the Schrödinger equation is given, and the relation to the properties of the spectrum and the dc-conductivity is outlined. In the papers by ABRIKOSOV and RYZHKIN [2] and by GOGOLIN [3] the main emphasis is on the average conductivity of quasi-one-dimensional systems. The most recent review is by ERDÖS and HERNDON [4]. Here, the main interest concentrates on the statistical properties of the resistance, and its logarithm.

Although quite a number of models have been considered, I shall use only the "tight binding" Hamiltonian which is defined on a lattice (lattice spacing $a \equiv 1$):

Springer Series in Solid-State Sciences Vol. 83: **Physics and Technology of Submicron Structures**
Editors: H. Heinrich · G. Bauer · F. Kuchar

$$H = \sum_{j=1}^{N} \{\varepsilon_j a_j^+ a_j + V(a_j^+ a_{j+1} + a_{j-1} a_j^+)\} , \tag{1}$$

V (= 1) is the constant hopping probability amplitude between the states a_j, $a_{j\pm1}$ associated with nearest neighbors. ε_j are the energies associated with the lattice sites. They are usually taken as random variables obeying some distribution function P (ε_1.....ε_N). For statistically independent site energies one can write

$$P(\varepsilon_1.....\varepsilon_N) = \prod_{j=1}^{N} P(\varepsilon_j) . \tag{2}$$

Particularly convenient distributions of site energies are the Gaussian, and the box distribution. A measure of the disorder is then given by the width W of the distribution function.

The diagonal part of the above Hamiltonian corresponds to the potential energy, whereas the off-diagonal part represents the kinetic energy of the particle. All necessary ingredients for a nontrivial description of localisation are incorporated in this extremely simple model.

Usually, when dealing with a random system which is replaced by a statistical ensemble via the introduction of a distribution function, a physical quantity denoted by, say, A is assigned to an ensemble average, i. e.

$$\langle A \rangle = \int d[\varepsilon]\, P(\varepsilon_1.....\varepsilon_N)\, A(\varepsilon_1.....\varepsilon_N) . \tag{3}$$

This is, however, only meaningful if A obeys the "central limit theorem", i. e. if the relative statistical fluctuations of A around the mean value become vanishingly small in the thermodynamic limit. This is in general not true, especially in a localised system, as will become clear later. We have then to investigate the distribution of A in order to decide whether or not it is a physically meaningful quantity.

2. THE LOCALISATION LENGTH

Localisation is commonly associated with the asymptotic behavior of the eigenstates of a quantum mechanical system at infinity. A wave function is called localised if its envelope decays exponentially away with a finite decay length γ^{-1}, i. e.

$$\lim_{j \to \infty} a_j = f_j e^{-\gamma |j|} , \tag{4}$$

where f_j is a random phase factor. $\gamma^{-1} = \lambda$ is called the localisation length. A numerical example is shown in Fig. 1.

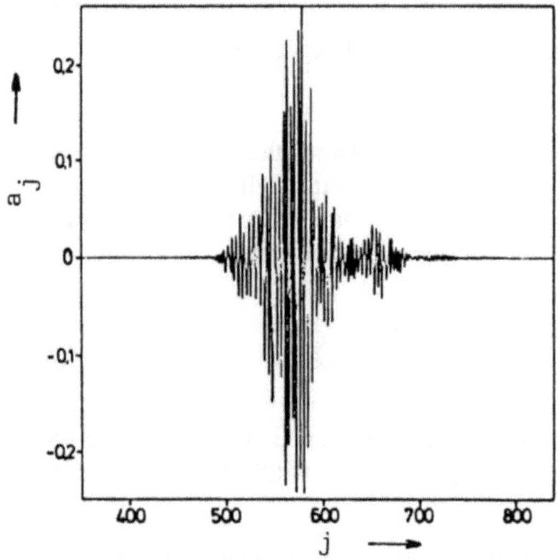

Figure 1: Localised state as obtained by numerical diagonalisation of the one-dimensional Anderson model (ref. [5], parameters: N = 1000, E = 0.089, box distribution with W = 2)

2.1 Asymptotic Behavior of Wave Functions

The inverse exponential decay length of the eigenstates, γ, may be calculated from

$$\lim_{j \to \infty} \frac{1}{j} \log (a_j^2 + a_{j+1}^2) = -2\gamma \ . \tag{5}$$

In general, it will depend on the particular realisation of the random potential, $\gamma = \gamma(\varepsilon_1 \ldots\ldots \varepsilon_N)$. However, as it is a self-averaging quantity (see below), the configurational average (cf. eq.(3)) agrees with the most probable value in the thermodynamic limit, $N \to \infty$.

The inverse localisation length may be calculated by considering the stationary Schrödinger equation

$$E a_j = \varepsilon_j a_j + a_{j-1} + a_{j+1} \tag{6}$$

as an initial value problem for a particular solution $a_j(E)$ corresponding to an energy E when the initial values a_0 and a_1 ($(a_0^2 + a_1^2) \neq 0$) are given. This can be written as

$$\begin{pmatrix} a_{j+1} \\ a_j \end{pmatrix} = \prod_{m=1}^{j} T_m \begin{pmatrix} a_1 \\ a_0 \end{pmatrix} \tag{7}$$

where the transfer matrix $\mathbf{T}_m$ is

$$\mathbf{T}_m = \begin{pmatrix} E - \varepsilon_m & -1 \\ 1 & 0 \end{pmatrix} . \tag{8}$$

Since ε_m are independent random variables we are concerned with the limiting properties of a product of random matrices when investigating the properties of a_j for $j \to \infty$.

The product matrix

$$\mathbf{M}_j = \prod_{m=1}^{j} \mathbf{T}_m \tag{9}$$

satisfies the theorem of Oseledec, i. e. there exists a limiting matrix

$$\mathbf{M} = \lim_{j \to \infty} (\mathbf{M}_j^+ \mathbf{M}_j)^{\frac{1}{2j}} \tag{10}$$

with eigenvalues $\exp(\pm\gamma)$ [6, 7]. The eigenvalues are inverse of each other since the $\mathbf{T}_m$ are symplectic. Thus, a_j is for $j \to \infty$ exponentially increasing with a characteristic length $\gamma(E)^{-1}$. Since this is again a self-averaging quantity one may assume continuity with E, although this assumption is not rigorously justified, and identify γ^{-1} with the inverse localisation length defined in eq. (5). The identification may also be justified by constructing the eigenstates by fitting particular solutions, which are exponentially increasing from the left and the right end of the system, to each other. The eigenvalues of H are then defined as those energies at which a continuous fit is possible [8].

The identification of the localisation length as a limiting property of a product of random matrices immediately provides also the proof that in a one dimensional stochastic system all eigenstates are localised [9-11], independent of the magnitude of the disorder W ($\neq 0$), via the theorem of Fürstenberg [1].

2.2 The Localisation Length

Many calculations of the localisation length have been performed using a variety of methods. Perturbational treatments [12-17] were very successful in the weak disorder limit. Numerical procedures [5, 18-20]were designed for treating even macroscopically large systems. I describe here briefly a recursive method which is very similar to the transfer matrix method described above. It is based on a definition that makes the connection between localisation and transport properties transparent. The starting point is the one-electron Green's function that is also very convenient for perturbation theory:

$$(E \pm i\eta - H)G^{\pm}(E) = \mathbf{1} \quad (\eta>0)\,. \tag{11}$$

The density of energy levels, n(E), is given by the imaginary part of the diagonal elements of G. The localisation properties of the states are incorporated in the non-diagonal elements, as one can see by using the spectral representation. It is straightforward to derive that the matrix element of $G^{\pm}(E)$ between the site states which correspond to the first and the j-th site of the system, $G^{\pm}_{1j}(E)$, are given by a_j^{-1}. Then

$$\gamma = -\lim_{\eta\to 0}\lim_{j\to\infty}\frac{1}{2j}\log|G^{+}_{1j}(E)|^2\;; \tag{12}$$

$T(j, E) = |G_{1j}(E)|^2$ is related to the transmission probability for a particle to go from the site 1 to the site j. The fact that γ is always finite means that an infinitely long one-dimensional disordered system cannot be transparent.

Writing

$$G^{+}_{1j}(E) = \prod_{m=1}^{j} g^{+}_{m}(E) \tag{13}$$

where

$$g^{+}_{m}(E) \equiv G^{+}_{mm}(E) = \frac{1}{E+i\eta-\varepsilon_m-g_{m-1}} \tag{14}$$

is the diagonal element of G at the end of a system of the length m, we obtain

$$\gamma_j = \frac{j-1}{j}\gamma_{j-1}+\frac{1}{j}\log|g_j|\,. \tag{15}$$

There is a relation between the spectral properties and the localisation length which was discovered by HERBERT and JONES [21]. It may readily be derived by calculating G_{1j} and using eq. (12):

$$\gamma(E) = \int_{-\infty}^{\infty} n(x)\log|E-x|\,dx = \mathrm{Re}\int_{-\infty}^{E}\lim_{N\to\infty}\frac{1}{N}\sum_{j=1}^{N}G_{jj}(x)dx\,. \tag{16}$$

The diagonal elements of the Green's function may be evaluated in second order perturbation theory [12]. In the limit of small disorder one obtains ($|E| < 2$, box distribution)

$$\gamma(E) = \frac{W^2}{24\,(4-E^2)}\;, \tag{17}$$

i. e. $\gamma(0) = W^2/96$.The numerical results which were obtained by using eq. (15) are consistent with $\gamma(0) = W^2/105$ [19,20]. The difference in the numerical factor is due to an anomaly in the band centre which originates in resonance effects leading to a breakdown of second order perturbation theory [13-15].

The result that the localisation length of a one-dimensional disordered system diverges at $W = 0$ as W^{-2} remains also true for the case of a spatially correlated potential as has been discussed by JOHNSTON and KRAMER [16] as long as the correlation length remains finite.

3. LOCALISATION AND TRANSPORT

3.1 The Conductivity

Localisation manifests itself in the transport properties of a system. The conductivity can be calculated by using the Kubo formula:

$$\sigma(T, \omega) = \frac{2\pi e^2}{m^2 \omega \Omega} \sum_{\alpha} \sum_{\alpha \neq \beta} |\langle\alpha| \mathbf{p}\, \mathbf{e} |\beta\rangle|^2 (f(E_\alpha) - f(E_\beta))\, \delta(E_\alpha - E_\beta - \hbar\omega); \quad (18)$$

$\mathbf{p}$ and $\mathbf{e}$ are the momentum operator and the direction of the electric field of frequency ω, respectively. E_α, E_β, are the energy eigenvalues corresponding to the eigenstates $|\alpha\rangle$, and $|\beta\rangle$. e and m are the electronic charge and mass, respectively, f(E) the Fermi distribution function. The volume of the system, Ω, is taken to infinity at the end of the calculation. Using the relation

$$\frac{\mathbf{p}}{m} = \frac{i}{\hbar}[H, \mathbf{r}] \quad (19)$$

one can rewrite eq. (18) in terms of the one-electron Green's function as

$$\sigma(T, \omega) = \frac{e^2 \omega}{\pi \Omega} \int_\Omega d\mathbf{r}_1 \int_\Omega d\mathbf{r}_2 [\mathbf{e}\cdot(\mathbf{r}_1 - \mathbf{r}_2)]^2 \int_{-\infty}^{\infty} dE\, [f(E + \tfrac{\hbar\omega}{2}) - f(E - \tfrac{\hbar\omega}{2})]$$

$$\times\, G^+(\mathbf{r}_1, \mathbf{r}_2; E + \tfrac{\hbar\omega}{2})\, G^-(\mathbf{r}_2, \mathbf{r}_1; E - \tfrac{\hbar\omega}{2}) \,. \quad (20)$$

For small ω ($\hbar\omega \ll kT$) one may replace the difference of the Fermi functions by the derivative with respect to the energy. This will be sharply peaked around the Fermi energy E_F at low temperature. Thus, the configurationally averaged dc-conductivity will vanish at the absolute zero of temperature if, on the average, the product $G^+(\mathbf{r}_1, \mathbf{r}_2; E) G^-(\mathbf{r}_2, \mathbf{r}_1; E)$ is asymptotically exponentially decreasing with increasing distance between the sites $\mathbf{r}_1$ and $\mathbf{r}_2$. This is the case when the condition eq. (12) is fulfilled, i. e. if the states near the Fermi energy are localised.

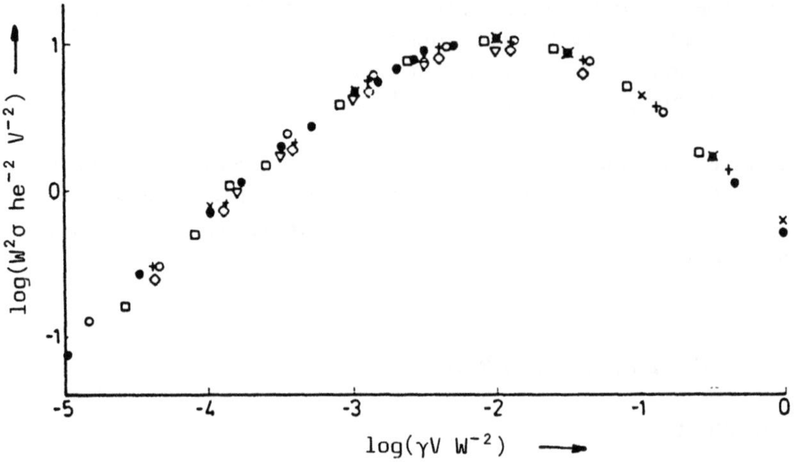

Figure 2: The dc-conductivity of the one-dimensional Anderson model with rectangular distribution of the site energies as a function of the disorder, W, and the imaginary part of the frequency, γ (from [18])

That this is indeed the case has been demonstrated numerically for the one-dimensional Anderson model with rectangular distribution of the site energies by MACKINNON [18] and THOULESS and KIRKPATRICK [22] by using the dc-version of eq. (20) at the absolute zero of temperature in a recursive procedure (Fig. 2).

Within the accuracy of the calculation the conductivity scales as

$$\sigma(\eta, W) = W^{-2}\sigma(\eta W^{-2}, 1)\ . \tag{21}$$

It is only for large disorder (W>5) that one observes deviations from this behavior. The scaling law allows for the necessary extrapolation with $\eta \rightarrow 0$ for any fixed W. For large η one has $\sigma \sim \eta^{-1}$ ("classical", Drude-like behavior) whereas for small η the conductivity tends to zero linearly with η.

3.2 Averages of Resistance and Conductance

Classically, the conductance S(L) of a d-dimensional hypercube of volume L^d is related to the conductivity by

$$S(L) = \sigma L^{d-2}\ . \tag{22}$$

Since the conductivity vanishes in the localised regime, it is no longer a useful parameter when considering the transport properties of a macroscopic, but finite, sample. One should have a theory of the conductance (or, equivalently, of its inverse, the resistance) without referring to the conductivity.

A very simple equation for the resistance R(L) of a one-dimensional wire of length L has been given by LANDAUER [23]. The basic idea is to consider the charge transport through a disordered wire, which is thought to be connected to ideally conducting wires on the left and on the right, as a quantum mechanical transmission problem. The voltage is generated by a difference in the charge densities on the left and the right hand side of the wire. The current is given by the total number of particles transmitted at the Fermi velocity. The result is

$$R(L) = \frac{h}{e^2}\,\frac{1-t(L)}{t(L)} \;; \tag{23}$$

t(L) is the transmission coefficient. An elementary derivation of this formula is given in [4]. Generalisations to quasi-one-dimensional systems (many transmission channels) have been discussed by several authors [24-31]. For a general discussion of its validity see the paper by Stone in this volume.

A dimensionless resistance may be defined by $r(L) = R(L)\,e^2/h = t(L)^{-1} - 1$. As the logarithm of t(L) is statistically a well-behaved quantity (it obeys the central limit theorem, see below), and its configurational average is asymptotically equal to $-2\gamma L$, one may expect that the resistance is an exponentially increasing function of L, on the average, although its probability distribution will not fulfill the central limit theorem, as I will discuss in the following section. The exponential increase of the resistance has been proven for a variety of models by several authors [32-38]. An elementary derivation for the Anderson model (E = 0) was given by ABRAHAMS and STEPHEN [32].

In order to evaluate the average resistance of a system of length N one has to consider the configurational average

$$\mathbf{U}_N = \langle \mathbf{M}_N^+ \mathbf{M}_N \rangle = \langle \mathbf{T}_N^+ \ldots\ldots \mathbf{T}_1^+ \mathbf{T}_1 \ldots\ldots \mathbf{T}_N \rangle \;. \tag{24}$$

Since the transfer matrices are statistically independent this yields a recursion relation

$$\mathbf{U}_{N+1} = \langle \mathbf{T}_{N+1}^+ \mathbf{U}_N \mathbf{T}_{N+1} \rangle \;. \tag{25}$$

If one takes $\langle \varepsilon_j \rangle = 0$ then $\mathbf{U}_N$ is diagonal, and can be cast into the form $\mathbf{U}_N = A_N \tau_z + B_N$, where τ_z is the Pauli matrix. The largest eigenvalue of the recursion relation eq. (25) is

$$\lambda_1 = \sigma_2/2 + (1 + \sigma_2^2/4)^{1/2} \tag{26}$$

with $\sigma_2 = \langle \varepsilon_j^2 \rangle$ ($\equiv W^2/12$ for the box distribution). Thus, asymptotically one has

$$\langle r(N) \rangle \propto \exp\{N \log\lambda_1\} \;; \tag{27}$$

$2\gamma_r = \log \lambda_1$ is the inverse of the "localisation length" which is characteristic for the exponential increase of the average of the resistance.

The dimensionless conductance, s(L), is defined as the inverse of the dimensionless resistance, r(L). The calculation of the corresponding configurational average is much more complicated than in the case of the resistance. It has been performed by ABRIKOSOV and RYZHKIN [2] in the limit of weak disorder. KIRKMAN and PENDRY have treated the general case [39]. The result for weak disorder, and in the center of the band (E = 0) is

$$\langle s(N)\rangle \propto \sigma_2^{-3/2} N^{-3/2} \exp\{-\sigma_2 N/16\} \tag{28}$$

in the limit of large N. Thus, $2\gamma_s = \sigma_2/16$ is the inverse of the "localisation length" which is characteristic for the exponential decrease of the average conductance.

By comparison with eqs. (17), (26), and (28) one observes that in the weak disorder limit there are relationships between the various localisation lengths, namely

$$\begin{aligned} \log\langle g\rangle &= \frac{1}{4}\langle \log g\rangle \\ \log\langle r\rangle &= 2\,\langle \log r\rangle \\ \langle \log r\rangle &= -\langle \log g\rangle \,. \end{aligned} \tag{29}$$

That the three quantities do not agree with each other reflects the fact that the resistance and the conductance are *not* self-averaging in one-dimensional disordered systems.

4. THE STATISTICAL PROPERTIES

4.1 Distribution of the Localisation Length

The central limit theorem for the localisation length has been shown to be valid for the disordered harmonic chain by O'CONNOR [40]. Approximate treatments for the electronic problem have been given by ANDERSON et al. [24] using a Landauer-type of approach, and by MEL'NIKOV [38] by estimating the distribution function of the resistance, and calculating from that the distribution function of its logarithm. The case of a Gaussian white noise potential has been treated by ABRIKOSOV [41], and by KREE and SCHMID [35]. Numerical results have been obtained by ANDERECK and ABRAHAMS [33], SAK and KRAMER [36], and by KANTOR and KAPITULNIK [42]. TANKEI and TAKANO [43] present somewhat different results in the limit of large disorder. The method of O'Connor can be directly applied to the Anderson model discussed here when identifying

$$\log(a_{j+1}^2 + a_j^2) \equiv x_{j+1}$$

$$a_j/a_{j-1} = \tan\theta_j \equiv y_j \tag{30}$$

which obey the recursion relation

$$x_{j+1} = x_j + \log\left[\frac{y_j^2}{1+y_j^2}(1+y_{j+1}^2)\right]$$

$$y_{j+1} = (E-\varepsilon_j) - y_j^{-1} . \tag{31}$$

This yields the most important result that the logarithm of the resistance (or of the conductance, altenatively) is a statistically well-behaved quantity, its limiting distribution function being a Gaussian with a positive variance.

The corresponding relative average fluctuations can be calculated explicitly by using approximative methods [35, 38, 41, 43], and by numerical procedures [33, 36, 42]. They *decrease* with increasing length of the system, i. e.

$$\frac{\langle(\Delta\log r)^2\rangle^{1/2}}{\langle\log r\rangle} = 2(\gamma N)^{-1/2} \qquad (\gamma N) >> 1 . \tag{32}$$

This holds for small γ [36, 41, 43]. In the limit of large γ, however,there seem to be deviations from this behavior [36, 43]. The reasons for these are not yet understood.

4.2 Resistance Fluctuations

As the distribution of log r is asymptotically well behaved, it is intuitively clear that the resistance as well as the conductance must have statistical distributions which yield asymptotically divergent fluctuations. This can be verified by considering the average of the square of the resistance $\langle r(N)^2\rangle$ which is given by

$$\langle r(N+1)^2\rangle \propto \langle(\mathrm{Tr}U_{N+1})^2\rangle \equiv \langle \mathrm{Tr}\,U_{N+1}^{(2)}\rangle . \tag{33}$$

$U_{N+1}^{(2)}$ is given by the recursion relation

$$\langle U_{N+1}^{(2)}\rangle = \langle(T_{N+1}^{+} \otimes T_{N+1}^{+})\langle U_N^{(2)}\rangle(T_{N+1}\otimes T_{N+1})\rangle . \tag{34}$$

The asymptotic behavior of the average square of the resistance is, as in the case of the average resistance, determined by the largest eigenvalue of this recursion relation, λ_2, i. e.

$$\langle r(N)^2\rangle \propto \exp(N\log\lambda_2) . \tag{35}$$

For small disorder a straightforward calculation gives

$$\lambda_2 = 1 + \sqrt{3}\,\sigma_2 \,. \tag{36}$$

Thus, the second moment of the resistance grows more rapidly with the length of the system than the average resistance itself [32,34]. Similar conclusions can be drawn for the higher moments of the resistance, and in the limit of large disorder. A full account of the asymptotic behavior of all of the moments of the resistance and the conductance has been given by KIRKMAN and PENDRY [37, 39], and by ABRIKOSOV [41], and MEL'NIKOV[38] by using approximative methods.

The result eq. (35) together with eq. (36) implies that the root-mean-square (rms) fluctuations of the resistance grow exponentially with the length of the system. The resistance is not a self-averaging quantity. Strictly speaking, it does not have any physical meaning. Similar statements hold also for the conductance.

5. CONCERNING EXPERIMENTAL CONSEQUENCES

If a physical quantity is self-averaging its relative rms fluctuations vanish in the thermodynamic limit. As a consequence, the experimentally determined value, which is usually obtained by investigating a single *macroscopic* sample, is well described by an ensemble average. This is nothing else but the statement that its most probable value agrees with the average value in the thermodynamic limit. The relative statistical error of the experimental value is proportional to the inverse of the root of the volume of the system (i. e. $N^{-1/2}$ in our case), and can be reduced by increasing the size of the system.

The considerations of the preceding sections imply that that the resistance and the conductance of one-dimensional disordered systems at zero temperature do not have this property. Measuring the resistances of a set of thin wires of equal length should yield considerable statistical scatter if the temperature is low enough. This scatter cannot be reduced by increasing the length of the wire. As we have seen, the rms fluctuations of the resistance diverge faster than the resistance itself in the thermodynamic limit.

At finite temperatures some of these effects might be obscured due to inelastic processes. In this case one can think of the mean distance between two inelastic scattering events, $L_i(T)$, as playing the role of an effective length of the system. $L_i(T)$ will be increasing with decreasing temperature T because the inelastic processes will be frozen out. The statistical resistance fluctuations can be expected to become more pronounced with decreasing temperature.The experimental work is reviewed in the article by WEBB in this volume

It is extremely difficult to prepare macroscopically identical wires of different length in order to investigate the length dependence of the fluctuations. It is easier to investigate the fluctuations in a given sample as a function of some parameter as, for instance, the temperature, the density of the charge carriers, the magnetic field, and the frequency, and to interpret the results in terms of an effective length L_{eff}. The question of the scaling laws for the averages of the conductance and the resistance taken over a *finite* ensemble has been addressed by STONE and JOANNOPOULOS [44]. These scaling laws are different from those for ⟨s⟩ and ⟨r⟩. They are most important for the interpretation of numerical data.

In the *metallic* limit, i. e. when the localisation length is large compared to the diameter of the system, the conductance fluctuations apparently do not diverge but become a universal constant of the order of e^2/h. In this regime standard perturbation theory can be used. This case has been reviewed by LEE et al. [45] and is treated in this volume in the article by STONE.

ACKNOWLEDGEMENTS

The author wishes to thank Jan Masek for many useful remarks and fruitful discussions during the preparation of this paper, especially concerning the clarification of the connection between the exponential decay of the transmission probability and the conductivity (cf. eq (20)).

REFERENCES

1. K. Ishii: Progr. Theor. Phys. (Suppl.) **53**, 77 (1973)
2. A. A. Abrikosov and I. A. Ryzhkin: Adv. Phys. **27**, 147 (1978)
3. A. A. Gogolin: Phys. Reports **86**, 2 (1982)
4. P. Erdös, R. C. Herndon: Adv. Phys. **31**, 65 (1982)
5. G. Czycholl, B. Kramer: Z. Phys. B **39**, 193 (1980)
6. V. I. Oseledec: Trans. Moscow Math. Soc. **19**, 197 (1968)
7. J. L. Pichard, G. Sarma: J. Phys. C: Solid State Phys. **14**, L127 (1981)
8. N. F. Mott, W. D. Twose: Adv. Phys. **10**, 107 (1961)
9. S. A. Molcanov: Math. USSR Izvestija **12**, 69 (1978)
10. H. Kunz, B. Souillard: Commun. Math. Phys. **78**, 201 (1980)
11. F. Delyon, H. Kunz, B. Souillard: J. Phys. A: Math. Gen. **16**, 25 (1983)
12. D. J. Thouless: in Ill-condensed Matter, eds. R. Balian, R. Maynard, G. Toulouse, 1 (North-Holland 1979)
13. M. Kappus, F. Wegner: Z. Phys. B**45**, 15 (1981)
14. B. Derrida, E. Gardner: J. Physique **45**, 1283 (1984)
15. C. J. Lambert: J. Phys. C: Solid State Phys. **17**, 2401 (1984)
16. R. Johnston, B. Kramer: Z. Phys. B **63**, 273 (1986)
17. M. Kasner, W. Weller: phys. stat. sol. (b) **134**, 731 (1986)
18. A. MacKinnon: J. Phys. C: Solid State Phys. **13**, L1031 (1980)
19. G. Czycholl, B. Kramer, A. MacKinnon: Z. Phys. B **43**, 5 (1981)
20. J. L. Pichard: J. Phys. C: Solid State Phys. **19**, 1519 (1986)

21. D. C. Herbert, R. Jones: J. Phys. C: Solid State Phys. **4**, 1145 (1971)
22. D. J. Thouless, S. Kirkpatrick: J. Phys. C: Solid State Phys. **14**, 235 (1981)
23. R. Landauer: Phil. Mag. **21**, 863 (1970)
24. P. W. Anderson, D. J. Thouless, E. Abrahams, D. S. Fisher: Phys. Rev. B **22**, 3519 (1980)
25. D. C. Langreth, E. Abrahams: Phys. Rev. B **31**, 2978 (1981)
26. D. S. Fisher, P. A. Lee: Phys. Rev. B **23**, 6851 (1981)
27. E. N. Economou, C. M. Soukoulis: Phys. Rev. Letters , 618 (1981); ibid. **47**, 972 (1981)
28. D. J. Thouless: Phys. Rev. Letters, **47**, 92 (1981)
29. P. W. Anderson: Phys. Rev. **23**, 4828 (1981)
30. R. Landauer: Springer Series in Solid-State Sciences **61** (Proceedings of LIT-PIM), 38 (Springer-Verlag 1985)
31. M. Büttiker, Y. Imry, R. Landauer, S. Pinhas: Phys. Rev. B **31**, 6207 (1985)
32. E. Abrahams, M. Stephen: J. Phys. C: Solid State Phys. **13**, L377 (1980)
33. B. Andereck, E. Abrahams: J. Phys. C: Solid State Phys. **13**, L383 (1980)
34. A. D. Stone, J. D. Joannopoulos, D. J. Chadi: Phys. Rev B **24**, 5583 (1981)
35. R. Kree, A. Schmid: Z. Phys. B **42**, 297 (1981)
36. J. Sak, B. Kramer: Phys. Rev. B **24**, 1761 (1981)
37. P. D. Kirkman, J. B. Pendry: J. Phys. C: Solid State Phys. **17**, 4327 (1984)
38. V. I. Mel'nikov: JETP Letters **32**, 225 (1981)
39. P. D. Kirkman, J. B. Pendry: J. Phys. C: Solid State Phys. **17**, 5707 (1984)
40. A. J. O'Connor: Commun. Math. Phys. , **45**, 63 (1975)
41. A. A. Abrikosov: Solid State Commun. **37**, 997 (1981)
42. Y. Kantor, A. Kapitulnik: Sol. State Commun. **42**, 161 (1982)
43. K. Tankei, F. Takano: J. Phys. Soc. Japan, **55**, 3516 (1986)
44. A. D. Stone, J. D. Joannopoulos: Phys. Rev. B **25**, 2400 (1982)
45. P. A. Lee, A. D. Stone, H. Fukuyama: Phys. Rev. B. **35**, 1039 (1987-II)

Universal Fluctuations and Conductance Asymmetry in Mesoscopic Silicon MOSFETs

S.B. Kaplan

IBM Research Division, T.J. Watson Research Center,
P.O. Box 218, Yorktown Heights, NY 10598, USA

The conductance of very small silicon metal-oxide-semiconductor field-effect transistors (MOSFETs) contains aperiodic structure known as universal conductance fluctuations. Fluctuations of nearly the same amplitude are observed when either the magnetic field or the gate voltage is varied. Structure in the magnetoconductance of inversion and accumulation layers is correlated for closely spaced values of the Fermi energy. The period of magnetoconductance fluctuations for samples tilted in a magnetic field is dependent on the perpendicular component of the applied field. When the source-drain voltage across the device is increased, the conductance becomes increasingly asymmetric until the fluctuation amplitude is decreased by electron heating. The role of quantum interference and disorder in causing these phenomena will be discussed.

1. Introduction

The transport properties of small electronic conductors are affected by the interference of electronic states describing different trajectories [1,2]. The simplest representation of the interference process is described in Fig. 1(a), in which an electron beam is split in two and then rejoined, thus forming an Aharonov-Bohm loop [3]. The application of a magnetic field perpendicular to the loop shifts the electronic phases, thereby changing the interference conditions that affect the conductance. This results in conductance oscillations with a period of h/e in the magnetic flux through the loop [4-6].

In a disordered metallic loop, electrons suffer many elastic scattering events as they diffuse through the sample, as shown schematicallly in Fig. 1(b). Aharonov-Bohm loops of various areas are formed in addition to the lithographically defined loop. The interference conditions are influenced by the area of each such Aharonov-Bohm loop. The conductance is sample-specific [1], because it reflects the spatial distribution of scattering sites. The magnetoconductance contains aperiodic fluctuations [7,8] due to the penetration of flux through the random array of loops [2,9]. Under certain conditions, these fluctuations are predicted to have a universal magnitude ($\sim G_0 \equiv e^2/h$), irrespective of sample material, shape, or the degree of disorder [9-12]. The interference that gives rise to these universal fluctuations can only take place if quantum mechanical coherence is preserved. The electronic phase is disturbed by inelastic and spin-flip processes for trajectories exceeding the phase coherence length, L_ϕ. Fluctuations are easily observed when the spatial extent of the sample is of the same order as L_ϕ, (which in the mesoscopic range [11], intermediate between microscopic and macroscopic). At

Springer Series in Solid-State Sciences Vol. 83: **Physics and Technology of Submicron Structures**
Editors: H. Heinrich · G. Bauer · F. Kuchar

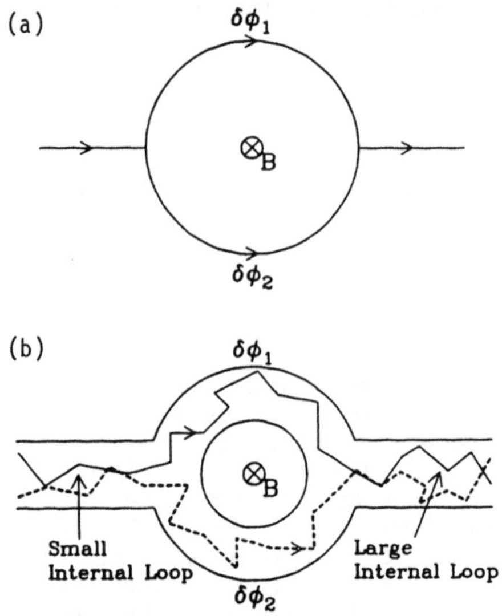

Figure 1. (a) An Aharonov-Bohm loop, in which an electron beam is split in two and rejoined. The magnetic field B results in the shifts of the electronic phases $\delta\phi_1$ and $\delta\phi_2$, resulting in a periodically varying current. (b) A schematic diagram of an Aharonov-Bohm loop made from a disordered metallic conductor. Two of the possible electronic paths are shown. Two ring-like pairs of trajectories which lead to aperiodic structure are identified.

a temperature of T~0.5K, L_ϕ is of order 1 μm for Si MOSFET accumulation or inversion layers, as well as for metallic samples. The fluctuations in macroscopic samples are not easily observed, since any measurement averages the universal contribution from small coherent subregions of the sample [10-12].

The purpose of this lecture is to describe two experiments that lend support to the theory of universal conductance fluctuations. The first experiment is a test of the orbital nature of the interference process [13,14]. The second is a demonstration of the nonlinear effects [15] predicted by theory [10] for small metallic samples.

2. The Orbital Nature of Universal Conductance Fluctuations

Aperiodic conductance fluctuations had been observed in many experiments on small structures, but remained unexplained. A representative example is the observation of fluctuations in the conductance of narrow Si MOSFETs [16] by A. Fowler et al. The main focus of their experiment was to study the transition from 1- to 2-dimensional conduction in a variable-width electron accumulation layer. However, the presence of conductance fluctuations widened their focus of study, and stimulated our interest.

The device layout is shown in Fig. 2(a). As the gate voltage, V_G, is increased above the device threshold voltage, a narrow conducting channel is formed in n-type Si between the depletion region of two p^+ control electrodes. (These electrodes were shorted to the substrate during most experiments.) A side view is shown in Fig. 2(b). The channel length is approximately 10 μm. The width of the narrow accumulation layer varies with the gate voltage, and cannot be measured microscopically.

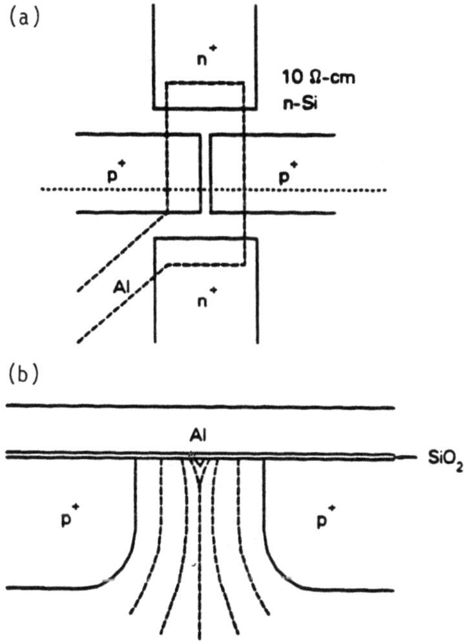

Figure 2. (a) The layout of a narrow Si accumulation layer device used in [13,14,16] is shown. The two p$^+$ control electrodes define the channel. The width between the controls is approximately 1 μm. The length of the narrow region is approximately 10 μm. The two n$^+$ regions are the source and drain. A cross-section taken along the dotted line is shown in (b). The diffusions are about 0.5 μm deep and the oxide is 30 nm thick.

Aperiodic fluctuations were observed in the conductance of these pinched accumulation layer samples as the gate voltage was varied [16]. At the lower values of V_G, the localization (or decay) length of the electronic wave function was found to be much shorter than the phase-coherence length, indicating that the electrons were strongly localized. The fluctuations in this regime were identified with changes in the variable-range hopping paths as the Fermi level was varied [17]. The conductance was observed to fluctuate aperiodically with magnetic field as well as gate voltage. These conductance fluctuations were found to persist up to the highest values of gate voltage, where the localization length is longer than L_ϕ, and variable-range hopping gives way to diffusive transport. This is the weakly-localized regime, where the theory of universal conductance fluctuations should apply.

The presence of fluctuations with variation of either gate voltage or magnetic field in the pinched accumulation layers is consistent with the theory of universal fluctuations [2,9]. As V_G is varied, the Fermi level is shifted, and the electronic wavelengths are therefore changed. This change in the interference conditions produces fluctuations of the same magnitude (to within a factor $\sqrt{2}$ [12]) as changing the magnetic field, as observed by Licini et al. in Si inversion layer MOSFETs [18]. They illustrated the coherent nature of the fluctuations by measuring the magnetoconductance at various values of V_G. The fluctuations in the magnetoconductance were observed to be similar for closely spaced values of V_G, but were very different for larger gate voltage spacings. Although these results are consistent with the theory of universal conductance fluctuations, a definitive exhibition of the orbital nature of the causative effects was lacking.

If the magnetoconductance fluctuations are due to electronic interference, the fluctuations should depend on the amount of magnetic flux threading the sample. Kaplan and Hartstein [13,14] exploited the two-dimensional properties of the Si accumulation layer samples described above to test this hypothesis. The electronic trajectories in such a system are essentially planar, even when the magnetic field vector is not normal to the sample surface. In order to determine whether the perpendicular component of magnetic field is the important parameter, the angle θ between the normal vector to the sample surface and the magnetic field was varied by tilting the sample with a gear set. The magnetoconductance of these devices was measured for gate voltages of from 10 to 12 V. The conductance in this gate voltage range is of order 10^{-4}S, which is in the metallic regime addressed by the universal fluctuation theory. The magnetic field, B, was varied between 0 and 1.5 T, and was oriented perpendicular to the channel. The magnetoconductance data are shown in Fig. 3(a) for various tilt angles.

The two-dimensional nature of the electronic density of states is evident by the presence of the large oscillations at the largest magnetic fields, which are found to be periodic in $(B\cos\theta)^{-1}$. These Shubnikov-de Haas oscillations arise from the quantizing effect of a magnetic field [19]. The quantum mechanical treatment of cyclotron motion for a two-dimensional electron gas in a magnetic field results in a set of energy levels (Landau levels) separated by an energy proportional to $(B\cos\theta)^{-1}$. The channel conductance is affected by the change in the number of electrons contributing to the transport as the magnetic field causes these levels to move through the Fermi level.

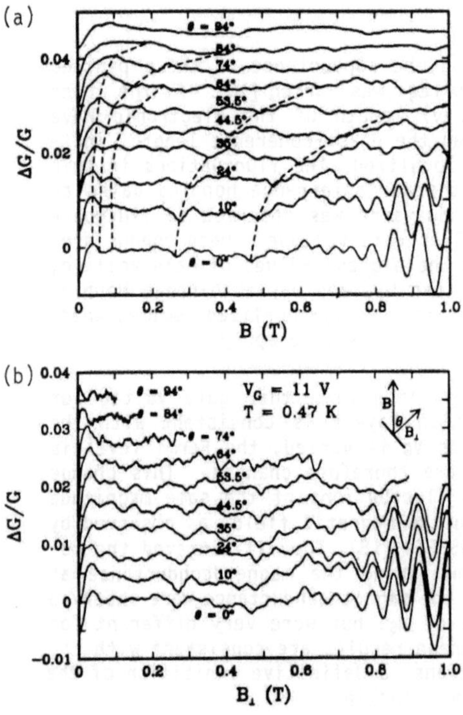

Figure 3. (a) Fractional change in the magnetoconductance of a pinched Si accumulation layer sample for various values of the angle θ between the magnetic field axis and the vector normal to the sample surface [13]. A slowly varying background and part of the low-field magnetoconductance have been subtracted. $G\sim2.3\times10^{-4}$ at $V_G=11$V, $T=0.47$K$\pm$0.02K, and B=0. The dotted lines are used to track several structures from curve to curve as θ is varied. (b) The same data plotted versus the perpendicular component of the magnetic field.

The dependence of the Landau level spacing on the perpendicular component of the applied field allows an independent check of the tilt angle.

The aperiodic fluctuations in the magnetoconductance data are shown in Fig. 3(a). The peaks and dips are observed to appear at larger values of B as θ increases. The data are replotted vs. $B\cos\theta$ in Fig. 3(b). One can see that the fluctuations lie at nearly the same value along the abscissa. The magnetoconductance data taken in a parallel magnetic field [14] do not show fluctuations above the noise level of the measurement system. These results indicate that universal conductance fluctuations are caused by an orbital effect, and that coherent electronic states are involved. The mechanism is the same type of Aharonov-Bohm interference that leads to oscillations with a flux period of h/e in small metallic loops.

It was observed in [13,14] that the measured fluctuation amplitudes were much smaller than e^2/h. This would appear to disagree with the universal fluctuation theory, but the prediction of a universal fluctuation amplitude is based on the assumption that phase-breaking processes are unimportant. This is certainly not true when the sample size exceeds L_ϕ. Each coherent subsection of the sample has conductance fluctuations of order e^2/h. Theoretical calculations show that the fluctuation amplitude for a long, narrow sample is equivalent to the contribution of a series string of resistors, each resistor representing a subsection of length $\sim L_\phi$ [10-12]. The root-mean-square (rms) fluctuation amplitude under these conditions is given by [12]

$$\delta G \sim G_0 \times (L_\phi/L)^{3/2} . \qquad (1)$$

L_ϕ for the pinched MOSFETs was estimated [13] to be $0.3\pm0.1\mu m$ for $T \sim 0.5$ K and $V_G = 11V$, as compared with the channel length $\sim 9.5\mu m$. The width was estimated [13] to be ~ 0.1 μm. The expected fluctuation amplitude of these narrow samples was estimated using Eq. (1) to be $\sim 2.2\times10^{-7}S$, in (perhaps fortuitous) agreement with the measured value of $2\pm0.5\times10^{-7}S$. The length dependence for long samples described by Eq. (1) has been verified in detail [20]. The fluctuation amplitudes are in only fair agreement with theoretical estimates, perhaps due to uncertainties in the estimation of L_ϕ.

3. Asymmetric Conductance and Nonlinear Effects

Al'tshuler and Khmel'nitskiĭ [10] predicted that quantum interference in disordered samples can produce a nonlinear and asymmetric conductance. The details of this sample-specific effect are dependent on the scattering potential due to the randomly placed elastic scattering sites. For example, when a source-drain bias is applied to a Si MOSFET, the disordered scattering potential becomes tilted by the bias energy eV (see Fig. 4). The excess energy, $E_F - V_{scat}$, determines the electronic wavelength. These energies are randomly distributed, because of the disordered nature of the potential. When the random potential is tilted, the wavelengths, and therefore the interference conditions, are changed. A significant (but random) phase change is possible along a coherent electronic trajectory when the voltage across it is large enough to change the number of wavelengths by one wavelength. A simple calculation shows that this occurs when the energy changes by $E_\phi \sim hD/L_\phi^2$, where D is the diffusion constant. This coherence energy is the same energy dif-

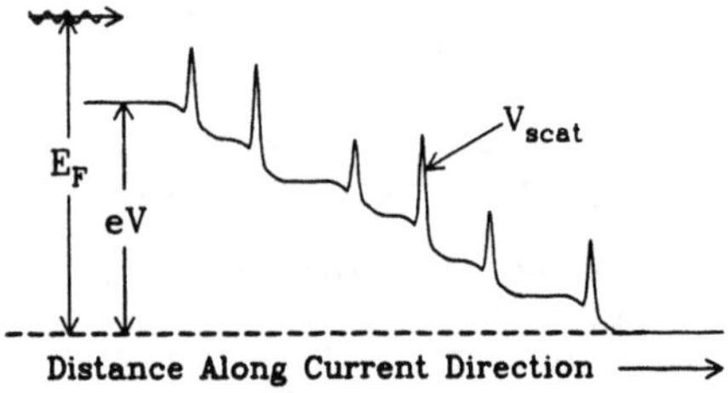

Figure 4. A schematic energy diagram for a disordered metallic sample under a voltage bias V (from [15]). The net current is towards the right. Electrons of energy E_F scatter from the random potential V_{scat}. If the bias is large enough, the electronic wavelength may be substantially shifted.

ference needed to change the fluctuation pattern by varying the gate voltage [9,18]. However, there is a difference: varying the gate voltage changes the wavelengths uniformly. Figure 4 shows that the when a source-drain bias is applied, the wavelengths on one side are influenced more than on the other. The random variation of interference conditions with source-drain bias leads to a nonlinear conductance. When the bias is reversed, the interference conditions are substantially different than for the original bias polarity, because disordered samples lack inversion symmetry with respect to exchange of source and drain. This leads to an asymmetric conductance that should fluctuate with gate voltage and magnetic field.

In order to observe these effects, I have measured the dependence of the conductance of Si MOSFETs with submicron dimensions as a function of the source-drain voltage, V_{SD} [15]. These devices were fabricated using a poly-Si gate and a self-aligned source and drain. The conductance of one such device, with length $L \sim 0.9\mu m$ and width $W \sim 0.5\mu m$, was first studied using $V_{SD} = 5\mu V$. This value of source-drain voltage is much less than either E_ϕ or k_BT, thus insuring that heating and nonlinear effects are negligible. The magnetoconductance fluctuations of this device were compared to that of a larger device with $L \sim 2.3$ and $W \sim 3.1 \pm 0.1\mu m$, in order to estimate the important parameters L_ϕ and $E_\phi \sim hD/L_\phi^2$. L_ϕ was estimated to be $0.3 \pm 0.15\mu m$.

When the gate voltage in these devices was varied, the magnetoconductance fluctuation pattern began to change shape over an energy scale of roughly 0.2 meV. This is in good agreement with the estimate of $E_\phi \sim hD/L_\phi^2 \sim 0.17$meV. One expects the fluctuation patterns to change when the voltage across a sample length equal to L_ϕ is of order E_ϕ. Since the sample is $\sim 3L_\phi$ in length, this should occur when $V_{SD} \sim 0.5$ mV. Nonlinear effects should become important on the same energy scale.

When the source-drain voltage, V_{SD}, was increased, the conductance was observed to become nonlinear and asymmetric. The shape of the conductance-voltage characteristic varied aperiodically with the applied magnetic field, providing a first indication that this is an interference effect. In order to observe many fluctuations without changing the source-drain voltage, the conductance was studied as the magnetic field was varied. The data for the antisymmetric magnetoconductance $G_A \equiv [G(V_{SD}) - G(-V_{SD})]$ are shown in Fig. 5, for several values of V_{SD}. The fluctuation patterns are shown to be similar to one another until V_{SD} approaches ~ 0.3meV. This is in fair agreement with our estimate of 0.5meV.

The magnitudes of the fluctuation patterns shown in Fig. 5 increase with V_{SD}. This is expected, since the asymmetry of the dephasing process shown in Fig. 4 becomes more pronounced as the potential becomes more tilted. For a

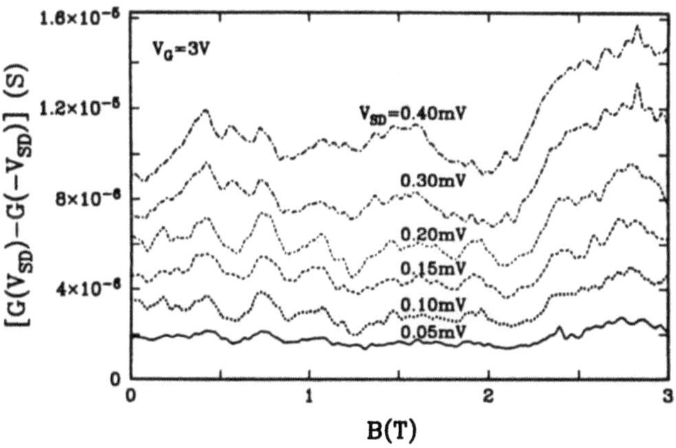

Figure 5. The fluctuations in the antisymmetric magnetoconductance for several values of source-drain voltage V_{SD} [15]. Each curve has been arbitrarily displaced along the vertical scale. Note that the fluctuation patterns change shape when V_{SD} changes by more than 0.2mV.

conductor with maximum spatial extent of order L_ϕ, the antisymmetric rms fluctuation amplitude at temperatures $T < E_\phi/k_B$ is predicted to be [10]

$$\delta G_A \equiv \langle [G(V) - G(-V)]^2 \rangle^{1/2} \sim G_0 \times (eV_{SD}/E_\phi). \qquad (2)$$

The fluctuations in a device larger than L_ϕ must decrease due to the averaging of the contributions of coherent device subregions [21]. Furthermore, there are additional complications when $eV_{SD}L_\phi/L$ approaches E_ϕ. Electron scattering may result in an electron temperature of order 3 or 4 K at some of the largest values of V_{SD} used in this experiment. L_ϕ decreases with increasing temperature. This changes the number of coherent sample subregions, thus decreasing the fluctuation amplitude. In addition, when k_BT exceeds E_ϕ, a number of fluctuation patterns equal to k_BT/E_ϕ is thermally averaged [2,9]. It is therefore not possible to vary V_{SD} without causing other changes. An attempt was made to (at least partially) separate these effects from the asymmetric behavior to be studied by comparing G_A to the symmetric amplitude $G_S \equiv [G(V_{SD}) + G(-V_{SD})]$, which is also affected by heating effects.

Both G_A and G_S are plotted in Fig. 6(a). The decrease in the symmetric fluctuation amplitude with V_{SD} graphically shows the effects of electron heating. G_A increases rather linearly with V_{SD} until V_{SD}~0.3mV. G_A then decreases, most likely for the same reason that G_S decreases. The ratio $\delta G_A/\delta G_S$ is plotted in Fig. 6(b). This gives a rough idea how G_A behaves in the absence of electron heating. We see that there is a linear increase up to V_{SD}~0.3mV, followed by a sublinear increase. The fact that the change in slope occurs near where $\delta G_A = \delta G_S$, and at the same value of V_{SD} that causes a change in the fluctuation pattern, lends support to the argument that this effect is due to quantum interference.

The change of the ratio $\delta G_A/\delta G_S$ in Fig. 6(b) to a nonlinear slope occurs when $eV_{SD} \gtrsim E_\phi$. This behavior is explained in [21]. The number of coherent energy

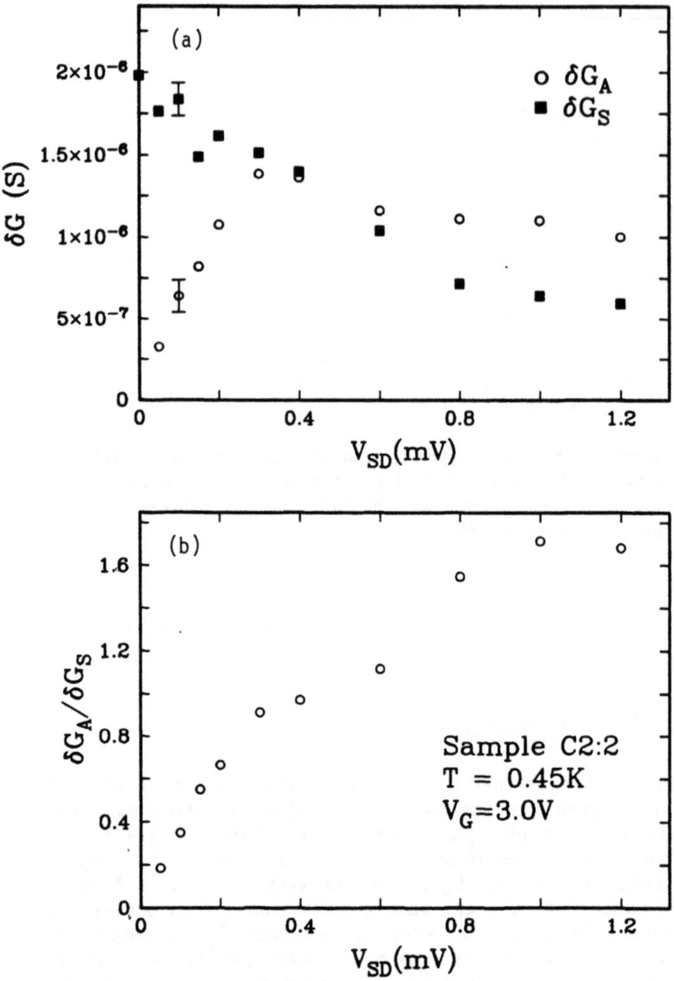

Figure 6(a). The rms amplitudes of the antisymmetric (δG_A) and the symmetric (δG_S) components of the conductance are plotted vs. V_{SD} [15]. The symmetric fluctuation amplitude decreases with source-drain voltage. The antisymmetric amplitude increases up to $V_{SD} \sim 0.3$mV, then decreases. (b) The ratio $\delta G_A/\delta G_S$ is plotted vs. V_{SD}. The data increase linearly up to $V_{SD} \sim 0.2$mV. Beyond this point, the increase is sublinear.

levels spanned by the source-drain voltage is given by $N = eV_{SD}/E_\phi$. Each of these levels contributes randomly to the conductance, resulting in a $\sqrt{N}$ dependence of the fluctuation amplitude. (This result is similar to that obtained by adding N incoherent conductors in parallel.) The antisymmetric conductance in the limit $N >> 1$ is given by [21]

$$\delta G_A \sim G_0 \times (eV/E_\phi)^{1/2}. \tag{3}$$

This behavior is in qualitative agreement with the data. A more detailed reconciliation between theory and experiment is not easily accomplished, because G_S also depends somewhat on V_{SD}. The ideal situation would be to observe G_A in the absence of heating effects. The effects of electron heating can be alleviated in shorter samples, where electrons may traverse the sample without scattering inelastically. Such work is being carried out. In fact, careful measurements on a range of channel lengths may reveal some of the details of nonequilibrium effects in short MOSFETs.

4. Conclusions

Aperiodic fluctuations in the conductance of small Si MOSFETs were observed when the gate voltage and the perpendicular magnetic field were varied. These fluctuations are similar to those seen by others using metal and GaAs-based heterostructure samples. The fluctuations of a sample tilted with respect to the magnetic field were shown to be dependent on the perpendicular component of that field. This demonstrates that the fluctuations are due to an orbital effect. In addition, the energy over which the fluctuations are correlated is close to theoretical estimates. It is concluded that the fluctuations arise from the same quantum interference mechanism that gives rise to the periodic oscillations with flux period h/e in metallic rings.

It has also been shown that the conductance of a mesoscopic sample is a nonlinear function of the voltage across it. The rectifying behavior predicted by theory is observed to vary with magnetic field and gate voltage. The field and energy scales show that these phenomena are also due to quantum interference. The data imply that second harmonic generation is possible, as predicted [10]. In the absence of electron heating, these effects should increase with the voltage across the sample. In real devices, electron heating does take place, but heating does not totally obscure the asymmetry or the fluctuations. This relative insensitivity to temperature is a striking property of mesoscopic coherence effects.

Many of the experiments in this field were done using specially designed samples, and at temperatures of tens of mK. It is important to note that the demonstration of fluctuating rectification made use of a quite ordinary device fabricated for technological purposes. These effects persisted above liquid helium temperatures.

I am grateful to M. Buttiker, Y. Imry, R. Landauer, P. A. Lee, P. Santhanum, A. Hartstein and A. Williams for interesting discussions. I would like to thank J. Sun for providing samples. The technical assistance of N. Albert and J. Tornello is also acknowledged.

References

1. B. L. Al'tshuler, Pis'ma Zh. Eskp. Teor. Fiz. 41, 530 (1985) [JETP Lett. 41, 648 (1985)]
2. A. D. Stone, Phys. Rev. Lett. 54, 2692 (1985).
3. Y. Aharonov and D. Bohm, Phys Rev. 115, 485 (1959); ibid, 123, 1511 (1961).
4. L. Gunther and Y. Imry, Sol. St. Comm. 7, 1391 (1969).
5. M. Buttiker, Y. Imry and R. Landauer, Phys. Lett. 96A, 365 (1983).

6. R. A. Webb, S. Washburn, C. P. Umbach and R. B. Laibowitz, Phys. Rev. Lett. 54, 2696 (1985).
7. C. P. Umbach, S. Washburn, R. B. Laibowitz and R. A. Webb, Phys. Rev. B30, 4048 (1984); R. A. Webb, S. Washburn, C. P. Umbach and R. B. Laibowitz, in Localization, Interaction and Transport Phenomena in Impure Metals, edited by B. Kramer, G. Bergmann and Y. Bruynseraede (Springer-Verlag, New York, 1985).
8. G. Blonder, Bull. Am. Phys. Soc. 29, 535 (1984).
9. P. A. Lee and A. D. Stone, Phys. Rev. Lett. 55, 1622 (1985).
10. B. L. Al'tshuler and D. E. Khmel'nitskiĭ, Pis'ma Zh. Eskp. Teor. Fiz. 42, 291 (1985); [JETP Lett. 42, 360 (1985)].
11. Y. Imry, Europhys. Lett. 1, 249 (1986).
12. P. A. Lee, A. D. Stone and H. Fukuyama, Phys. Rev. B 35, 1039 (1987).
13. S. B. Kaplan and A. Hartstein, Phys. Rev. Lett. 56, 2403 (1986).
14. S. B. Kaplan and A. Hartstein, in Proceedings of the 18th Int. Conf. on the Physics of Semiconductors, (World Scientific, Singapore, 1987), p. 1499.
15. S. B. Kaplan, to appear in Surface Science.
16. R. A. Webb, A. Hartstein, J. J. Wainer and A. B. Fowler, Phys. Rev. Lett. 54, 1577 (1985).
17. P. A. Lee, Phys. Rev. Lett. 53, 2042-2045, (1984).
18. J. C. Licini, D. J. Bishop, M. A. Kastner and J. Melngailis, Phys. Rev. Lett. 55, 2987 (1985).
19. T. Ando, A. B. Fowler and F. Stern, Rev. Mod. Phys., 54, 437 (1982).
20. W. J. Skocpol, P. M. Mankiewich, R. E. Howard, L. D. Jackel, D. M. Tennant and A. Douglas Stone, Phys. Rev. Lett. 56, 2865 (1986).
21. A. I. Larkin and D. E. Khmel'nitskiĭ, Zh. Eksp. Teor. Fiz. 91, 1815 (1986).

Part IV

From Two Dimensions to One: Conductance Phenomena and Optical Properties

Groundstate Properties of an Electron Gas in a Magnetic Field at the Cross-over from Two- to One-Dimensional Behaviour

U. Wulf and R.R. Gerhardts

Max-Planck-Institut für Festkörperforschung,
Heisenbergstr. 1, D-7000 Stuttgart 80, Fed. Rep. of Germany

The groundstate of a 2DEG with a lateral density modulation in a perpendicular magnetic field is calculated within the Hartree approximation. Emphasis is put on the non-linear screening properties of this system.

1. Introduction

Currently there is a great interest in the fabrication and the experimental and theoretical investigation of low dimensional electron systems. Employing modern microlithography techniques, very successfully mirostructures on Si-MOSFETs and AlGaAs-heterostructures have been produced which allow lateral confinement of the originally two-dimensional electron gas (2DEG) in these systems. Apart from split-gate structures with isolated quasi-1D electron channels, devices with a periodic microstructured gate have been produced, which allow a tunable periodic density modulation of the 2DEG [1,2]. The density can be varied from a nearly homogeneous 2DEG at weak modulation to a quasi one-dimensional situation where the electron system splits into an array of weakly coupled channels. Transport and optical experiments have been performed in a perpendicular magnetic field to study for example Shubnikov-de Haas oscillations, cyclotron resonance, and magnetoplasmons near the cross-over between one- and two-dimensional behaviour [1].

For the interpretation of these measurements a sound knowledge of the energy spectrum is inevitable. Previous work on the groundstate properties in microstructures has covered interacting electrons in zero magnetic field [3-5] and non-interacting electrons in the presence of a magnetic field [6]. Especially in a strong magnetic field the mutual Coulomb interaction between the electrons is, however, expected to lead to peculiar screening effects related to the oscillatory dependence of the singular Landau density of states in the magnetic field [7].

The purpose of the present paper is to discuss the dependence of the groundstate properties of the electron gas on the modulation potential and on the magnetic field under due consideration of the screening effects. The mutual Coulomb interaction will be treated in the Hartree approximation, which allows us to calculate the screening of the external potentials beyond the familiar linear approximation. After introducing our model for a microstructured heterostructure in Sect.2, we consider several typical situations and demonstrate that non-linear screening properties lead to interesting effects which apparently can be observed in very different situations.

2. Model and self-consistent equations

A typical example for the microstructures we want to consider is sketched in Fig.1. The 2DEG is located near the interface between p-doped GaAs and an n-doped AlGaAs layer wich carries a cap layer with a linear grating produced by some kind of etching

Springer Series in Solid-State Sciences Vol. 83: **Physics and Technology of Submicron Structures**
Editors: H. Heinrich · G. Bauer · F. Kuchar

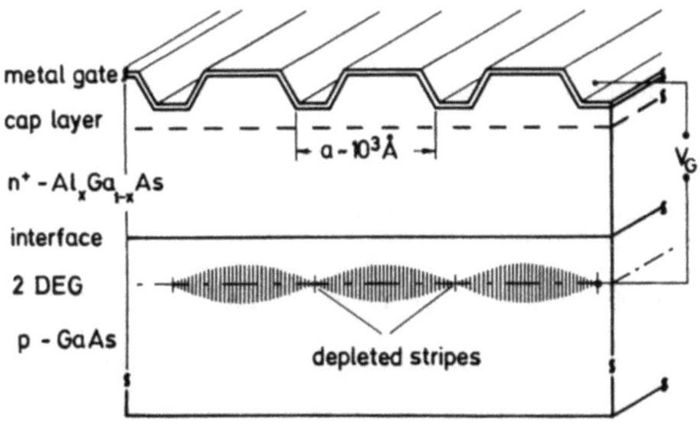

Fig.1: Schematic diagram of a laterally microstructured GaAs-AlGaAs heterostructure

procedure. The grating is covered with a metal gate, so that a gate voltage can be applied between the structured gate and the 2DEG.

We choose coordinates so that the z-direction is perpendicular to the plane of the 2DEG, and points from the interface at $z = 0$ into the GaAs. We assume translational invariance in the y-direction and a periodic modulation in the x-direction with a period a of the grating ($a \sim 100nm$). Donor and acceptor charges are simulated by a background charge density $\rho(z)$ independent of x and y. If a gate voltage is applied in order to deplete electrons from the 2DEG, the electron density decreases first in the stripes closest to the metal gate, where the repelling forces are largest. In the equilibrium state the density of the 2DEG $n_s(x,z)$ is modulated in such a manner that these forces are counterbalanced by the mutual Coulomb repulsion of the electrons, which we treat in the Hartree approximation. Thus, the electrostatic part $V(x,z)$ of the potential energy of each electron is determined by Poisson's equation,

$$\Delta V(x,z) = -\frac{4\pi e^2}{\kappa(z)}\left[n_s(x,z) + n_d(z)\right], \tag{2.1}$$

and suitable boundary conditions at interface and gate. The background charge density has been expressed as $\rho(z) = -en_d(z)$, and $\kappa(z)$ refers to the static dielectric constant of $GaAs$ ($\kappa(z) = \kappa_{sc}$ for $z > 0$) and of $Al_xGa_{1-x}As$ ($\kappa(z) = \kappa_{ins}$ for $z < 0$), respectively. The boundary conditions are as usual,

$$\partial V/\partial x \quad \text{and} \quad \kappa\partial V/\partial z \ \text{continuous at} \ z = 0, \tag{2.2}$$

and $V(x,z) = V_G$ on the metal gate, where V_G is the gate voltage .

The electron density in turn is determined by the energy eigenfunctions $\Psi_\lambda(\vec{r})$ of the single-particle Schrödinger equation

$$\left[\frac{1}{2m}\left(\frac{\hbar}{i}\vec{\nabla} + \frac{e}{c}\vec{A}(\vec{r})\right)^2 + V_{tot}(\vec{r}) - E_\lambda\right]\Psi_\lambda(\vec{r}) = 0 \tag{2.3}$$

and their occupation numbers,

$$n_s(\vec{r}) = g\sum_\lambda |\Psi_\lambda(\vec{r})|^2 f(E_\lambda - \mu), \tag{2.4}$$

where $f(E) = [\exp(E/k_BT)+1]^{-1}$ is the Fermi-function, g a degeneracy factor ($g = 2$ for the spin degeneracy in GaAs), $V_{tot}(\vec{r}) = V(x,z) + \Delta E_c\ \Theta(-z)$ is the sum of electrostatic potential energy and conduction band offset, and

$$\vec{A} = (0, Bx, 0) \qquad (2.5)$$

is the vector potential in the Landau gauge. μ is the electrochemical potential of the 2DEG. The symmetry of the system is exploited by the ansatz

$$\Psi_\lambda(\vec{r}) = L_y^{-1/2}\exp(iyk_y)\Phi_\lambda(x,z), \qquad k_yL_y/2\pi = 0, \pm 1, ..., \qquad (2.6)$$

where periodic boundary conditions in the y-direction with a period L_y have been imposed, and by the Fourier expansions

$$n_s(x,z) = \sum_{r=-\infty}^{\infty} n_{s,r}(z)\exp(iq_rx), \quad V(x,z) = \sum_{r=-\infty}^{\infty} V_r(z)\exp(iq_rx), \qquad q_r = r\ 2\pi/a \quad (2.7)$$

for electron density and potential. From Poisson's equation (2.1) we see that a potential contribution $V_r(z)$ originating from a source at $z = z'$ decays exponentially, $V_r(z) \sim exp(-|q_r||z-z'|)$. Thus, in sufficiently large distance from the gate only the fundamental harmonic ($|r| = 1$) will survive, even if higher harmonics of the potential are created at the gate. Taking advantage of this fact, we avoid the solution of the actual problem with a complicated $V(x,z)$ near the gate by a simple boundary condition

$$V(x,-D) = V_G + V_M\cos(k_0x), \qquad k_0 = 2\pi/a, \qquad (2.8)$$

where $z = -D$ is the average position of the gate.

Our aim is to achieve a good understanding of the response of the 2DEG to applied modulations in the x-direction. We are here not interested in accompanying effects such as an average shift or a fluctuating spread of the wave fuctions in the z-direction. Thus we choose the model parameters so that these additional effects are small. For convenience, we require for the electron density $n_s(x,z) = 0$ for $z < 0$, i.e. we assume $\Delta E_c \to \infty$. Then we solve (2.1) with boundary conditions (2.2) and (2.8). For $r \neq 0$ and $z > 0$ we get [5]

$$V_r(z) = I_r(z) + \exp(-|q_r|z)\left[\frac{1}{2}\epsilon V_M\delta_{|r|,1} + a_r^- I_r(0)\right]/a_r^+, \qquad (2.9)$$

where $\epsilon = \kappa_{ins}/\kappa_{sc}$, $a_r^\pm = \sinh(|q_r|D) \pm \epsilon\cosh(|q_r|D)$, and

$$I_r(z) = \frac{2\pi e^2}{\kappa_{sc}|q_r|}\int_0^\infty dz' n_{s,r}(z')\exp(-|q_r||z-z'|).$$

Furthermore, we assume a depletion layer with $N_{dep} = n_dd$ ionized acceptors per cm^2 distributed homogeneously over the thickness d which is much larger than the spread of the 2DEG. Then the $r = 0$ component of the potential for $z > 0$ can be written as [5]

$$V_0(z) = \frac{4\pi e^2}{\kappa_{sc}}\{z\left[N_{dep}(1-\frac{z}{2d}) + \frac{1}{2}N_s\right] - \frac{1}{2}\int_0^\infty dz' n_s(z')[|z-z'| - z']\}, \qquad (2.10)$$

where N_s is the average number of 2D electrons per cm^2, and the term $z/2d$ may be neglected.

If we neglect the modulation, $V_r(z) \equiv 0$ for $r \neq 0$, and keep only the linear-in-z term of $V_0(z)$, the eigenfunctions are products of the form $\Phi(x,z) = \phi(x)\, u(z)$ where the $u(z)$ are shifted Airy functions, whereas $\phi(x)$ are plane waves for zero magnetic field and Landau functions for finite B. In the general case, we expand the exact eigenfunctions into a sum of such product functions. In order to discuss screening properties in a strong magnetic field, we assume a large depletion charge, $N_{dep} \gg N_s$, which leads to a strong confinement of the 2DEG in the z-direction and a large splitting of the electrical subband energies, so that the coupling of the electric subbands by the periodic modulation potential can be neglected. Then, the wavefunctions of the lowest electric subband, which is partly occuppied, can be well approximated by $\Phi(x,z) = \phi(x)u_o(z)$ where $u_o(z)$ is an Airy function and ϕ satisfies the reduced Schrödinger equation

$$[-\frac{\hbar^2}{2m}\frac{\partial^2}{\partial x^2} + \frac{1}{2}m\omega_c^2(x-x_o)^2 + V(x) - \epsilon_\lambda]\phi_\lambda(x) = 0 \tag{2.11}$$

with

$$\omega_c = \frac{eB}{mc}, \quad x_o = -l^2 k_y, \quad l = (\frac{\hbar c}{eB})^{1/2} \tag{2.12}$$

being the cyclotron frequency, center coordinate of the cyclotron motion, and the magnetic length, respectively. The effective potential follows from

$$V(x) = \sum_r \exp(iq_r x) \int_0^\infty dz V_r(z)|u_o(z)|^2. \tag{2.13}$$

The Fourier coefficients of the electron density in this approximation are $n_{s,r}(z) = N_{s,r}|u_o(z)|^2$. The magnetic field introduces the parabolic confinement potential into (2.11). In the rest of this paper we consider typical situations which may occur depending on the relative strength of this magnetic potential and the modulation potential $V(x)$. We start in Sect.3 with the limit of zero magnetic field. In Sect.4 we consider a weak modulation which allows us to study magnetic field dependent screening properties of the 2DEG. Finally, we discuss in Sect.5 the case of a strong periodic modulation, i.e. the quasi 1D case of weakly coupled channels. We will demonstrate that in all these cases interesting screening effects occur.

3. Zero magnetic field

In the limit of vanishing magnetic field the free motion in the y-direction is decoupled from that in the x-direction. The magnetic term in (2.11) reduces to the free electron energy dispersion in the y-direction,

$$\lim_{B\to 0} \frac{1}{2}m\omega_c^2(x + l^2k_y)^2 = \frac{1}{2m}\hbar^2 k_y^2.$$

The periodic modulation in the x-direction leads to the well known Bloch band structure. Here we are interested in screening effects on the band structure which, e.g., show up in the dependence of the lowest Bloch energy E_o on the amplitude V_c of the external potential V_{ext} in the plane $\bar{z}$, where the mean electron density $n_{r,0}(z)$ is maximum, $V_C = \epsilon V_M \exp(-2\pi\bar{z}/a)/a^+_{r=1}$, cf.(2.9). The upper insert of Fig.2 visualizes these quantities. The external potential V_{ext} (solid line) is taken to oscillate around the energy zero. The resulting screened potential V, which determines the band structure, oscillates around an energy offset ΔV given by the average Hartree energy of the system without modulation.

The solid lines in Fig.2 show the V_c-dependence of the bottom of the Bloch bands for the lowest three electric subbands. The calculations are for a (100)-Si-MOSFET

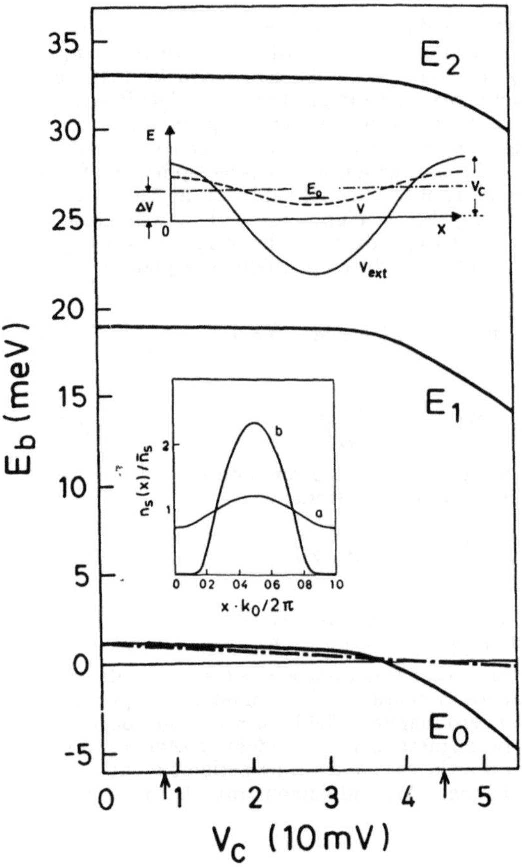

Fig.2: Minimum energies of the three lowest electric subbands calculated within the Hartree approximation (solid line), and minima of the screened potential calculated within linear screening theory (dash-dotted line) versus V_c. Upper insert: Schematic graph of the external potential (solid line) and screened potential (dashed line). Lower insert: Normalized electron density corresponding to (a) $V_c = 9meV$ and (b) $V_c = 44meV$.

($\kappa_{sc} = 11.5, \kappa_{ins} = 3.9, m_z = 0.92m_o, m_x = m_y = m = 0.19m_o, g = 4, D = 26nm, a = 200nm, N_{dep} = 3 \cdot 10^{11}cm^{-2}, N_s = 10^{11}cm^{-2}$) and have been published previously [5]. The fact that the curves are nearly parallel indicates that the electric subband splitting is nearly unaffected by the modulation in the x-direction.

For small amplitude V_c of the external cosine potential, E_o decreases linearly with increasing V_c, until, at a rather well defined V_c-value $V_c^o(\approx 35meV)$ a much stronger decrease sets in. The numerical results show that this happens when the 2DEG splits into an array of quasi-1D channels. For $V_c < V_c^o$, the electron density shows only a weak modulation with rather large values at the potential maxima (cf. curve a of lower insert of Fig.2, and left arrow at the V_c-scale) for $V_c > V_c^o$, on the other hand, the density at the potential maxima is very low (cf. curve b of the insert, right arrow) and the electrons arrange themselves in weakly coupled channels. The transition between the two situations occurs in a rather small interval of V_c-values and can be attributed to the breakdown of linear screening of the external potential. This is demonstrated by the dash-dotted straight line in Fig.2, which indicates the minimum of the screened potential calculated within the linear Thomas-Fermi theory of screening (see (3.2)-(3.6) below). For $V_c < V_c^o$ this minimum is slightly lower than the minimum energy eigenvalue E_o, as expected. However, near V_c^o the linear approximation obviously starts to fail. The numerical data show that there the amplitude of the screened potential becomes equal to the Fermi energy of the homogeneous, unmodulated 2DEG.

In order to understand this, we now briefly sketch the Thomas-Fermi theory of screening in the 2DEG. One starts with the solution

$$V(\vec{r}) = \frac{e^2}{\kappa} \int d^3r' \, \frac{n(x',y')\delta(z'-\bar{z})}{|\vec{r}-\vec{r}\,'|} \tag{3.1}$$

of the 3D Poisson equation for the electrostatic potential energy originating from the 2D electron density $n(x,y)$ in the plane $z = \bar{z}$. Restricting the potentials to this plane and taking 2D Fourier transforms, one obtains for the response of the 2DEG on the external potential V_{ext}:

$$V(\vec{k}) = V_{ext}(\vec{k}) + \frac{2\pi e^2}{\kappa|\vec{k}|}\delta n_s(\vec{k}), \tag{3.2}$$

i.e. the potential $V(\vec{k})$ is the sum of the external potential and the induced potential, expressed in terms of the induced density fluctuation $\delta n_s(\vec{k})$, which, in turn, is determined by the total potential V. If V varies slowly in space, it can be treated like a local thermodynamic variable, and the induced density is given by

$$\delta n_s(x,y) = n_s(x,y) - \bar{n}_s = \int dE D(E) \left[f(E + V(x,y) - \mu) - f(E-\mu) \right], \tag{3.3}$$

with $D(E)$ the density of states (DOS) of the 2DEG. If V is, furthermore, sufficiently small, one can linearize,

$$\delta n_s(x,y) = -D_T \, V(x,y), \tag{3.4}$$

where

$$D_T = \frac{\partial \bar{n}_s}{\partial \mu} = \int dE \; D(E) \left[-\frac{d}{dE} f(E-\mu) \right] \tag{3.5}$$

is the thermodynamic density of states at the Fermi level. Inserting (3.4) into (3.2), we read off the 2D dielectric constant

$$\epsilon(\vec{k}) = 1 + Q/|\vec{k}|, \qquad Q = 2\pi D_T e^2/\kappa. \tag{3.6}$$

The corresponding 2D Thomas-Fermi screening length is $2\pi/Q$. For the 2DEG in zero magnetic field the DOS is constant, $D(E) = D_o\Theta(E)$, with $D_o = g \; m/2\pi\hbar^2$. For low temperatures (3.3) yields

$$n_s(x) = (\mu - V(x)) \; D_o\Theta(\mu - V(x)) \,. \tag{3.7}$$

Thus, for $V(x) \leq \mu$ the linearization is exact. For $V(x) > \mu$, however, the linear approximation is no longer justified. According to (3.7), the electron density vanishes where the total screened potential is larger than the Fermi energy. That means that screening is less effective and the sreened potential follows the external potential more closely than predicted by the linear approximation.

These considerations nicely explain the numerical results. As the linear screening breaks down, a more or less abrupt transition from the 2D to 1D behaviour occurs. Such a sudden formation of 1D channels at a critical value of the modulation amplitude is also observed in the experiments [1]. There the effect is even more drastic, since with increasing modulation, i.e., gate voltage, the average electron density becomes smaller, so that screening is reduced even more. Within a linear screening approximation or even a strict non-interacting electron picture such a sudden change to 1D behaviour can not be understood.

4. Weak modulation and strong magnetic field

In this section we discuss peculiar screening properties of the 2DEG in a strong magnetic field. We will consider only the screening of external potentials which vary slowly on scale of the magnetic length, so that the energy spectrum is dominated by the Landau quantization. First, however, some general remarks are necessary.

In finite magnetic field the periodic modulation again leads to a band structure which, however, is different from the familiar Bloch band structure. Owing to the translation invariance in the y-direction, k_y remains a good quantum number, which enters the Schrödinger equation (2.11) for the x-direction as center coordinate x_o. For fixed x_o, the parabolic magnetic potential term destroys the periodicity and leads to discrete energy eigenvalues $\epsilon_{x_o,n}$ with bounded eigenfunctions $\phi_{x_o,n}(x)$. Considered as a function of x_o, the eigenvalue problem becomes periodic with the period a of the modulation. From (2.11) one immediately obtains

$$\epsilon_{x_o+a,n} = \epsilon_{x_o,n}, \qquad \phi_{x_o+a,n}(x) = \phi_{x_o,n}(x-a). \tag{4.1}$$

Thus the whole information on the energy spectrum is contained in $\epsilon_{x_o,n}$ with $|x_o| \leq a/2$. We see that, owing to the coupling of the motion in the x- and y-direction by the Lorentz force, the periodic modulation in the x-direction leads to a band structure for the k_y-dependence of the energy spectrum with a unit cell of width a/l^2. For $B = 0$ one has, on the other hand, the usual Bloch bands depending on the continuous quantum number k_x, which varies in the Brillouin zone of width $2\pi/a$ and has no analogue in the case $B \neq 0$. Furthermore for $B = 0$ the x- and y- direction are decoupled, with the free electron energy dispersion in the y-direction.

We now consider a modulation potential $V(x)$ varying slowly on the scale of the magnetic length, $l|dV/dx| \ll \hbar\omega_c$. Locally this leads only to a small perturbation of the Landau wavefunctions. For a fixed x_o, the matrix elements of $V(x)$ between different Landau wavefunctions are small, so that the level mixing owing to $V(x)$ is not important. The expectation value of $V(x)$ in states with small Landau quantum number n is well approximated by $V(x_o)$, since their spatial extent is of the order $\sqrt{n+1}\, l$. We, therefore, obtain, at least for small n,

$$\epsilon_{x_o,n} = \hbar\omega_c(n+\frac{1}{2}) + V(x_o) \tag{4.2}$$

as a reasonable approximation for the energy spectrum, i.e. the unperturbed Landau energies with an offset given by the potential at the center of the oscillator functions. The modulation lifts the high degeneracy of the Landau energies and leads to Landau bands which reflect closely the spatial variation of the modulation potential $V(x)$.

The simple result (4.2) tells us a lot about the screening properties of the system. Let e.g., a Landau level be half filled in the absence of modulation and apply (4.2) to an external modulation potential V_{ext} with a total variation of much more than the thermal energy k_BT. Then only states with center coordinates x_o near the minima of V_{ext} will be occupied while those near the maxima will remain empty. A large spatial variation of the electron density $n_s(x)$ follows which leads to an electrostatic field screening the external modulation field. As a result, the external potential V_{ext} leads to a total screened potential $V(x)$ which varies only by amounts of the order k_BT. Here we used the weak modulation assumption in the form that the 2DEG will not split into 1D channels with essentially zero electron density near the potential maxima.

The same result in a more quantitative version can be obtained from the linearized Thomas-Fermi result (3.6) if one calculates D_T from the Landau DOS

$$D(E) = \frac{g}{2\pi l^2}\sum_{n=0}^{\infty}\delta(E-\hbar\omega_c(n+\frac{1}{2})), \tag{4.3}$$

where $g = 2$ includes the spin degeneracy. Defining the total filling factor ν and the filling factor ν_n of the n-th Landau level according to

$$\nu = 2\pi l^2 N_s = \sum_n g\ f(\hbar\omega_c(n+\frac{1}{2}) - \mu) = \sum_n \nu_n, \tag{4.4}$$

we obtain

$$\epsilon(\vec{k}) = 1 + \frac{\hbar\omega_c}{k_B T}\sum_n \frac{\nu_n}{g}(1-\frac{\nu_n}{g})\frac{2}{|\vec{k}|a_B}, \tag{4.5}$$

where $a_B = \kappa\hbar^2/(me^2)$ is the effective Bohr radius. For $\hbar\omega_c \gg k_B T$ only the Landau level containing the Fermi energy contributes considerably to the sum in (4.5), the others give $\nu_n = g$ or $\nu_n = 0$. In addition to the nearly perfect screening for half-filled Landau levels, (4.5) predicts that screening becomes very poor if a Landau level is nearly full or empty. This prediction is compatible with the result (4.2) only if the total variation of the external potential is less than $\hbar\omega_c$. If it is larger, and that is possible for large periods $a \gg l$ even with $l|dV/dx| \ll \hbar\omega_c$ we can argue with (4.2) and the weak modulation assumption, that the total variation of the screened potential, within an accuracy of the order of $k_B T$ must be $\hbar\omega_c$. Thus the linear screening approach breaks down in this case. This is indeed easily understood from a close inspection of the non-linearized expression (3.3).

With these qualitative considerations our numerical results are easily understood. Parameters corresponding to an AlGaAs-heterostructure have been chosen ($\kappa_{sc} = 12.4$, $\kappa_{ins} = 11.6, m^* = 0.067m_o, D = 26nm, a = 180nm, N_s = 2.25 \cdot 10^{11}cm^{-2}$, $N_{dep} = 3 \cdot 10^{11}cm^{-2}$) and the amplitude V_c of the external cosine potential is taken as $V_c = 7.8meV$. Figure 3 shows for $T = 1K$ the three lowest Landau bands and the Fermi level for three values of the filling factor ($\nu = 1.6, 1.7$, and 2.0). By taking into account up to eight Landau levels for the expansion of the eigenfunctions, we have numerically checked that only three Landau levels are necessary to calculate the two lowest Landau bands with sufficient accuracy. This is a consequence of the small level mixing at fixed x_o.

As is seen from Fig.3, the width of the Landau bands and thus, according to (4.2), the total variation of the screened potential is very small for filling factor $\nu = 1.6$, considerably larger for $\nu = 1.7$, and close to $\hbar\omega_c$ at $\nu = 2.0$. For $\nu = 1.7$ one clearly observes pinning of the lowest Landau band to the Fermi energy. This pinning effect can

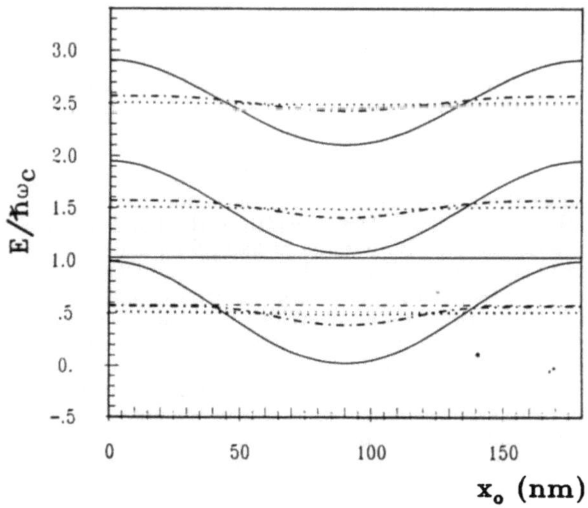

Fig.3: Landau bands (thick lines) and corresponding Fermi levels (thin horizontal lines) for $V_c = 7.8meV$, $T = 1K$, and different filling factors: $\nu = 2.0$ (solid), $\nu = 1.7$ (dash-dotted) and 1.6 (dotted).

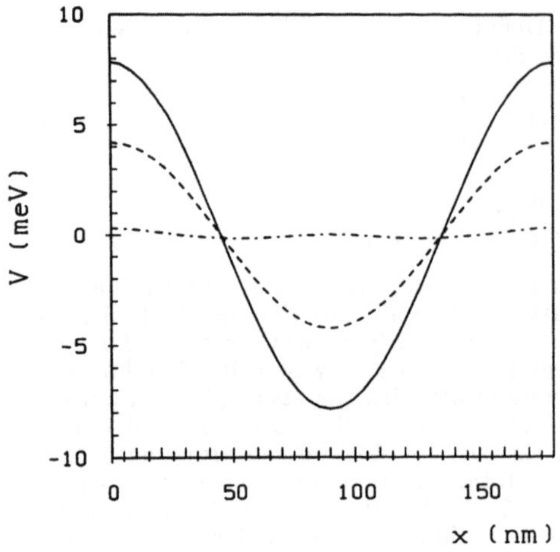

Fig.4: External potential (solid line) for $V_c = 7.8meV$ and screened potential at $\nu = 2.0$ (dashed line) and $\nu = 2.4$ (dash-dotted line).

also be understood from (4.2) and the weak modulation assumption, which prohibits regions of nearly vanishing $n_s(x)$. We have also calculated the energy spectrum for $\nu > 2$ and we find that screening improves and the Landau bands become smaller again as the filling of the $n = 1$ Landau level increases. For $2 \leq \nu < 3$, we observe pinning to the Fermi level of the $n = 1$ Landau band near its minimum. In this way an accumulation of high electron density near the potential minimum, which would give rise to large electrostatic fields is avoided.

Figure 4 shows the external potential, corresponding to the data of Fig.3, together with the screened potential for $\nu = 2$ and $\nu = 2.4$. In contrast to the linear approximation, a considerable screening is observed at $\nu = 2$, so that the total variation of the screened potential is $\hbar\omega_c \approx 8.4meV < 2V_c$. For $\nu = 2.4$ an oscillation with period $a/2$ is seen in the screened potential which also indicates non-linear effects. We note that the density modulation which produces the screening also depends on the filling factor. But in the cases considered here this dependence is not dramatic. We obtained a modulation which looks cosine-like with an amplitude of about 18% for $\nu = 1.6$ and $\nu = 2.4$, and of 9% for $\nu = 2.0$. Further details are given elsewhere [8].

We also performed calculations for stronger modulation [$V_c = 19meV$] and obtained similar results. Especially for $\nu = 2.0$ the total variation of the screened potential was again $\hbar\omega_c$, as expected.

As the results of this section demonstrate, screening properties strongly affect the single-particle spectrum if long-range fluctuations of the electrostatic potential of sufficiently large amplitude occur in the sample. The effective width of the Landau levels will oscillate as a function of the filling factor, with minima at half filling and maxima of order $\hbar\omega_c$ at total filling of the Landau levels. Due to the pinning, the DOS at the Fermi energy is always considerable, at least within an energy interval of width k_BT, even if the Fermi energy is well between two Landau energies. One can give simple statistical arguments that in real samples such potential fluctuations must occur on the length scale of about10^3nm [8,9]. Thus, the calculations of this section provide a qualitative understanding of many experiments [10] which revealed an unexpectedly high DOS between the Landau levels in systems under conditions of the quantum Hall effect. Moreover, they give a microscopic justification of a statistical inhomogeneity model developed recently by Gerhardts and Gudmundsson [9] in order to explain these experiments. This model includes, on a phenomenological level, in addition to the nonlinear screening of long-range potential fluctuations the level broadening effects of short range fluctuations, and yields excellent agreement with the experiments for reasonable values of the material parameters [11].

Peculiar screening effects and pinning of the energy spectrum in the presence of long-range potential fluctuations have been predicted by Luryi [7] on the basis of qualitative arguments. An analytical non-linear screening approach on the basis of the Hartree approximation has recently been formulated by Labbe [12]. He made, however, no attempt to solve his non-linear set of equations for the screened potential explicitly, and he did not discuss the effects we have emphasized. His linearized dielectric constant reduces in leading order of $|\vec{k}|l \ll 1$ to (4.5) with $g = 1$ (no spin).

Similar oscillations of the Landau level width as a function of the filling factor have previously been calculated within a perturbative approach [13]. Low order Bornian scattering by randomly distributed charged impurities leads in these theories to the level broadening, and non-linear effects occur since the level broadening is included in a self-consistent way in the dielectric function which screens the impurity potentials. The results for level broadening and DOS obtained from these theories, especially in the case of completely filles Landau levels, are, however, not convincing.

5. Strong modulation and finite magnetic field

For strong modulation the electron density at the potential maxima becomes very small, and weakly coupled 1D channels are formed. We discuss two complementary cases, first that of a weak magnetic field and then that of a strong magnetic field where the Landau splitting exceeds the barrier height of the modulation potential.

5.1. Weak magnetic field

For a weak magnetic field the magnetic length is comparable with or larger than the scale on which the modulation potential varies. Then the approximations leading to (4.2) are not valid. The potential mixes the Landau levels considerably and a complicated energy band structure may result. A typical example is shown in Fig.5 for $B = 1T$. Again

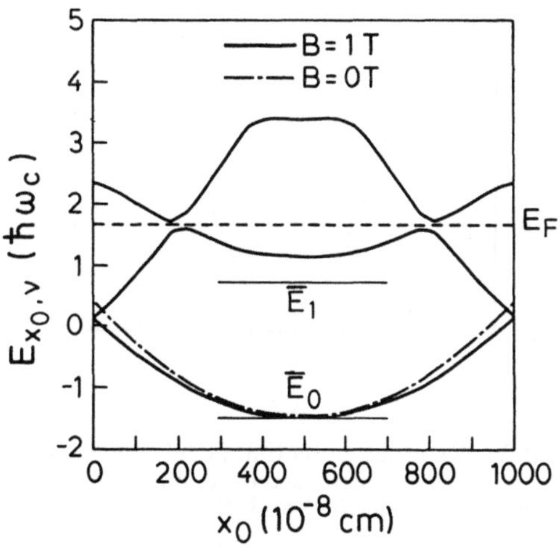

Fig.5: Energy spectrum at $V_c = 18.5meV$ and $B = 1T$ (solid line) with the Fermi level (dashed line). For comparison the lowest energy levels $\bar{E}_o$ and $\bar{E}_1$ of the quantization in the x-direction at $B = 0T$ and the free energy dispersion in the y-direction (dash-dotted line) are plotted.

a AlGaAs-heterostructure is assumed, but now with $N_s = 10^{11} cm^{-2}, a = 100nm$ and $V_c = 18.5meV$. The energy spectrum closely resembles the energy spectrum for $B = 0T$. To demonstrate this, we have, for $B = 0T$, indicated the energy levels $\bar{E}_0$ and $\bar{E}_1$ due to the confinement in the x-direction and the free dispersion $\hbar^2 k_y^2/2m$ adding to $\bar{E}_0$. Here k_y is expressed in terms of x_o for $B = 1T$. For a qualitative understanding of the whole bandstructure, it is useful to approximate the modulation potential at its minimum by a parabola, say $V(x) = \frac{1}{2}\omega_o^2 x^2$. Then, the total potential in (2.11) is parabolic. The energy eigenvalues are

$$E_n(x_o) = \hbar\Omega(n + \frac{1}{2}) + \frac{1}{2}m\omega_c^2 \frac{\omega_o^2}{\Omega^2} x_o^2 \tag{6.1}$$

with $\Omega = (\omega_o^2 + \omega_c^2)^{1/2}$ and the eigenfunctions are rescaled oscillator functions centered around $\tilde{x}_o = x_o\omega_c^2/\Omega^2$. For $\omega_c \ll \omega_o$ the energy spectrum (5.1) is a set of flat parabolas. Plotting the corresponding parabolas for each valley of the periodic modulation potential, one gets in the center zone $|x_o| < a/2$ a kind of backfolded band scheme which is a reasonable approximation of the correct one for low energies. The wavefunctions corresponding to the different parabolas which cross at a point x_o inside the zone belong to different valleys. For $\omega_c \ll \omega_o$ their centers $\tilde{x}_0$ are far apart, their overlap is small, and the correct bandstructure has a small gap at that x_o-value.

With increasing magnetic field the parabolas $E_n(x_o + ma)$ become steeper, the crossing points x_o move towards the zone boundaries $\pm a/2$, the corresponding $\tilde{x}_o$-values come closer, the overlap of the corresponding wavefunctions increases, the gaps become larger, and the bands become flatter. Finally for strong magnetic fields the simple bandstructure of Fig.6 results.

The complicated bandstructure is, of course, a single-particle effect. Similar results have been obtained by Maan for non-interacting electrons in superlattices with a magnetic field parallel to the lattice planes [6]. It is interesting to note that the vertical distance between adjacent bands changes drastically for small B-values, exhibiting maxima as well as minima within the elementary cell. This should be observable in optical experiments with an electric field vector polarized in the x-direction.

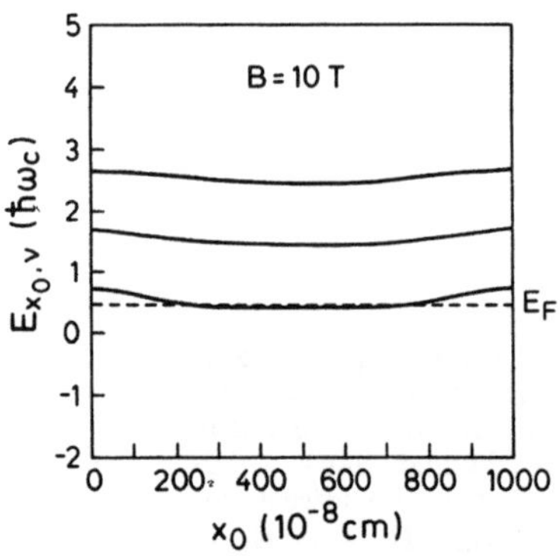

Fig.6: Energy spectrum at $V_c = 18.5meV$ and $B = 10T$

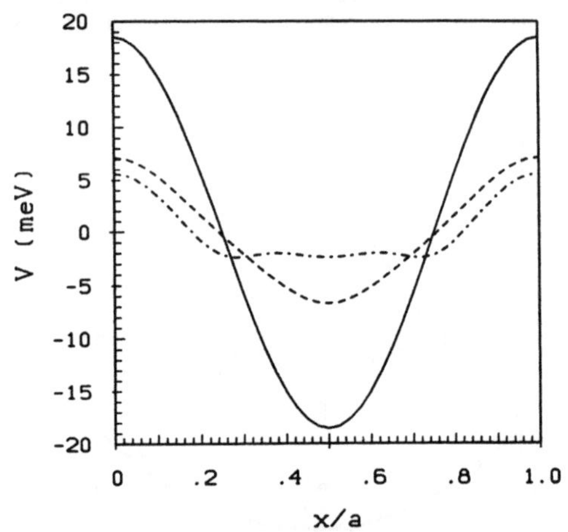

Fig.7: External potential (solid line) for $V_c = 18.5meV$ with screened potential at $B = 1T$ (dashed line) and $B = 10T$ (dash-dotted line).

5.2. Strong magnetic field

The energy spectrum arising at a magnetic field of $10T$ is depicted in Fig.6. The cyclotron energy, $\hbar\omega_c \approx 17meV$, is much larger than the confinement energy $\hbar\omega_o \approx 4meV$ corresponding to $V_c = 18.5meV$ in the approximation discussed in Sect.5.1. No backfolding is left and the band distance is very close to $\hbar\omega_c$ as we expect from the simple parabolic model. But we see that over a large central region the bands are flat without any indication of a parabolic curvature. This is a screening effect, resembling the pinning which was discussed in the weak modulation case. It is seen more clearly in the plot of the screened potential shown in Fig.7. At weak magnetic fields the screened potential follows closely the external potential. In a strong magnetic field, however, the screening becomes strongly non-linear and the effective potential develops a flat central region, where the confinement potential is completely screened out. The electrons in this region behave very much like free 2D electrons in a magnetic field.

It is interesting to investigate the width W of the potential plateau as a function of the magnetic field (Fig.8). W is defined as $W = (x^+ - x^-)$, where $x^+(x^-)$ is the largest (lowest) x-value within one microstructure cell with $V(x^+) = V(x^-) = V(\bar{x})$, where $\bar{x}$ is the center of the cell. For small B we have essentially linear screening, the screened potential is cosine-like, and we have no such plateau. Then, in a certain narrow range of magnetic fields the screening properties change drastically, the plateau region develops and the nearly perfect screening discussed in the previous section is observed. The change in the screening properties can also be seen in the electron density (Fig.9), although here the effects are not so drastic. For weak magnetic fields the density at the potential maxima between the valleys are small but still noticable. For strong fields the electrons are essentially localized in the perfect screening region in the center of the wells and the overlap between the neighbouring wells vanishes.

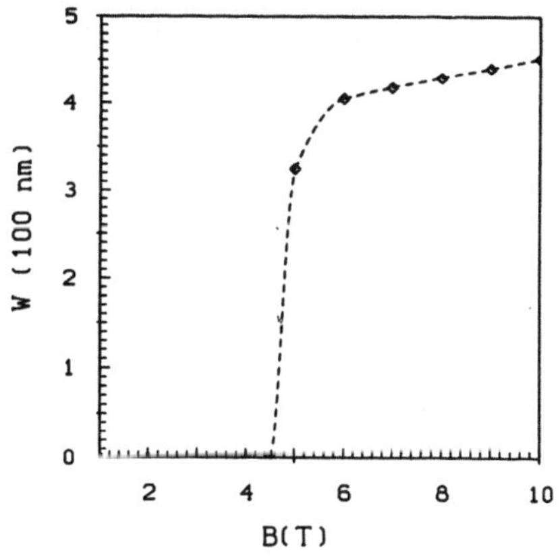

Fig.8: Potential plateau width W (defined in the text) versus the magnetic field.

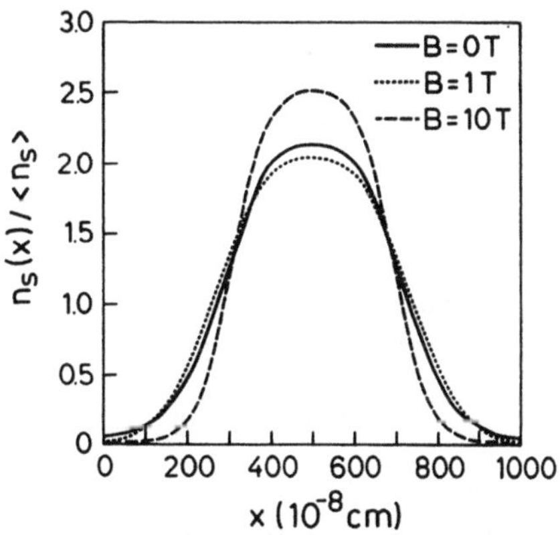

Fig.9: Electron density corresponding to the energy spectrum of Fig.5 (dotted line) and Fig.6 (dashed line) for $T = 1K$. The corresponding distribution at $B = 0T$ and $T = 0K$ in solid line.

6. Summary

A laterally modulated 2DEG in a perpendicular magnetic field has been considered emphasizing the importance of non-linear screening effects which were treated within the Hartree approximation. The properties of the system depend on the relative strength of the cyclotron energy and the modulation potential. Typical situations were discussed.

For zero magnetic field we considered the transition from 2D to 1D behaviour with increasing amplitude of the external modulation potential. Owing to the break-down of linear screening, a rather sharply defined transition was obtained. This seems to be in agrrement with the experimental observations by Hansen et al. [1].

For weak modulation and strong magnetic field interesting non-linear screening effects were obtained: pinning of the energy levels to the Fermi energy and a strong variation of the band width with the filling factor, for instance. These results allow a simple explanation of the apparent constant background density of states between Landau levels, seen in many experiments on the 2DEG in strong magnetic fields. These results also have interesting consequences for the localization mechanism in the integer quantum Hall effect which could not be discussed here.

For strongly modulated systems we have obtained a complicated band structure for weak and moderate magnetic fields. From these results an interesting line splitting of intrasubband resonances for weak magnetic fields is expected, which is absent in zero as well as in strong magnetic fields. At present, the available microstructures are probably not narrow and not clean enough to study these effects experimentally. It is, however, possible - although not yet clear - that strong splittings of the cyclotron resonance lines which recently have been observed have their origin in a similar interaction of electric confinement potentials with magnetic forces. For strong magnetic fields we find a transition to a 2D behaviour of the electron gas within a quasi 1D-channel of the modulated system. This transition is again accompanied with a break-down of the linear screening.

Acknowledgement

We gratefully acknowledge the pleasant and close cooperation with Vidar Gudmundsson who performed calculations similar to those of Sect. 4 for a strictly 2D laterally bounded system. We are also indebted to D. Heitmann and E. Böckenhoff for valuable discussions. The work was supported by the Bundesministerium für Forschung und Technologie, West Germany (Grant No.NT-2718-C).

References

[1] W.Hansen, M.Horst, J.P.Kotthaus, U.Merkt, Ch.Sikorski, and K.Ploog, Phys. Rev. Lett. 58, 2586 (1987)

[2] T.P.Smith III, H.Arnot, J.M.Hong, C.M.Knoedler, S.E.Laux, and H.Schmid, Phys. Rev. Lett. 59, 2802 (1987)

[3] S.E.Laux and F.Stern, Appl. Phys. Lett. 49, 91 (1986)

[4] W.Y. Lai and S.Das Sarma, Phys. Rev. B33, 8874 (1986)

[5] U.Wulf, Phys.Rev.B35, 9754 (1987)
(All values for V_M given in this paper are to be multiplied by the factor 2 due to a trivial error. See Erratum submitted to Phys.Rev.B)

[6] J.C.Maan, in *'Two-Dimensional Systems, Heterostructures and Superlattices'*, Vol53 Springer Series in Solid-State Sciences, ed. by G.Bauer, F.Kuchar, and H.Heinrich, (Springer 1984), p.183

[7] S.Luryi,in *'High Magnetic Fields in Semiconductor Physics'*, Vol71 Springer Series in Solid-State Sciences, ed. G.Landwehr (Springer 1986), p.16

[8] U.Wulf, V.Gudmundsson, and R.R.Gerhardts, submitted to Phys.Rev.B

[9] R.R.Gerhardts and V.Gudmundsson, Phys. Rev. B34, 2999 (1986); V.Gudmundsson and R.R.Gerhardts, Phys. Rev. B35, 8005 (1987)

[10] D.Weiss, K.v.Klitzing, and V.Mosser, in *'Two-Dimensional Systems:Physics and new Devices'*, Vol67 Springer Series in Solid State Sciences, ed. by G.Bauer, F.Kuchar and H.Heinrich (Springer,1986), p.204; R.Lassnig and E.Gornik, ibid p.218

[11] D.Weiss, V.Mosser, V.Gudmundsson, R.R.Gerhardts, and K.v.Klitzing, Solid State Commun. 62, 89 (1987); V.Gudmundsson and R.R.Gerhardts, Phys. Rev. B, 1988 in press.

[12] J.Labbe, Phys. Rev. B35, 1373 (1987)

[13] R.Lassnig and E.Gornik, Solid State Commun. 47, 959, (1983); T.Ando and Y.Murayama, J. Phys. Soc. Jpn 54, 1519 (1985); W.Cai and T.S.Ting, Phys. Rev. B33, 3967 (1986)

Visualization and Theoretical Modelling of the Atomistic Structure of Semiconductor Quantum Well Interfaces

J. Christen[1], *D. Bimberg*[1], *T. Fukunaga*[2], *H. Nakashima*[3], *D.E. Mars*[4], and *J.N. Miller*[4]

[1]Institut für Festkörperphysik der Technischen Universität, Hardenbergstr. 36, D-1000 Berlin 12, Germany
[2] Oki Electric Industry Co., Tokyo, Japan
[3]Institute of Industrial and Scientific Research, Osaka University, Japan
[4]Hewlett-Packard Laboratories, 3500 Deer Creek Road, Palo Alto, CA 94304, USA

Abstract

The atomic scale crystallographic and chemical properties of interfaces between semiconductors are of decisive importance for the performance of novel generations of electronic and photonic devices and are in addition of large fundamental interest. Optical methods like luminescence and absorption have recently emerged to yield quantitative information on these properties, if the corresponding lineshape are carefully analyzed. We emphasize here luminescence. The natural lineshape of luminescence from a quantum well shows Gaussian broadening if its interfaces are not ideally abrupt. A detailed lineshape theory is outlined, allowing for a quantitative determination of the interface roughness distribution function. We find this function to depend in a delicate way on growth rates, temperature, interruption time and chemical compositon of the growth surface. The results of an experimental study of the model quantum well system AlGaAs/GaAs/AlGaAs grown by molecular beam epitaxy with and without interruption of the growth at the interfaces is presented. Roughness reduction upon growth interruption is analyzed in detail. For specific growth conditions and interruptions of 2 min at both interfaces formation of up to 7 µm large interface islands differing by a one monolayer step (2.8 A) are observed. Consequently such quantum wells have a columnar structure, which can be directly visualized using cathodoluminescence imaging. Strong reduction of island size indicating transition from planar growth to three-dimensional growth is observed by CLI upon an increase of growth temperature from T_g = 600° C to 660° C.

1. Introduction

Our present knowledge of structural, chemical and electronic properties of interfaces between semiconductors is inversely proportional to their importance for the design and the understanding of a whole generation of novel microstructured electronic and photonic devices. These devices such as double barrier diodes consist wholly or partly of interfaces and regions close to them. Two of the main difficulties which can be presently indentified to cause this situation are:

(i) Many properties of interfaces and of structures and devices consisting of multiples of interfaces depend on seemingly minor details of the technology used to fabricate them, as well as on fundamental physical properties of the materials used. Frequently it is extremely difficult to analyze and separate in an unambiguous manner the various causes giving rise to a certain property.
(ii) Clear-cut experimental information on interfaces is scarce, since some of the few existing methods are only qualitative or semiquantitative of nature, like reflection high-energy electron diffraction (RHEED), and for others, like some electrical methods, the evaluation procedures are dubious.

In addition, interfaces are only accessible to the presently existing surface characterization methods if the structure is grown by molecular-beam epitaxy (MBE). During the brief time interval before the second material is grown on top of the first one the future interface can be inspected in the form of an open surface e.g., by RHEED.

Detailed mathematical analysis of QW luminescence lineshapes is demonstrated to provide quantitative information on the statistics of structural/chemical disorder at semiconductor interfaces with sub-Angstrom depth resolution. A model, recently proposed by some of us /1/, established the qualitative relationship between quantum well luminescence lineshapes and interface disorder. An extension of this model is presented here, allowing for a quantification of the information gained from lineshape analysis. Our results /2/ are found to be in general agreement with recent predictions /3-5/ based upon observations of the dynamics of RHEED oscillations demonstrating the reduction of surface disorder upon interruption of MBE growth for a few seconds. In contrast to RHEED observations, however, the luminescence results can now be interpreted quantitatively. Thus, structural parameters of interfaces are correlated with optical properties of quantum wells. In addition we present a cathodoluminescence imaging system enabling us to visualize for the first time directly monoatomic steps at semiconductor interfaces and the columnar structure of single quantum wells (SQW's) grown by growth interrupted MBE. Shrinkage of the lateral extension of these islands for increasing growth temperature at constant growth rate is observed.

This paper is organized as follows: Section 2 contains a brief description of the samples used for our experiments and of the MBE growth procedure. The cathodoluminescence imaging (CLI) and the photoluminescence set-ups are described in Sect. 3. The theoretical background necessary for an understanding of our derivation of interface roughness data from luminescence lineshapes is given in Sect. 4. Photoluminescence results of quantum wells grown at various growth rates, temperatures and interruption times are presented and analysed in Sect. 5. Conditions of carrier thermalization and nonthermalization are discovered. Direct images of growth islands at interfaces are presented in Sect. 6. Sect. 7 contains a summary.

2. Samples

In order to eliminate effects of interlayer well size fluctuations produced e.g. by fluctuations of the substrate or effusion cell temperature during MBE growth, predominantly single quantum wells are investigated. All samples are grown on GaAs substrates oriented within 0.1° to (100). The As_4/Ga flux ratio is about 4. Results from three series of samples grown at largely varying temperatures and rates will be presented here. For growth details see Ref. 2.

3. Experimental Methods

a) Photoluminescence

The main features are: excitation wavelength 514 nm or 632 nm, typical excitation density 90 mW/cm^2, temperature range 1.5 K - 300 K spectral resolution 0.06 - 0.3 nm, being largest at low temperature. Details of the method were described recently /1/.

b) Cathodoluminescence Imaging - CLI

Features essential for the present system /2/ are the following: The miniature continous-flow cryostate is fitted to a standard specimen stage and allows a variation of the temperature in the range 5 K $<$ T $<$ 300 K. The kinetic energy of the electrons used for excitation in the CLI experiments is reduced to 3 keV, as compared to 30 keV in our standard CL work /6,7/. At 3 keV the center of excitation is 0.04 µm below the surface and the diameter of the generation volume of secondary electrons and final-state electron-hole pairs, the Bethe-range, is thus largely reduced to 0.2 µm. This value, and not the much smaller spot size of the electron beam, limits the lateral resolution if no deconvolution procedure is applied. The emitted light is dispersed by a 30 cm vacuum monochromator having a linear dispersion of 6 nm/mm. Spectral resolution in the temperature range 5 K -100 K is $\approx$ 0.6 nm. The electron beam is digitally scanned over an rectangular area of typical dimensions 32 µm x 18 µm. This area is equally divided into 128 x 100 pixels. (maximum value is 512 x 400 pixels). On each pixel the luminescence intensity is measured at a fixed wavelength with a dynamic range of 10^3. The emission wavelength from quantum well areas of different widths is different. Thus information on structural properties of a QW can be translated into information on optical properties. A CLI discloses directly the width distribution of the QW under investigation (its columnar structure) and thus the variation of the z-coordinate of the interface positions in the x-y-plane.

Figure 1 shows schematically the operating principle of CLI for an interface showing an island-like structure with single monolayer steps. Details of the point focus CL system have been described recently[7].

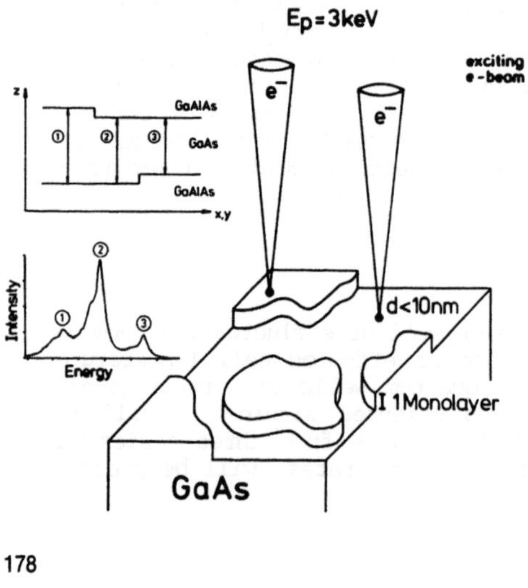

Fig. 1 Principle of cathodoluminescence imaging: the lower half of an interface is shown. Triplet splitting of recombination results from single monolayer steps at each interface if the island area is larger than the projection of the recombination volume.

4. Correlation of Interface Roughness with QW-Luminescence Lineshapes: Theoretical Background

A complete model for the lineshape of the luminescence from a quantum well is established, extending considerably our previous approach /1/ of a simple Gaussian or Lorentzian shape. A detailed mathematical procedure is given, which demonstrates how to calculate interface roughness from the observed lineshape broadening.

The interface of a heterostructure consists of two adjacent lattice planes of different chemical composition. Growth of the second material usually commences on a plane which is not monoatomically flat but shows a structure of monolayer height steps due to incomplete coverage. The position of the interface in growth direction z thus depends on the x-y-coordinate. This effect is sometimes called interface roughness. Interdiffusion of atoms at the usually rather high growth temperatures additionally contributes to roughness.

A quantum well consists of a narrow, chemically uniform layer limited by two interfaces to the barrier material. The well is truly chemically uniform if it is made up out of an element or a binary compound. In all other cases there exists a mean composition which is defined as an average over an information volume /8/. The information volume might contain a statistical or a nonstatistical distribution (e.g. due to ordering effects /9/) of the compound. In what follows we will consider only wells of binary compounds like GaAs. The roughness of each of the two interfaces contributes to a x, y-dependence of L_z, the QW width. In general, the contributions from the top and the bottom interface are different, since the growth surface of the bottom interface consists of the barrier material, whereas the growth surface of the top interface consists of the well material. We have demonstrated recently /10/ that the surface kinetics of such rather similar materials as GaAs and AlGaAs are still far from identical.

The lineshape function I(E) of the luminescence of an electron-hole pair from a quantum well as a function of photon energy E is given by

$$I(E) \sim f(E) \int_0^{\infty} D(E,E')\ B(E',E_n)\ dE' \qquad (1)$$

if momentum conservation is taken care of by a separate mechanism. D(E) is the two-dimensional density of states, B (E) is a broadening function, which, in the case of discrete quantization energies E_n, is a Lorentzian broadening due to final state interaction /11/. It is largest for small energies close to E_n and can be neglected if the statistical contribution to the lineshape broadening is much larger.

f (E) is the thermal distribution function. In general, Fermi functions with different Fermi energies for electrons and holes have to be inserted. Under the present circumstances of excitonic recombination after low to medium excitation a Maxwell-Boltzmann function presents an excellent approximation. The low energy onset of recombination occurs at

$$E_n = E_g + E_e (L_z) + E_h (L_z) - E_x^{2D} (L_z) \qquad (2)$$

with E_g the 3-dimensional gap, E_e the energy of the lowest two-dimensional electron subband, E_h the energy of the lowest two-dimensional hole subband and E_x the binding energy of the two-dimensional exciton, if the recombination is excitonic, as it is here /12/. Otherwise $E_x = 0$. E_e, E_h are the well known solutions of the two-dimensional "particle in the box" Schrödinger equation /13/ which are L_z -dependent. E_x also depends on L_z, but its absolute value is much smaller. If L_z is statistically distributed over a certain "information volume", then E_e, E_h and thus E_n are also statistically distributed. E_n in Eq.(1) and Eq.(2) has then to be replaced by its probability distribution $P(E_n, \langle E_n \rangle)$ and its expectation value $\langle E_n \rangle$ respectively. $B(E', E_n)$ is no longer given by a Lorentzian but its convolution with the probability distribution $P(E_n, \langle E_n \rangle)$. The information volume V_I is equal to the exciton volume if the recombination is excitonic. Otherwise it is equal to the recombination volume which can be calculated from the mean e-h distance.

In a luminescence experiment recombination of many electron-hole pairs from different, although closely neighbouring areas are detected. The experimentally observed lineshape I_{exp} thus presents a statistical average over the observation volume of all "single event" lineshapes Eq. (1):

$$I_{exp} = \int_V I \left[E\,(V) \right] \, d\,V. \qquad (3)$$

This additional broadening effect has to be taken into account for larger lateral inhomogenities. Then the variance $\bar{\sigma}$ of the continuous distribution of the mean values of the width $\bar{L}_z$ cannot be neglected as compared to the variance σ of the width L_z within one single recombination volume. Temperature dependent luminescence taken at varying positions of the samples shows that this additional broadening is observable, but small, and can be neglected for the present set of samples.

With increasing growth interruption time the lateral size of the islands at the interfaces becomes larger than the recombination volume. Then discrete mean width values $L_{z\nu}$ exist, and Eq. (3) is replaced by

$$I_{exp.} = \sum_{\nu=1}^{\infty} I_\nu (E) \qquad (4)$$

with n = 3, if exactly one monolayer step occurs at each interface and the islands at both interfaces are larger than the projection of V_I onto them. Fig. 1 depicts in a schematic way this correlation of structural and optical properties.

A practical example of the impact of quantum well width variation, and thus of interface roughness, on the luminescence broadening is given in Fig. 2. The L_z-dependence of the spectral broadening is given for a fixed halfwidth equal to a, the lattice constant, of the L_z -distribution func-

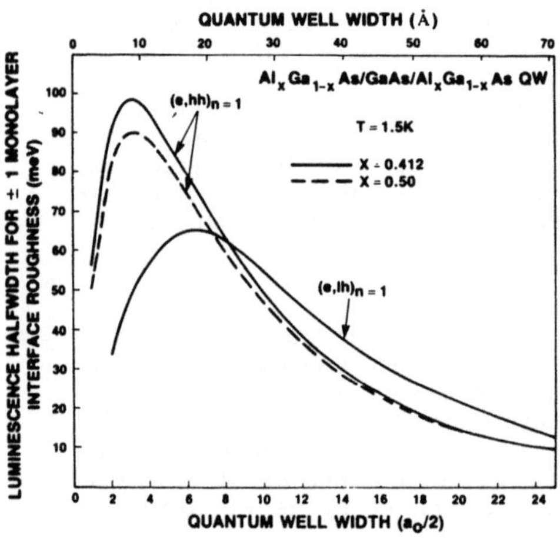

Fig. 2 L_z-dependence of energy broadening of the luminescence for a fixed FWHM of one lattice constant of the L_z -distribution function. Broadening values are given for the $(e, hh)_{n=1}$ and the $(e, lh)_{n=1}$ transitions for $[Al]$ = 41.2 %, and for the $(e,hh)_{n=1}$ transition for $[Al]$ = 50 %.

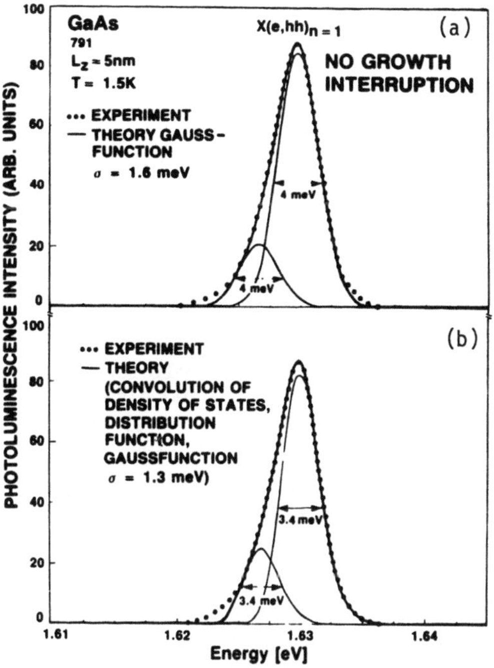

Fig. 3 T = 1.5 K luminescence spectrum of a 5 nm QW and theoretical fit according to Eq. (1) neglecting Lorentzian broadening (b). A perfect agreement is achieved. Part (a) shows a fit with a simple Gaussian for comparison. The origin of the doublet structure is discussed in Ref.[14].

tion. Fig. 2 is calculated for a finite square well potential, (Al = 41.2 % and 50 %, $\Delta E_c = 0.62\ E_g$), assuming continuity of the envelope function $\psi(z)$ and of $\psi'(z)/m^*$ at the interface and energy dependent masses[13]. The effective masses which we use for this calculation are

for GaAs:
$m_e^* = 0.0665$, $m_{lh}^* = 0.091$, $m_{hh}^* = 0.377$
for $Al_{0.41}Ga_{0.59}As$:
$m_e^* = 0.0895$, $m_{lh}^* = 0.137$, $m_{hh}^* = 0.424$
for $Al_{0.5}Ga_{0.5}As$:
$m_e^* = 0.0895$, $m_{lh}^* = 0.148$, $m_{hh}^* = 0.435$

Fig. 3 b presents a fit of Eq. (1) to an experimental luminescence line obtained from a 4.2 nm QW, grown according to Sec. 2 without interruption of the growth. The fit is perfect, although only one fitting parameter (the variance of the $P(E_n)$ distribution) is used. The temperature is directly taken from a semilogarithmic plot of the luminescence line. The variance of the theoretical spectral broadening function of Fig. 4 b (σ= 1.3 meV) together with Fig. 3 yields a variance of the quantum well width distribution function σ= 0.33 Å, which is approximately 20 % less than the value we derived earlier /1/ from a fit with a simple Gaussian, shown in Fig. 3 a. This value has now to be distributed among the inequivalent interfaces. For equivalent interfaces each of them would show a roughness of σ = 0.23 Å.

Details are given elsewhere /10/. The origin of the doublet character of the luminescence line in Fig. 4 is not related to structural effects and is discussed elsewhere /14/.

5. Photoluminescence Spectra of Growth Interrupted Quantum Wells: Evidence for Suppression of Carrier Thermalization

Fig. 4 a shows T = 100 K spectra of $\approx$ 5 nm wide QW's grown at 620° C at a rate of 2.8 Å/s with interruption of the growth Δt at both interfaces of 1 s, 10 s, 100 s, respectively (see Sect. 2). Obviously the statistical contribution to the width shrinks appreciably with increasing growth interruption time. A theoretical analysis according to Eq. (1) yields variances of the energy broadening function of 2 meV, 1.4 meV and 0.9 meV, respectively, yielding variances of the L_z distribution function of 0.62 Å, 0.44 Å and 0.28 Å, respectively, using Fig. 2. The results are shown in Fig. 4 b. The roughness of one interface would be 0.7 of these values for equivalent interfaces, which is usually not the case.

The Δt = 100 s spectrum shows a doublet. A similar spectrum taken at lower temperature shows a single line (we ignore here for simplicity the "intrinsic" doublet splitting of the heavy hole exciton reported elsewhere /14/. At higher temperatures no third line on the high energy side emerges. Apparently the islands at the GaAs surface (top interface) are larger than the exciton diameter /1,10/ whereas the islands at the AlGaAs surface (bottom interface) are smaller. Figure 5 shows this situation schematically. The disappearance with decreasing temperature of the second line is indicative of thermalization in a two-level-system and is not at all surprising for given lifetimes of the order of 0.5 ns /12/.

Fig. 6, in contrast, presents a very surprising result. The 2 K photoluminescence spectra of two L_z 5 - 6 nm QW's grown with almost the same interruption time of 100 s (# 792) and 120 s (# NAK1) at the interfaces are compared (see Sect. 2). We recognize one component (with doublet finestructure /14/) and a triplet (with doublet finestructure /14/). Apparently the excitons do not thermalize any more in the quantum wells

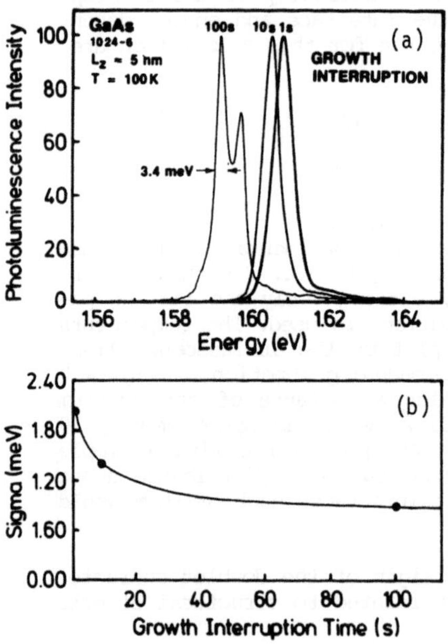

Fig. 4 (a) Comparison of T = 100 K photoluminescence spectra of 5 nm wide QW's grown with growth interruptions at both interfaces of 1 s, 10 s, 100 s, respectively. Part (b) shows the variance of the energy broadening function versus growth interruption time.

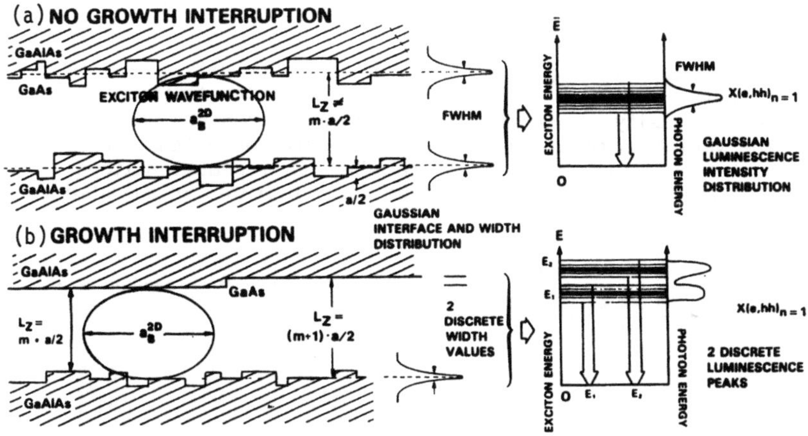

Fig. 5 Model for the interface disorder of a AlGaAs/GaAs QW and for the energy state(s) of the X(e, hh)-exciton in such a well grown (a) without and (b) with interruption of the growth. The GaAs surface smoothes more rapidly than the AlGaAs one.

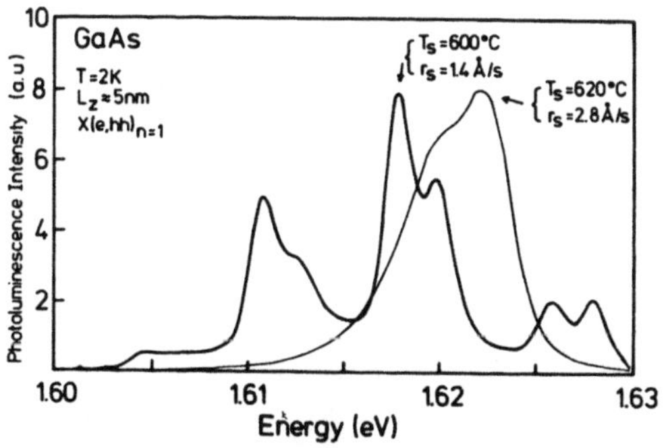

Fig. 6 Comparison of T = 2 K photoluminescence spectra of two samples grown with similar growth interruption times but at different growth rates and temperatures: a) T_S = 600° C, r_S = 1.4 A/s, b) T_S = 620° C, r_S = 2.8 A/s

of sample NAK1. Similar results are obtained for the other three SQW's of that sample (see Fig. 1 of Ref. 14). Apparently the lateral extension of islands at both interfaces of that sample is equal to or larger than the diffusion length L_D:

$$L_D = (\mu kT \tau /e)^{1/2}. \qquad (5)$$

Here T is the temperature, τ is the lifetime and μ is the mobility. For T = 2 K, τ = 0.5 ns /12/ and $\mu \approx 1 - 2 \times 10^5$ cm²/Vs, as limiting value of the ambipolar mobility, which is taken to be equal to some of the best known hole mobilities /15/, we get $L_D \lesssim$ 0.9 -1.3 µm. The extension of these very smooth islands must be thus larger than $\approx$ 1 µm. Spot CL spectra showing single peaks at different photon energies for different spots corroborate this conclusion. Consequently, the direct observation of such interface islands by CLI should be possible. The next section proves that we indeed are able to obtain such images. In that section we will also discuss the influence of the various growth parameters on island size.

6. Images of Growth Islands at Heterointerfaces

Fig. 7 shows CLI images of sample NAK 1 grown at 600° C at r_s = 0.14 nm/s. The spectrometer is set at a fixed wavelength given by the wavelength of maximum intensity of one of the PL triplet components shown in Fig. 6. Then the variation of CL-luminescence intensity as a function of position of excitation is recorded. This procedure is repeated alltogether three times. The black/white scale of the three pictures is adjusted

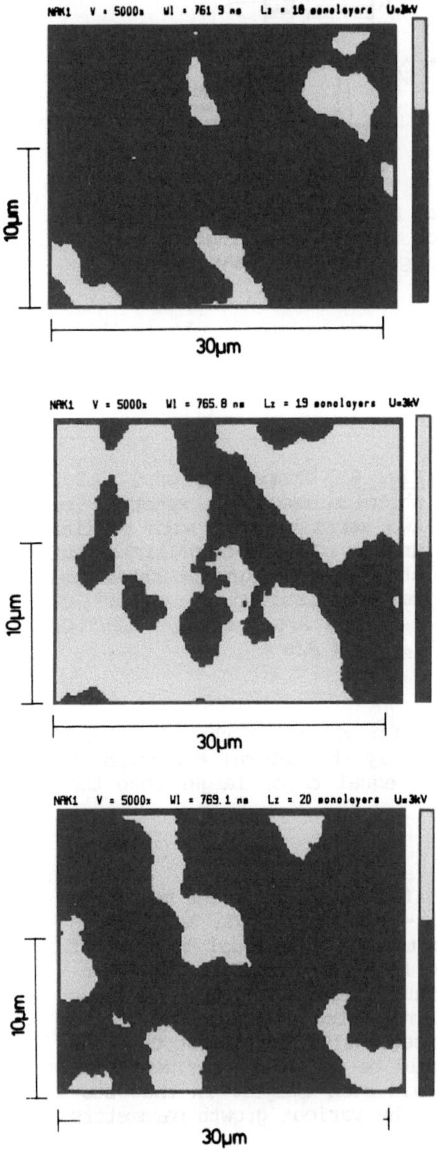

Fig. 7 CLI images of sample NAK 1 grown at 600° C (Δt = 120 s) taken at the three different wavelengths of the peaks of the triplet (for a spectrum see Fig. 6).

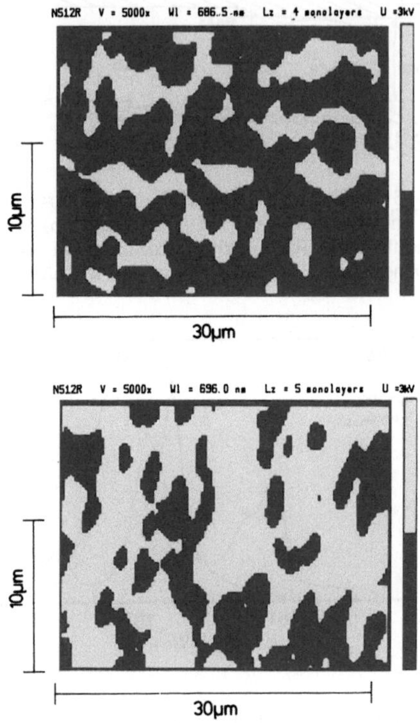

Fig. 8 CLI images of sample 512 grown at 660° C (Δt = 120 s) taken in the same way as for Fig. 7.

such that the sum of all white areas covers or sligthly overcovers the total surface. Then the bright areas in Fig. 7 show the extension of quantum well columns of thickness close to 18, 19 and 20 monolayers, respectively. At the edge of each bright area a step of 2.8 A height occurs at one of the interfaces. From a larger number of such experiments we find that the mean size of the islands is 6 - 8 μm. Diffusion effects influence the size of the islands reported here only in a minor way, as shown by time resolved experiments. Details of these experiments are given elsewhere.

Fig. 8 shows CLI images of sample 512 grown at 660° C at r_s = 0.17 nm/s. Obviously the islands are much smaller. Their mean size is determined to 2 μm. An increase of the growth rate and an increase of the temperature lead to a reduction of the interface island size and to an increase of interface roughness indicating the transition from planar to three-dimensional growth. The influence of growth rate is particularly important at low substrate temperatures of the order of 600° C - 620° C as illustrated by Fig. 7. A detailed discussion of the separate influence of both growth parameters in terms of surface kinetics will be given elsewhere.

Time resolved cathodoluminescence images directly visualize the lateral diffusion of excitons. Their lateral diffusion velocity was determined in a separate study /16/ for the samples of Fig. 7 as 10^5 cm/s.

7. Conclusions

Quantitative information on the reduction of roughness of the quantum well interfaces with increasing interruption time of the growth process is obtained from detailed analysis of luminescence lineshapes. A full theory of the shape of luminescence from quantum wells and its dependence on structural properties of the interfaces is given. For the specific example of GaAs/AlGaAs heterostructures grown at 620° C at a rate of 2.8 A/s a reduction of interface roughness from 0.31 A to 0.14 A is observed for an increase of interruption time from 1 s to 100 s. Earlier reports by us on the inequivalence of the top/bottom interfaces are corroborated.

Direct images of growth islands differing by 2.8 A (one monolayer) height at GaAs/AlGaAs heterointerfaces and of the columnar structure of GaAs quantum wells grown by MBE with growth interruption at the interfaces are reported. The method used to obtain these images is scanning cathodoluminescence of quantum wells. Atomic scale structural information is expressed in terms of spectral properties. An outline of this method is given.

The dependence of the lateral extension of such islands on the parameters of MBE growth is investigated. For fixed GaAs growth rate r_s = 0.5 monolayer/s the mean island size decreases from 6 μm - 8 μm to 2 μm upon an increase of growth temperature from T_g = 600° C to 660° C, indicating the transition from 2-dimensional to 3-dimensional growth.

The lateral extension of the islands can thus be much larger than the ambipolar diffusion length L_D $<$ 1 μm and inter-island charge carrier transfer due to thermalization can be almost completely suppressed. Experimental evidence for this suppression of thermalization is presented.

Acknowledgements

The work at TUB is supported by DFG in the framework of SFB 6. We are very much indebted to R. Bauer and D. Oertel for supplying the photoluminescence data.

References

/1/ D. Bimberg, D. Mars, J.N. Miller, R. Bauer, D. Oertel, J.Vac.Sci.Technol. B 4, 1014 (1986)

/2/ D. Bimberg, J. Christen, T. Fukunaga, H. Nakashima, D. Mars, J.N. Miller, J.Vac.Sci.Technol. B 5, 1191 (1987)

/3/ J.H. Neave, B.A. Joyce, P.J. Dobson, N. Norton, Appl.Phys. A 31, 1 (1983)

/4/ T. Sakamato, H. Funabashi, K. Ohta, T. Nakagawa, N.J. Kowai, T. Kojima, Y. Bando, Superl. and Microstr., 1, 347 (1985)

/5/ B.F. Lewis, F.J. Grunthamer, A. Madhukar, T.C. Lee, R. Fernandez, J.Vac.Sci.Technol. B 3, 1317 (1985)

/6/ A. Steckenborn, H. Münzel, D. Bimberg, J.Luminescence 24/25, 351 (1981) and Inst.Phys.Conf., Ser.60, 185 (1981)

/7/ D. Bimberg, H. Münzel, A. Steckenborn, J. Christen, Phys.Rev. B 31, 7788 (1985)

/8/ J. Singh, K.K. Bajaj, S. Chaudhuri, Appl.Phys.Lett. 44, 805 (1984) and J. Singh, K.K. Bajaj, J.Appl.Phys. 57, 5433 (1985)

/9/ R. Hull, K.W. Carey, J.E. Fouquet, G.A. Reid, S.J. Rosner, D. Bimberg, D. Oertel, Proc.Int.Symposium on GaAs and Related Compounds, Las Vegas 1986, in print

/10/ D. Bimberg, D. Mars, J.N. Miller, R. Bauer, D. Oertel, J. Christen, Superl.and Synth.Microstructures 3, 79 (1987) and Proc. MSS III, J.de Physique, in print

/11/ P.T. Landsberg, Proc.Phys.Soc., A 62, 806 (1949) and Phys.Stat.Sol. 15, 623 (1966)

/12/ J. Christen, D. Bimberg, A. Steckenborn, G. Weimann, Appl.Phys.Lett. 44, 84 (1984) and D. Bimberg, J. Chrsiten, A. Werner, M. Kunst, G. Weimann, W. Schlapp, Appl.Phys.Lett. 49, 76 (1986)

/13/ G. Bastard, Phys.Rev. B 24, 5693 (1981)

/14/ R. Bauer, D. Bimberg, J. Christen, D. Oertel, D. Mars, J.N. Miller,T. Fukinaga, H. Nakashima, Proc. 18th Conf. Phys. Semic., Stockholm 1986, in print

/15/ G. Weimann, private communication

/16/ D. Bimberg, J. Christen, T. Fukunaga, H. Nakashima, D.E. Mars and J.N. Miller, Superlatt. and Microstructures 4 (1988), in print

Quasi-One-Dimensional Electron Channels in AlGaAs/GaAs Heterojunctions

W. Hansen and J.P. Kotthaus

Institut für Angewandte Physik, Universität Hamburg,
Jungiusstr. 11, D-2000 Hamburg 36, Fed. Rep. of Germany

Abstract. With submicron lithography it is possible to prepare narrow inversion channels at semiconductor interfaces, in which electrons can move freely only in one direction. For experiments on such quasi-one-dimensional (1D) systems we employ AlGaAs/GaAs heterojunctions with a laterally microstructured gate electrode. A negative gate bias is used to deplete parallel channels in the originally homogeneous two-dimensional (2D) inversion layer and thus cause a transition from 2D to 1D electronic behavior. Quantum oscillatons of the magnetoresistance as well as far infrared spectroscopy of the electronic excitations are used to characterize the transition from 2D to 1D behavior. From such experiments and with simple models we can derive 1D subband spacings, 1D electron densities and effective channel widths. Also, we can estimate the importance of many-body effects on the directly observed intersubband resonance transition between 1D subbands.

1. Introduction

The boundaries of a spatially constricted electron system influence the electron energy dispersion increasingly when the width of the system becomes smaller. The electronic energies become quantized for motion in the direction of confinement, if the width approaches the Fermi wavelength of the electron system. This is well known from investigations of two-dimensional (2D) electron systems confined at a semiconductor interface by an electric field normal to the interface [1]. There the energy dispersion is split into so called 2D subbands: continuous 2D energy bands from free motion parallel to the interface add to the quantized energies that arise from confinement at the interface.

Preparation of quasi-one-dimensional (1D) or even zero-dimensional electron systems in semiconductors usually starts from a 2D electron system. With lithographically defined photoresist patterns and various subsequent preparation techniques additional potentials are imposed on the electron system that vary in a lateral dimension. If these potentials confine the electron system to widths below the electronic mean free path and approaching the Fermi wavelength, the electron energies are modified into 1D subbands or even completely quantized in the zero-dimensional case. Different experiments have been reported to verify lateral quantization: Oscillations in gate voltage dependence of the conductance in multichannel Si-MOS devices [2] or of the capacitance in gated heterojunctions [3], deviations from 2D behavior of magnetoconductance oscillations in single channel GaAs heterojunctions [4,5], and fine structure in the cathodoluminescence [6] of quantum well wires and in the tunneling current through zero-dimensional quantum dots [7] have been attributed to lateral quantization. Here we summarize investigations of the static (DC) as well as

Springer Series in Solid-State Sciences Vol. 83: **Physics and Technology of Submicron Structures**
Editors: H. Heinrich • G. Bauer • F. Kuchar

dynamic conductivity at far infrared (FIR) frequencies in laterally microstructured AlGaAs/GaAs heterojunctions [8-11]. Whereas the static conductivity perpendicular to the direction of free motion vanishes in a 1D system, the dynamic conductivity exhibits a resonance at finite frequency. In a one-electron model this resonance energy would be equal to the energy spacing between the 1D subbands, since the FIR absorption in this polarization arises from resonant transitions between these subbands. In a many-electron system, however, the resonance energy is affected by many-body interactions [12-14]. Comparison of the DC-transport results and the spectroscopic measurements yields valuable information about many-body contributions to the intersubband resonance frequency.

In the systems studied here application of a gate voltage between a laterally microstructured gate and the electron system results in a laterally periodic potential with strength controlled by the gate voltage. Just by changing the gate voltage a transition from a 2D electron system in a periodic potential "superlattice" to an array of 1D electron waveguides is observed. The properties of such an electron system in both gate voltage regimes are not only interesting for basic research but also may reveal novel device applications.

2. Sample Preparation

In preparing our samples we start from conventional AlGaAs/GaAs heterojunctions grown by molecular beam epitaxy [15]. A typical sample consists of a 1μm thick GaAs buffer layer on top of a semi-insulating GaAs substrate, a 100nm thick undoped AlGaAs spacer, a 400nm thick Si doped perpendicular to the sample surface the density of states oscillations at the

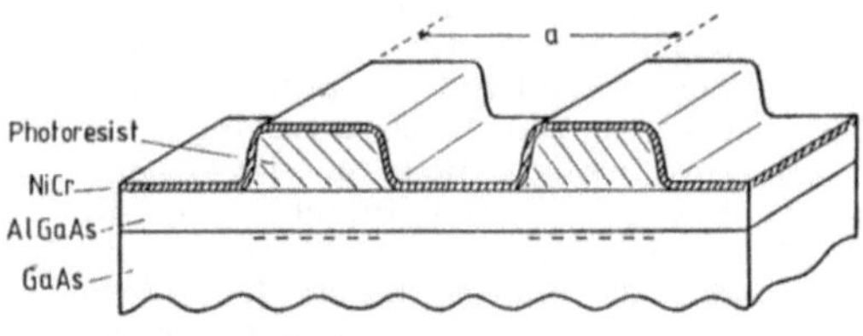

(a)

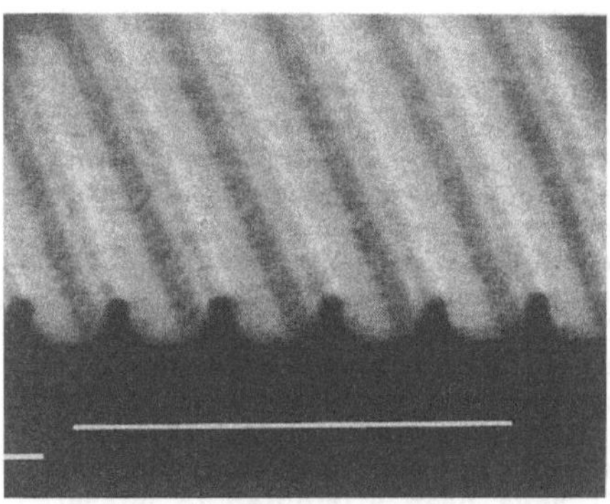

(b)

Fig. 1:

AlGaAs/GaAs heterojunction with laterally periodic modulated gate. (a) Sketch of the cross section, (b) SEM photograph of the photoresist profile with 1μm marker. The sample is slightly tilted, so that a large area of the sample surface is surveyed at a glancing angle.

AlGaAs layer, and a 5nm thick undoped GaAs cap layer. The substrate is wedged at an angle of about 3° to avoid Fabry Pérot resonances in FIR experiments. Wet etching is used to define an active mesa of desired geometry and ohmic contacts to the inversion channel are alloyed with Indium. We use holographic lithography to pattern the laterally micro-structured gate electrode as shown in Fig. 1 homogeneously onto a large active area (~6mm²) that is necessary for FIR transmission experiments [16]. A typically 120nm thick photoresist layer on the sample is exposed by two laser beams of wavelength λ=458nm, that interfere at an angle δ and thus generate a light grating of period $a=\lambda/[(2\sin(\delta)]$. Exposure of such grating with a scanning e-beam would typically take 1000x longer. After development of the periodic array of photoresist stripes (grating constant a≤500nm) a NiCr gate (~7nm thick) is evaporated. The above desribed sample preparation has presumably less effect on the mobility in the 2D channel than e-beam writing or etching procedures.

3. Sample Characterisation

The generation of an array of isolated electron channels is deduced from measurements of capacitance-gate voltage traces as demonstrated in Fig. 2. The differential capacitance almost remains constant in the gate voltage regime $-0.5V<V_g<0V$ and then drops rapidly. This is easily understood by considering the effective area that contributes to the sample capacitance. If the negative gate bias is sufficiently low to completely deplete the inversion channels between the photoresist stripes, the corresponding area no longer contributes to the capacitance. Consequently at the gate voltage $V_g=V_D$, where the capacitance drops abruptly, screening of the applied potential by the 2D electron system breaks down and isolated electron channels remain below the photoresist stripes. The capacitance below this threshold voltage V_D exhibits sample dependent structure. This structure cannot be easily interpreted as depopulation of 1D subbands as in Ref. 3 and remains to be explained.

Lateral quantization in the narrow inversion channels is deduced from transport experiments along the channels [4]. In a magnetic field B applied Fermi energy are visible as quantum oscillations of the conductivity. Roughly, we expect that with increasing magnetic field the 1D density of

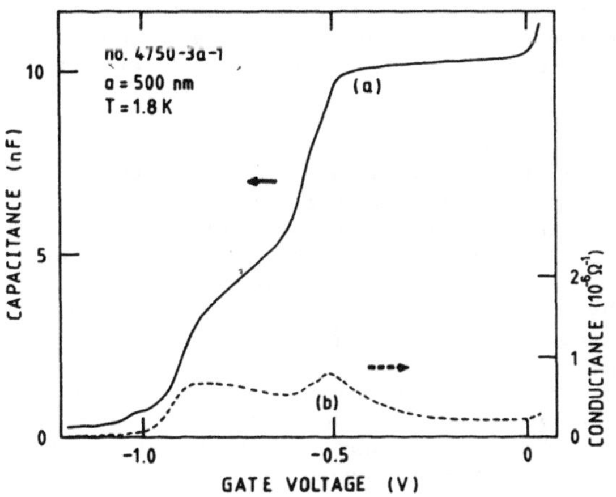

Fig. 2:

Capacitance-voltage trace (a) of a heterojunction with a laterally modulated gate. The capacitance signal is measured out of phase to a gate voltage modulation. In this sample the gate voltage at which isolated electron channels are formed is $V_g=V_D=-0.5V$. (b) conductance signal measured in phase to the gate voltage modulation [9]

states changes into the density of 2D Landau levels, if the Landau orbit diameter becomes less than the channel width. This is most easily modeled with a harmonic confinement potential $V_{(x)}=(m^*/2)\Omega_x^2 x^2$, where x denotes the direction of lateral confinement. Since the confinement perpendicular to the interface (z-direction) is much stronger, we assume z- and x-motion to be separable. Also, we consider here only the lowest 2D subband, since higher 2D subbands are not occupied in our samples. Neglecting spin splitting in a magnetic field the parabolic confinement model yields the energy dispersion

$$E_n(k_y)= E_o+ \hbar\omega \; (n + \frac{1}{2}) \; + \; (\frac{\Omega_x}{\omega})^2 \; \frac{\hbar^2 k_y^2}{2m^*} \; , \qquad (1)$$

where E_o is the bottom of the lowest 2D subband, $\hbar\omega=\hbar(\omega_c^2+\Omega^2)^{1/2}$ the 1D subband spacing, and $\omega_c=eB/m^*$. The density of states may be written as

$$D_{(E)} \; = \frac{2}{\pi\hbar} \sqrt{\frac{m^*}{2}} \; \frac{\omega}{\Omega_x} \; \sum_{E_n<E} [E-E_n]^{-1/2} \qquad (2)$$

with $E_n=\hbar\omega(n+1/2)$ describing the bottoms of the 1D subbands. Quantum oscillations arise, whenever the bottom of a 1D subband crosses the Fermi energy ($E_F(B)=E_n$). For comparison with experimental results we calculate the corresponding magnetic field values using the Fermi energy $E_F(B=0)$ and the subband spacing $\hbar\Omega$ at B=0 as fit parameters.

Fig.3 displays experimental traces of quantum oscillations at different gate voltages as observed in the gate voltage derivative of the two-terminal source-drain resistance [11]. In Fig.4 the running index ($n+\Phi$) of the oscillation maxima is plotted in a fan diagram vs their reciprocal magnetic field value. Here one is free to choose a suitable phase Φ, since one does not expect the oscillation maxima in dR/dV_g to coincide with integer quantum index n. We have adjusted the phase Φ so that the data extrapolate to n=0 at 1/B=0. At high magnetic fields a nearly linear behavior in 1/B is observed as is expected for Landau levels in a 2D system [1] and the high field slope may be used to determine the Fermi energy E_F. At low magnetic fields we observe deviations from linear behavior that directly reflect the confinement of the electron system. In Fig.4 the 1/B values at quantum index n calculated from the harmonic potential model with parameters E_F and $\hbar\Omega$ that best fit the magnetic field dependence of the measured quantum oscillations are also entered. Between $V_g=V_D=-0.5V$ and $V_g=-0.66V$ we find the subband spacing to rise from $\hbar\Omega=1.6meV$ to $\hbar\Omega=2.0meV$ and the Fermi energy to decrease from $E_F=14.5meV$ to $E_F=11.7meV$, respectively. If we vary the somewhat arbitrary phase Φ of the experimental data plotted in Fig.4, the values we obtain for Fermi energy and subband spacing vary by less than 20%. Surprisingly, we already observe deviations from linear behavior at gate voltages ($-0.5V<V_g<-0.2V$) where from capacitance measurements we expect the electron system to be still 2D. This indicates that an external potential already affects Landau levels in a 2D electron system, where the external potential is strongly screened by the modulated density. In this regime above $V_g=-0.5V$ we also use the harmonic potential model to derive a quantization energy. In lack of self-consistent calculations for the effective confining potential in our structures, it is not clear whether the simple harmonic model realistically describes the real effective potential. Self-consistent calculations for similar structures [17-20] reveal that the effective potential, which confines the electrons laterally, develops with increasing number of occupied 1D subbands from parabolic to square well shape. For the sake of simplicity we adhere to the parabolic potential model, but state qualitatively, that fits to a square well potential model would yield higher subband spacings at the Fermi level.

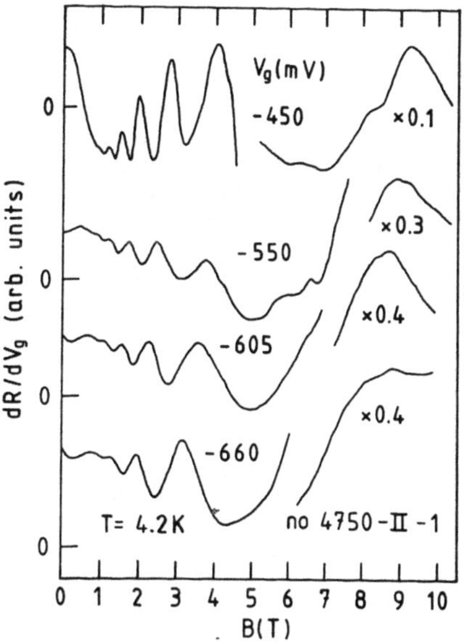

Fig. 3:

Gate voltage derivative of the source-drain resistance dR/dV_g vs magnetic field B of an AlGaAs/GaAs heterojunction with laterally periodic gate of period a=400nm [11]. At V_g=0V the the inversion channel has an electron density of $N_{so}=6\cdot10^{11}cm^{-2}$

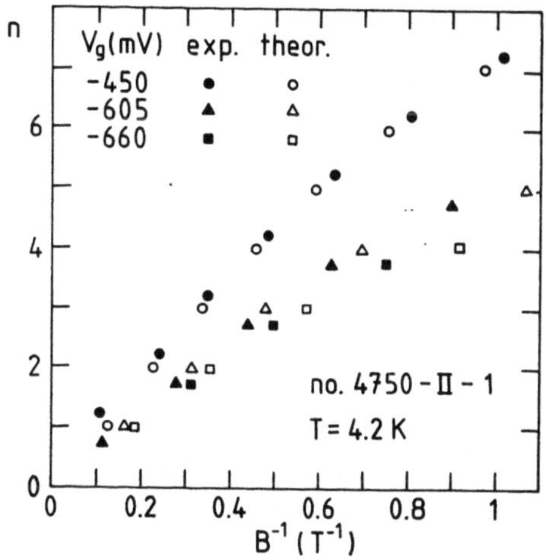

Fig. 4:

Quantum index n of the maxima in dR/dV_g from Fig. 3 vs reciprocal magnetic field. The open symbols are values derived from a fit to a parabolic confinement model. Experimental values (filled symbols) are phase shifted to the fit values

4. Far Infrared Excitations

Far infrared transmission through the microstructured AlGaAs/GaAs heterojunctions is investigated at low temperatures (T=2K) with a Fourier-transform spectrometer. The radiation is normally incident onto the sample and a linear polarizer defines the radiation field to be either parallel or perpendicular to the grating.

We evaluate the frequency dependent relative change of transmission

$$\frac{\Delta T}{T} = [T(V_{g1},B_1) - T(V_{g2},B_2)] / T(V_{g2},B_2) \qquad (3)$$

through the sample. The reference $T_2(V_{q2},B_2)$ is measured at gate voltage V_{g2} and magnetic field B_2 at which the conductivity of the electron system is essentially zero in the spectral regime shown. Thus the spectra $\Delta T/T$ directly reflect the dynamic conductivity of the electron system at V_{g1} and B_1. The FIR excitations of laterally microstructured electron systems are strong resonances with characteristic properties in two different gate voltage regimes above and below V_D. In Fig. 5 we display the frequency dependence of $-\Delta T/T$ at zero magnetic field and various gate voltages for perpendicularly polarized radiation. From capacitance measurements as described above, we infer that at gate voltages above V_D=-0.5V the electron system is 2D with modulated electron density. The resonances in this gate voltage regime are shown in the lower part of the figure. Here the micro-structured metal gate with periodically modulated distance to the electron system acts as a grating coupler for 2D-like plasmon excitations [16,21]. At V_g=0V the plasmon frequency at wave vector $q=2\pi/a$ is determined by the homogeneous electron density $N_{so}=6\cdot10^{11}cm^{-2}$ and an effective dielectric constant $\bar{\varepsilon}$ that takes into account screening by the metal gate and media surrounding the electron system: $\omega_p^2=e^2N_{so}q/(2\bar{\varepsilon}\varepsilon_0m^*)$. With decreasing V_g ($-0.5V<V_g<0V$) the resonance position in Fig.5 decreases. We now observe plasmon excitations in an electron system with spatially modulated charge density:

$$N_s(x) = \sum_m \bar{N}_{sm} \cos(m\, 2\pi x/a) . \qquad (4)$$

Plasmon excitations in charge density modulated 2D system have been investigated experimentally [16,21] and theoretically [22,23]. There it is found that the plasmon resonance at $q=2\pi/a$ does not split, if the second Fourier component $\bar{N}_{s2}$ of the charge density is small. The plasmon frequency then is determined by the average areal density $\bar{N}_{so}$:

$$\omega_p^2 = e^2\bar{N}_{so}q/(2\bar{\varepsilon}\varepsilon_0m^*) . \qquad (5)$$

Here the effective dielectric constant $\bar{\varepsilon}$ is derived from the measured plasmon frequency and electron density N_{so} at V_g=0V. Thus we can extract $\bar{N}_{so}$ at V_g<0V from measured plasmon frequencies and find an almost linear decrease of $\bar{N}_{so}$ with decreasing gate voltage. In the gate voltage regime $V_g<V_D$ we may also define an average areal density $\bar{N}_{so}^*=N_{1D}/a$ and calculate $\bar{N}_{so}^*$ from the 1D electron density N_{1D}, that is determined from magneto-transport measurements. For the sample shown in Fig.3 and at $V_g=V_D$=-0.5V we thus get $\bar{N}_{so}^*=1.7\cdot10^{11}cm^{-2}$, which compares well to the value $\bar{N}_{so}=1.4\cdot10^{11}cm^{-2}$ obtained from the plasmon frequency on the same sample.

The infrared excitations change character just at the gate voltage $V_g=V_D$, where an array of isolated electron channels is formed (upper part of Fig.5). With decreasing gate voltage $V_g<V_D$ the resonance frequency now increases. The oscillator strength that rises suddenly around $V_g=V_D$ decreases at $V_g<V_D$. The resonance linewidth is broad at the transition voltage V_D, becomes narrower below and then broadens again. The polarization and magnetic field dependence of the oscillator strength and the magnetic field dependence of the resonance frequency in this gate voltage regime are found to be typical for electron channels confined in a 2D potential [10,13,14]. These properties are well described by a quasi-classical model of the conductivity of a single electron channel harmonically bound by the potential $V_{(x,z)}=m^*/2(\Omega_x^2x^2+\Omega_z^2z^2)$ with $\Omega_z>>\Omega_x$. This model predicts for perpendicular polarization (in x-direction) a single resonance at finite frequency:

$$\omega_0^2 = \omega_c^2 + \Omega_x^2 + \omega_d^2 . \qquad (6)$$

The shift of the 1D intersubband resonance frequency from the subband spacing in (6) by the frequency ω_d arises, because the electrons in the channel feel an internal current driving field, that is different from the external radiation field. The transitions between the 1D subbands in our

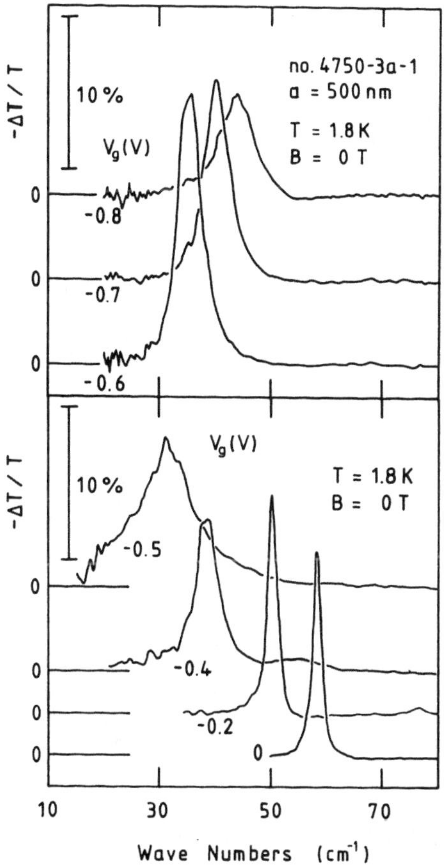

Fig. 5:

Infrared spectra of an AlGaAs/GaAs heterojunction with microstructured gate of period a=500nm. The lower part represents data in the 2D regime, the upper part in the 1D regime [10]

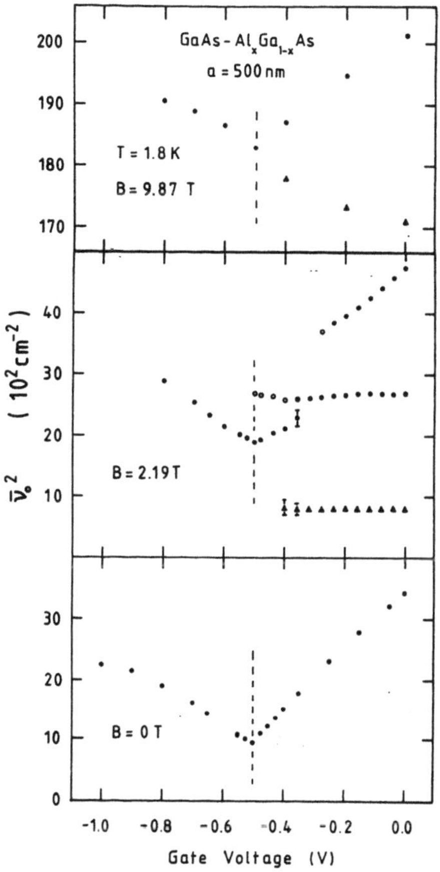

Fig. 6:

Square of the resonance frequency $\bar{\nu}_o$ of the FIR excitations vs gate voltage at different magnetic fields. In the 2D regime cyclotron resonances are marked by triangles, plasmon resonances by circles [10]

system are coherently excited and accompanied by a change in the lateral density distribution. Thus many-body interactions affect the resonance energy. This is well known from investigations of intersubband resonances in 2D systems [1]. For parallel polarization and magnetic field B=0 the model predicts a Drude like conductivity with maximum at ω=0. With increasing magnetic field the oscillator strength of this maximum is transferred to a resonance at the same frequency ω_0, where we find the resonance in perpendicular polarization. At high magnetic fields the resonances are predicted to be polarization independent and cyclotron resonance like. Experimentally, we find a magnetic field dependence of the resonance frequency that is well described by (6). In a magnetic field applied perpendicular to the sample surface a single resonance at polarization independent resonance frequency is observed with oscillator strengths, that behave qualitatively as predicted by the model.

Fig.6 shows the gate voltage dependence of the squared resonance frequencies at zero (bottom) and two finite magnetic fields. Independent of the magnetic field the resonances change character at the gate voltage $V_g=V_D=-0.5V$, which is marked with a dashed horizontal line. Below $V_g=V_D$ we always find a single resonance with increasing resonance frequency, if the gate voltage is decreased. Above the threshold voltage $V_g=V_D$ in a strong magnetic field we observe cyclotron resonance and the $q=2\pi/a$ magnetoplasmon, which may split (e.g., in Fig. 6 at B=2.2T) due to non-local interaction with the cyclotron harmonic $2\omega_c$ [24]. Whereas the magnetoplasmon resonance frequency decreases with decreasing gate voltage, the cyclotron resonance frequency is not affected by V_g at low magnetic fields. The oscillator strength of the cyclotron resonance decreases in proportion to the average electron density at first and suddenly vanishes slightly above $V_g=V_D$. In higher magnetic fields (B>6T) the cyclotron resonance shows a more gradual transition into the 1D intersubband excitation observed below $V_g=-0.5V$. This behaviour indicates in agreement with the DC-transport results, that the electron states in a magnetic field are affected by the lateral potential before screening breaks down and isolated electron channels are formed.

Information about the many-body contributions to the resonance frequency (6) is obtained by comparison of the far infrared resonance frequency with the subband spacings determined from magnetotransport measurements. In Fig.7 the resonance energies at B=0 are displayed together with the subband spacings extracted from magnetotransport measurements on the same sample vs the gate voltage. As the resonance energy is significantly higher than the subband spacings, we may conclude that the resonance frequency is dominantly determined by the many-body contributions. This is in contrast to recent results obtained on microstructured InSb-MOS systems [25]. There the collective contributions are found to be effectively screened by a Schottky grating adjacent to the

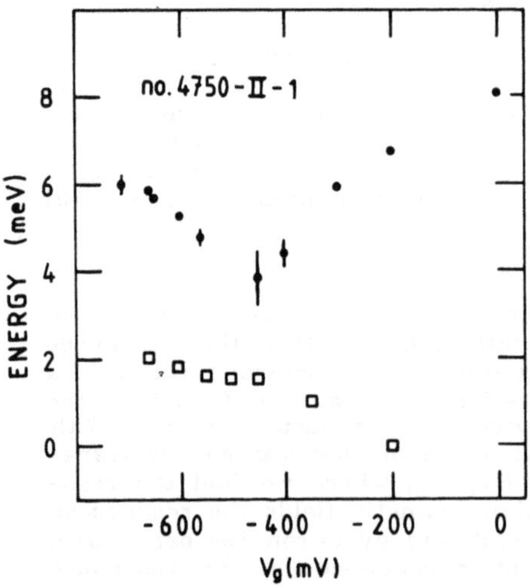

Fig. 7:

Subband spacings (open squares) determined from quantum oscillations of magnetotransport and resonance energies of the infrared excitations (filled circles) vs gate voltage [11]. The far infrared resonances are measured at B=0 in perpendicular polarization, the period of the gate structure is a=400nm.

electron channels and account for only about 20% of the intersubband resonance energies.

An exact quantum mechanical calculation of the many-body contributions to the 1D intersubband resonance in multi-channel structures with several occupied 1D subbands is complex [26]. It has to include the interactions of electrons in one channel as well as between adjacent channels and take into account properly the screening of the metal gate. For a rough estimate we use a classical approach, that only considers depolarization and screening by the metal gate in a single channel [12,14]. This yields for our geometry a frequency

$$\omega_d^2 = 2e^2 N_{1D}/(\bar{\varepsilon}\varepsilon_0 m^* W^2). \tag{7}$$

Screening by the metal gate is described by the same effective dielectric constant $\bar{\varepsilon}$ as in the 2D regime (eq. 5). The channel width W may be estimated from the 1D electron density N_{1D} derived from magnetotransport and the corresponding 2D electron density $N_{2D}=m^*E_F/(\pi\hbar^2)$ calculated from the Fermi energy as: $W=N_{1D}/N_{2D}$. For the data in Fig.7 this results in depolarization frequencies ω_d that are about a factor 2 higher than the measured resonance frequency at V_g=-0.5V. The effective channel width, that should be used in eq. (7) may be larger, since, e.g., interaction between adjacent channels may soften the depolarization effect. From the harmonic potential model we may define $W=[8E_F/(m^*\Omega^2)]^{1/2}$ as the width of the potential at the Fermi energy. With this W one calculates depolarization energies that rise for the sample in Fig.7 from $\hbar\omega_d$=5.7meV at V_g=-0.5V to $\hbar\omega_d$=6.3meV at V_g=-0.66V. These values are only slightly higher than the observed resonance energies. We note, however, that the increase of experimental resonance energies with decreasing gate voltage is more pronounced than the increase of both the calculated depolarization frequencies as well as the intersubband spacings derived from magnetotransport.

The resonance linewidth provide us with information about energy broadening and macroscopic channel width fluctuations. At high magnetic fields ($\omega_c^2 \gg \Omega_x^2+\omega_d^2$) the linewidth will be dominated by scattering processes that result in homogeneous broadening. From comparison of calculated cyclotron resonance line shapes with the experimental spectra we deduce typical values for scattering times of $\tau=2\cdot10^{-12}$sec corresponding to a mobility of μ=50000cm²/Vsec or an energy broadening of $\hbar/\tau$=0.3meV. At B=0 the resonance frequency is dependent on the channel width and therefore one expects inhomogeneous broadening, if macroscopic fluctuations of the channel width exist. From comparison of the resonance line shapes at B=0 to the ones calculated for a harmonically bound oscillator we obtain relaxation times, that are roughly a factor of 2 smaller than those at high magnetic fields: $\hbar/\tau$=0.7meV for the sample of Fig.7. Channel width fluctuations, that cause an inhomogeneous broadening of 0.4meV, do not destroy the far infrared excitations but seriously reduce conductance oscillations in static transport experiments. Since conductance oscillations depend on subband occupation, they may be averaged out completely, if width fluctuations result in largely inhomogeneous subband occupations. This may be one reason why, so far, without magnetic field no quantum oscillations are observed in the voltage dependent DC-resistance that can be linked to 1D subband depopulation.

5. Summary

AlGaAs/GaAs heterojunctions with laterally microstructured gates are an ideal system to study the transition from a 2D electron system with charge density modulation to an array of isolated 1D electron channels. DC-transport measurements, which verify quantization of the isolated electron channels, as well as far infrared spectroscopy provide valuable informations about the properties of this system. A comparison of the FIR intersubband resonance energy with DC-transport results indicates that the resonance position in heterojunctions studied so far may be dominated by many-body contributions. For more detailed interpretation of experimental results, self consistent calculations of effective potential and quantum mechanical calculations of the response including many-body interactions are highly desirable.

Acknowledgement We thank F. Brinkop and K. Ploog for their valuable contributions to the work summarized here and gratefully acknowledge financial support from the Stiftung Volkswagenwerk.

References

1. for a review see: T. Ando, A.B. Fowler, and F. Stern, Rev. Mod. Phys. 54, 437 (1982)
2. A. C. Warren, D. A. Antoniadis, and H. I. Smith Phys. Rev. Lett. 56, 1858 (1986)
3. T. P. Smith, III, H. Arnot, J. M. Hong, C. M. Knoedler, S. E. Laux and H. Schmid, Phys. Rev. Lett. 59, 2802 (1987)
4. K.-F. Berggren, T. J. Thornton, D. J. Newson, and M. Pepper Phys. Rev. Lett. 57, 1769 (1986)
5. H. van Houten, B. J. van Wees, J. E. Mooij, G. Roos, and K.-F. Berggren, Superlattices and Microstructures, in press
6. J. Cibert, P. M. Petroff, G. J. Dolan, S. J. Pearton, A. C. Gossard, and J. H. English, Appl. Phys. Lett. 49, 1275 (1986)
7. M. A. Reed, J. N. Randall, R. J. Aggarwal, R. J. Matyi, T. M. Moore, and A. E. Wetsel, Phys. Rev. Lett. 60, 535 (1988)
8. W. Hansen, M. Horst, J. P. Kotthaus, U. Merkt, Ch. Sikorksi, and K. Ploog, Phys. Rev. Lett. 58, 2586 (1987)
9. W. Hansen, Ph. D. thesis, Univ. Hamburg, 1987
10. J. P. Kotthaus, W. Hansen, H. Pohlmann, M. Wassermeier, and K. Ploog, Surf. Sci., in press (1988).
11. F. Brinkop, W. Hansen, J.P. Kotthaus, and K. Ploog, Phys.Rev. B in press
12. S. J. Allen, Jr., H. L. Störmer, and J.C.M. Hwang, Phys. Rev. B 28, 4875 (1983)
13. S. J. Allen, F. DeRosa, G. H. Dolan, and C. W. Tu: In Proceedings of the 17th Intern. Conf. on the Physics of Semiconductors, ed. by J. D. Chadi and W. A. Harrison, (Springer, New York 1985) p.313
14. W. Hansen, J. P. Kotthaus, A. V. Chaplik, and K. Ploog: In High Magnetic Fields in Semiconductor Physics, ed. by G. Landwehr, (Springer, Berlin 1987) p.266
15. K. Ploog, Ann. Rev. Mater. Sci. 11, 171 (1981).
16. D. Heitmann: In Two-Dimensional Systems: Physics and New Devices, Ed. by G. Bauer, F. Kuchar , and H. Heinrich (Springer Berlin 1986) p.285
17. W.-Y. Lai and S. Das Sarma, Phys. Rev. B 33, 8874 (1986).
18. S. E. Laux and F. Stern, Appl. Phys. Lett. 49, 91 (1986).
19. U. Wulf, Phys. Rev. B 35, 9754 (1987)
20. F. Stern, Surf. Sci., in press (1988)
21. U. Mackens, D. Heitmann, L. Prager, J. P. Kotthaus and W. Beinvogl, Phys. Rev. Lett. 53, 1485 (1984).
22. M. V. Krasheninnikov and A. V. Chaplik, Sov. Phys. Semicond. 15, 19 (1981).

23. W.-Y. Lai, A. Kobayashi, and S. Das Sarma, Phys. Rev. B 34, 7380 (1986).
24. E. Batke, D. Heitmann, J. P. Kotthaus, and K. Ploog, Phys. Rev. Lett. 54, 2367 (1985)
25. J. Alsmeier, Ch. Sikorski, and U. Merkt, Phys. Rev. B, in press
26. S. Das Sarma and W.-Y. Lai, Phys. Rev. B 32, 1401 (1985)

Quantum and Classical Ballistic Transport in Constricted Two-Dimensional Electron Gases

H. van Houten[1], *B.J. van Wees*[2], *and C.W.J. Beenakker*[1]

[1]Philips Research Laboratories, 5600 JA Eindhoven, The Netherlands
[2]Delft University for Technology, 2600 GA Delft, The Netherlands

An experimental and theoretical study of transport in constricted geometries in the two-dimensional electron gas in high mobility GaAs-AlGaAs heterostructures is described. A discussion is given of the influence of boundary scattering on the quantum interference corrections to the classical Drude conductivity in a quasi-ballistic regime, where the transport is ballistic over the channel width, but diffusive over its length. Additionally, results are presented of a recent study of fully ballistic transport through quantum point contacts of variable width defined in the two dimensional electron gas. The zero field conductance of the point contacts is found to be quantized at integer multiples of $2e^2/h$, and the injection of ballistic electrons in the two dimensional electron gas is demonstrated in a transverse electron focussing experiment.

1 Introduction

A variety of length scales governs the electron transport at low temperatures (see Table I). The mean free path l_e in a degenerate electron gas is determined by elastic scattering from stationary impurities. In homogeneous samples the classical transport can be described by the local Drude conductivity $\sigma_D = ne^2 l_e/m\mathrm{v}_F$, with n the electron gas density, v_F the Fermi velocity, and m the electron effective mass. In a two-dimensional electron gas (2-DEG) this classical expression can also be written as $\sigma_D = (e^2/h)\, k_F l_e$, with $k_F = m\mathrm{v}_F/\hbar$ the Fermi wave vector. The Drude conductance G_D of a 2-DEG channel of length L and width W is $G_D = (W/L)\sigma_D$. In two dimensions the smallest system for which a conductivity can be defined is a square of side of order l_e. The above expression for the conductance G_D of a larger channel can thus be seen as the result of the classical addition of such small squares in a (W/l_e) by (L/l_e) array (see Fig. 1). It is clear that any correlation between the squares is neglected in this approach.

It is now well known that such correlations are important because of quantum interference [1]. Constructive interference between time reversed backscattered electron waves leads to a conductivity decrease known as weak localization. Clearly, a prerequisite for this quantum interference effect is that the electrons maintain their phase coherence over a time τ_ϕ long compared to the elastic scattering time $\tau_e \equiv l_e/\mathrm{v}_F$.

Table I. Length scales for electron transport at low temperatures

elastic mean free path	$l_e = \mathrm{v}_F\tau_e$	Fermi wavelength	$\lambda_F = 2\pi/k_F$
phase coherence length	$l_\phi = (D\tau_\phi)^{1/2}$	thermal length	$l_T = (\hbar D/kT)^{1/2}$
magnetic length	$l_B = (\hbar/eB)^{1/2}$	cyclotron radius	$l_c = \hbar k_F/eB$

Springer Series in Solid-State Sciences Vol. 83: **Physics and Technology of Submicron Structures**
Editors: H. Heinrich • G. Bauer • F. Kuchar

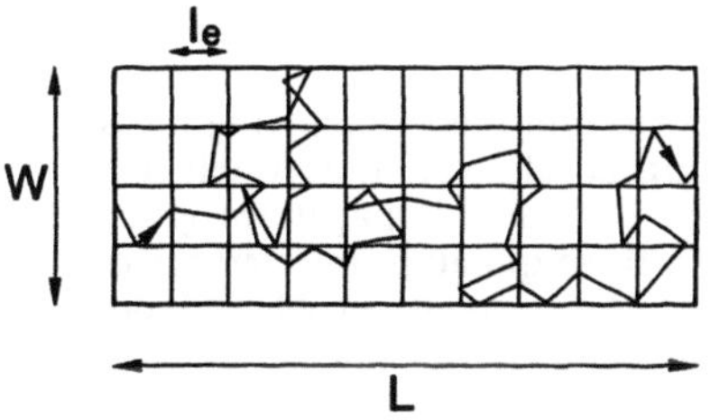

Fig. 1 Idealized picture of the Drude conductance; diffusive motion occurs on length scales large compared to l_e, and ballistic motion on shorter length scales.

The phase coherence length is a diffusion length given by $l_\phi \equiv (D\tau_\phi)^{1/2}$, with $D = v_F l_e/2$ the diffusion constant. The localization is one-dimensional (1D) if $l_\phi > W$. A second interference effect is the occurrence of universal (magneto) conductance fluctuations (UCF) in small samples, caused by interference of electrons on different trajectories [2]. This effect has two characteristic length scales, l_ϕ and the thermal length $l_T \equiv (\hbar D/kT)^{1/2}$. From a beautiful series of experiments performed in the diffusive regime in metal rings [3] and in silicon MOSFETs [4] the quantum interference corrections to the Drude conductivity are known to be non-local on length scales smaller than l_ϕ .

Classically, the simple scaling formula $G_D = (W/L)\sigma_D$ breaks down on length scales short compared to the mean free path. In this paper we distinguish a *quasi – ballistic* regime ($L \gg l_e > W$), where the electrons move ballistically over the sample width, but diffusively over its length, and a fully *ballistic* regime ($l_e \gg W,L$). The scattering of the electrons by typical smooth electron gas boundaries is specular, because of the large Fermi wave length ($\lambda_F \sim 40$ nm) . We note that the presence of voltage probes on the channel sides will in general have an inextricable influence on the electron transport [5]. We will here concentrate on channels which are smooth constrictions connected by broad 2-DEG regions to ohmic contacts (see Fig. 2). While in the quasi-ballistic regime the measured conductance is an intrinsic property of the narrow channel, this is no longer the case in the ballistic regime. Here the channel does not have an intrinsic resistance and, as we will discuss below, the conductance is determined by a geomet-

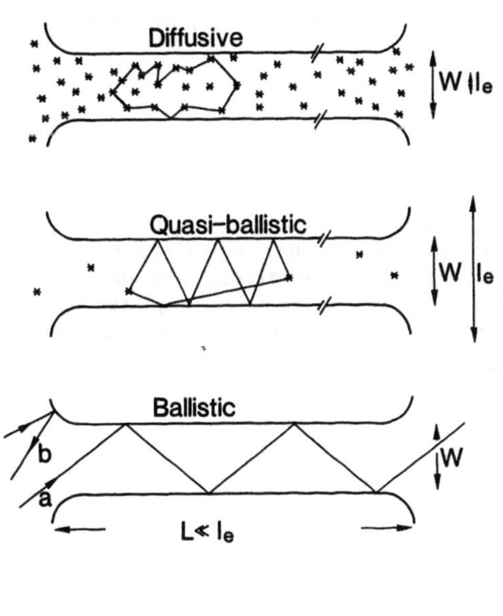

Fig. 2 Transport in constricted regions of length L and width W for three mobility regimes. Closed trajectories, necessary for weak localization, occur in the diffusive and quasi-ballistic regimes. Only open (a) trajectories occur in the ballistic regime, where the resistance is a geometrical constriction effect, associated with trajectories of type (b).

rical constriction effect. If λ_F is comparable to the constriction width it is possible to study quantum ballistic transport.

In the quasi-ballistic regime we have experimentally [6] and theoretically [7] investigated weak localization and UCF in long narrow channels defined in moderately high mobility GaAs-AlGaAs heterostructures. In the first part of this paper we give a qualitative account of the effect of a magnetic field on these quantum interference effects in narrow channels for various mobility regimes. An unsolved problem is presented by the breakdown of coherent diffusion on timescales shorter than τ_e, which is important if τ_ϕ and τ_e are comparable.

In the fully ballistic regime disorder related quantum interference effects are absent. In metal physics Sharvin [8] point contacts (i.e. constrictions much narrower and shorter than the mean free path) can ideally be used to study ballistic transport. We have very recently succeeded in fabricating such point contacts in the 2-DEG of high mobility GaAs-AlGaAs heterostructures [9,10]. A split-gate lateral depletion technique allows the width of the constrictions to be continuously varied (between 0 and 200 nm). We have found that the conductance of these quantum point contacts exhibits quantized plateaux at integer multiples of $2e^2/h$ if their width is varied. An explanation of this novel effect is given in terms of the quantization of the transverse electron momentum in the narrow part of the constriction. A semi-classical treatment of the quantized conductance in such ballistic point contacts is presented and, alternatively, it is discussed in the context of the Landauer formula [16].

Finally, the existence of skipping orbits of electrons in the 2-DEG is demonstrated in a transverse electron focussing experiment. Here electrons are injected by one point contact and, after deflection by a magnetic field and repeated reflections on a 2-DEG boundary, they are collected by a second point contact. Finestructure in the focussing spectra, attributed to interference of the ballistic electrons, is resolved at low temperatures.

2 Flux Cancellation Effect on Weak Localization and UCF

Weak localization originates in the constructive interference between a closed electron trajectory and its time reverse. A magnetic field destroys this effect, leading to a negative magnetoresistance. A semi-classical treatment in terms of electron trajectories is allowed for channels much wider than the Fermi wavelength. The 1D-weak localization conductance correction can be expressed as a time integral over the classical probability of return $C(t)$[12]

$$\delta G_{loc}(B) = -\frac{4e^2}{h}\frac{D}{L}\int_0^\infty dt\, C(t)\, e^{-t/\tau_\phi} < e^{i\phi(t)} > . \tag{1}$$

The factor $\exp(-t/\tau_\phi)$ accounts for the fact that only electrons which have retained their phase coherence over a time t can interfere constructively. In a magnetic field a phase difference $\phi(t)$ develops between time reversed trajectories. This effect is taken into account by the factor $< \exp i\phi >$. The brackets denote an average over impurity configurations. For weak magnetic fields this factor decays as $\exp(-t/\tau_B)$ (see Ref. [7]). The meaning of the relaxation time τ_B is that, in the presence of a magnetic field B, trajectories with a duration t exceeding τ_B no longer contribute to the localization. On length scales larger than l_e the probability density of return is given by $C(t) = (4\pi Dt)^{-1/2}$. One thus finds [1]

$$\delta G_{loc}(B) = -\frac{2e^2}{h}\frac{\sqrt{D}}{L}\left(\frac{1}{\tau_\phi} + \frac{1}{\tau_B}\right)^{-1/2} . \tag{2}$$

We define a critical field B^* for the suppression of weak localization as the field for which $\tau_\phi = \tau_B$. The relaxation time τ_B in a narrow channel with specular boundary scattering has been calculated in Ref. [7] , for the quasi-ballistic regime (see Refs. [13 ... 15] for other regimes). In Table II the main results of the analysis are summarized. In this section we present a simple physical interpretation of these results.

Table II. Magnetic field phase relaxation time τ_B and critical field B^* for 1D-weak localization [7,13] ($C_1 = 9.5$ and $C_2 = 24/5$). In the quasi-ballistic regime the critical field is enhanced as a consequence of the flux cancellation effect.

Diffusive ($l_e \ll W, L$)	Quasi-ballistic ($W \ll l_e \ll L$)	
1D $l_B^2 \gg W^2$	Weak field $l_B^2 \gg Wl_e$	Strong field $Wl_e \gg l_B^2 \gg W^2$
$\tau_B = \dfrac{6l_B^4}{W^2 v_F l_e}$	$\tau_B^{\text{weak}} = \dfrac{C_1 l_B^4}{W^3 v_F}$	$\tau_B^{\text{strong}} = \dfrac{C_2 l_B^2 l_e}{W^2 v_F}$
$B^* = \dfrac{\hbar}{e}\dfrac{\sqrt{3}}{Wl_\phi}$	$B^* = \dfrac{\hbar}{e}\dfrac{1}{Wl_\phi}\sqrt{\dfrac{C_1 l_e}{2W}}$	$B^* = \dfrac{\hbar}{e}\dfrac{1}{Wl_\phi}\dfrac{C_2 l_e^2}{2Wl_\phi}$

The effectiveness of a magnetic field in suppressing weak localization depends in an interesting way on the shape of the electron trajectories. Three regimes can be distinguished, depending on the relative magnitudes of l_e, W, and the magnetic length $l_B \equiv (\hbar/eB)^{1/2}$. This is illustrated in Fig. 3, where the relaxation time τ_B is plotted as a function of l_e/W for a fixed ratio l_B/W. The non-monotonous dependence of τ_B on l_e can be understood as follows.

In the diffusive regime ($l_e \ll W,L$) an increase of the mean free path is seen to lead to a decrease of τ_B . The phase change for a closed loop is given by the enclosed flux divided by $\hbar/e$, or equivalently, by its area divided by l_B^2. In the diffusive regime the enclosed area for a trajectory of duration τ_B is of order $W(D\tau_B)^{1/2}$. Setting this area equal to l_B^2 one finds (apart from numerical coefficients) the diffusive regime result for

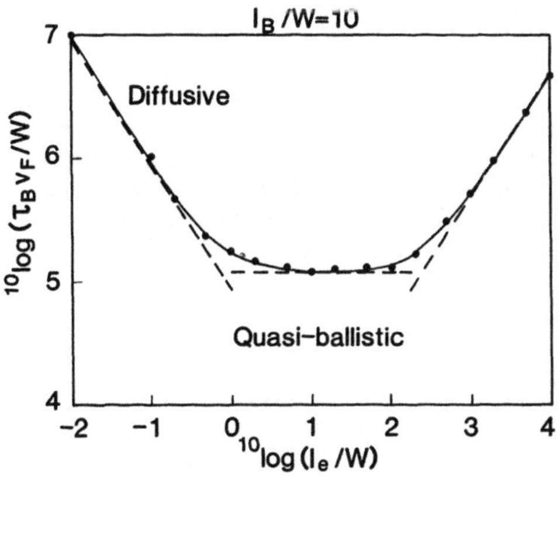

Fig. 3 Phase relaxation time τ_B as a function of the elastic mean free path l_e. The plot is obtained by a numerical calculation [7] with $l_B = 10W$. Dashed lines are analytic results (see Table II) The results show the characteristic dependence of τ_B on l_e in, respectively, the diffusive and quasi-ballistic regime. The latter regime can be subdivided in a weak ($Wl_e/l_B^2 \ll 1$) and a strong ($Wl_e/l_B^2 \gg 1$) field regime.

τ_B listed in Table II. The decrease of τ_B on increasing l_e is thus simply a consequence of the faster diffusion, leading to a larger enclosed area for a given duration ($D = v_F l_e/2$) .

On further increasing the mean free path the quasi-ballistic regime is reached. In this regime frequent collisions with the boundary lead to a crossing of the returning electron trajectories. Since the various closed parts of such trajectories are traversed in opposite directions the net enclosed flux is greatly reduced (*flux cancellation effect*)[16]. This effect is illustrated in the inset of Fig. 4. A weak and strong field regime have to be distinguished, depending on the ratio Wl_e/l_B^2. This ratio corresponds to the maximum phase change on a closed trajectory of linear extension l_e.

In the *weak field* regime ($Wl_e/l_B^2 \ll 1$) , many impurity collisions are needed before the electron loop encloses sufficient flux for a complete phase relaxation. In this regime a further increase of the mean free path does not lead to a larger enclosed flux, because the effect of the larger loop area is compensated by the increased flux cancellation. Consequently τ_B in the plot of Fig. 3 is approximately constant. On comparing the result for B in the weak field regime with the result for the diffusive regime, we see an enhancement of the critical field by a factor $(l_e/W)^{1/2}$ (see Table II).

The *strong field* regime is reached if $Wl_e/l_B^2 \gg 1$ (note that the theory only applies for fields limited by the condition $l_B \gg W$). Under these conditions, trajectories involving a single glancing angle reflection from the channel boundary enclose sufficient flux. In this regime the phase relaxation for trajectories of a given duration t is no longer determined by the average linear extension, but by the probability for these glancing angle reflections to occur. This probability is proportional to the number of impurity collisions, and thus to $1/l_e$. The relaxation time τ_B accordingly *increases* with l_e in this regime (see Fig. 3).

As seen in Fig. 3 the results given in Table II for the various asymptotic regimes agree very well with the results of a numerical calculation. For comparison with experiments in the quasi-ballistic regime the following simple interpolation formula can be used:

$$\tau_B = \tau_B^{weak} + \tau_B^{strong} \quad , \tag{3}$$

with τ_B^{weak} and τ_B^{strong} as given in Table II. So far we have assumed that the transport is diffusive on time scales corresponding to τ_ϕ. This will be a good approximation only if $\tau_\phi \gg \tau_e$. As we have discussed elsewhere [6], a modification of Eq. (1) is necessary to take the breakdown of coherent diffusion into account in the case that τ_ϕ and τ_e are comparable. We note that for very high mobility channels, with smooth boundaries, the weak localization effect vanishes, because the ballistic electron motion prevents the backscattering needed for this effect (see Fig. 2).

We now turn to the related problem of the magnetoconductance fluctuations observed in small samples. From the experimental data a conductance auto correlation function $F(\Delta B) = <\delta G(B)\delta G(B+\Delta B)>$ with $\delta G(B) = G(B) - <G(B)>$ can be determined. Here the average is over magnetic field values [2]. The field increment ΔB_c such that $F(\Delta B_c) = F(0)/2$ is by definition the correlation field, while $F(0)$ represents the variance. At $T = 0$ the variance $F(0)$ is of order $(e^2/h)^2$. At finite temperatures in quasi-one dimensional channels ($W \ll l_\phi \ll L$), $F(0)$ can be approximated by [7]

$$F(0) = 6\left(\frac{e^2}{h}\right)^2 \left(\frac{l_\phi}{L}\right)^3 \left[1 + \frac{9}{2\pi}\left(\frac{l_\phi}{l_T}\right)^2\right]^{-1} \quad . \tag{4}$$

Here l_T accounts for the non-monochromaticity of the electrons associated with the thermal smearing of the Fermi-Dirac distribution [2] . This effect is of importance only if $l_T < l_\phi$. On length scales L larger than l_ϕ the relative magnitude of the fluctuations $F(0)^{1/2}/G_D$ decreases as $L^{-1/2}$ as a consequence of the classical averaging over uncorrelated segments of length l_ϕ. (For comparison we note that $\delta G_{loc}/G_D$ does not depend on L if $L \gg l_\phi$). The expression (4) for the variance applies to the diffusive and quasi-ballistic regimes, under the assumption of coherent diffusion ($\tau_e \ll \tau_\phi$). As mentioned earlier in the context of weak localization, these disorder related quantum interference effects vanish for purely ballistic electron motion.

The correlation field ΔB_c depends on the shape of the trajectories (via the enclosed flux), and this is where boundary scattering plays an essential role. Even though the type of trajectories involved differs from the closed trajectories responsible for weak localization, the problem of determining ΔB_c is essentially the same as the one for B^* discussed before. As a typical example the theoretical results [7] for the correlation field are plotted in Fig. 4 for the diffusive and quasi-ballistic regimes (for the case $l_T \gg l_\phi$). This figure clearly shows the enhancement of the correlation field as a consequence of the flux cancellation effect in the quasi-ballistic regime.

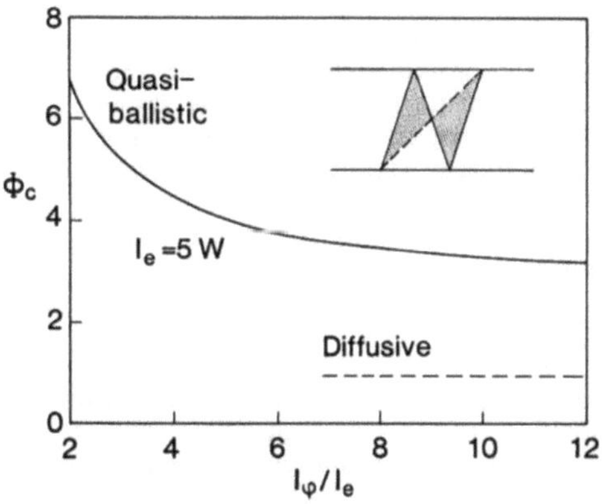

Fig. 4 Plot of the correlation flux $\phi_c = \Delta B_c/l_\phi W$ (in units of h/e) for Universal Conductance Fluctuations as a function of the phase coherence length l_ϕ (normalized by the elastic mean free path) in the quasi-ballistic and diffusive regime. The inset illustrates the characteristic flux cancellation: the shaded areas are equal and of opposite orientation.

3 Quantum Point Contacts

The conductance of ballistic, or Sharvin, point contacts is purely determined by its geometry, and by the electron momentum at the Fermi level $\hbar k_F$. For simplicity we consider the equivalent (by the Einstein relation) transport problem of uncharged particles [11], where the current arises as a consequence of the concentration gradient or chemical potential difference $\Delta\mu$ across the constriction. The electron distribution in **k**-space in the constricted region is schematically drawn in Fig. 5. The current I passing through a constriction of length L and width W equals the number of excess carriers present in the constriction $(WL)N_o\Delta\mu/2$, divided by the transit time $L/v_F \cos\phi$. Here $N_o = m/\pi\hbar^2$ is the 2-DEG areal density of states (assuming spin-degeneracy). The conductance $G \equiv Ie^2/\Delta\mu$ is thus $G = (1/2)e^2 N_o W < v_F \cos\phi >$, where the brackets denote an angular average over positive values of $\cos\phi$. Classically, all values of ϕ have equal weight, leading to a conductance [9]

$$G = \frac{2e^2}{h} \frac{k_F W}{\pi} , \tag{5}$$

proportional to the constriction width and the Fermi momentum.

We have recently fabricated the first ballistic point contacts in the 2-DEG in high mobility GaAs-AlGaAs heterostructures [9,10]. A split-gate lateral depletion tech-

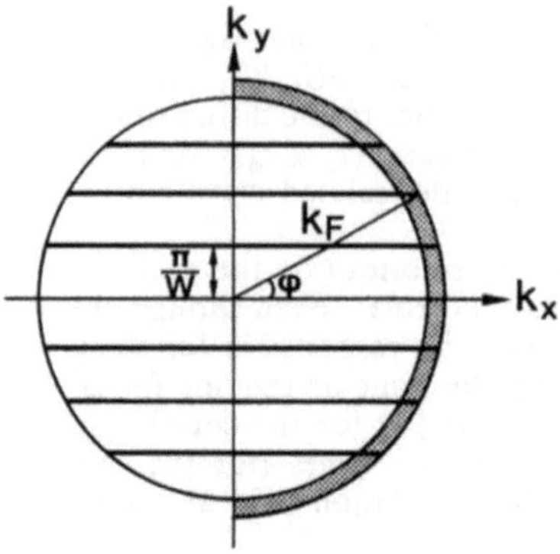

Fig. 5 Electron distribution in **k**-space in a constricted 2-DEG region. The current in the x-direction is carried by ballistic electrons in the shaded area. Quasi-one-dimensional electric subbands (horizontal lines), lead to the quantized conductance of Fig. 6.

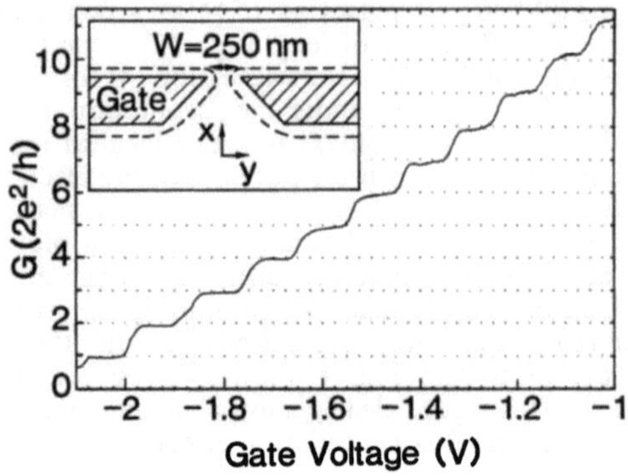

Fig. 6 Conductance of a ballistic quantum point contact at 0.6 K as a function of gate voltage in the absence of a magnetic field. The inset shows the sample geometry. The dashed line schematically indicates the 2-DEG boundary. Note that the point contact is actually a constriction of finite length.

nique [17] is used to vary the constriction width continuously. On varying the gate-voltage we have found the two-terminal conductance of these ballistic point contacts to exhibit quantized plateaux at integer multiples of $2e^2/h$ in the *absence* of a magnetic field (see Fig. 6) [18]. This novel effect can be understood as a consequence of the quantization of the transverse electron momentum in the narrow region, whereby only discrete values for the angle ϕ are allowed (see Fig. (5)). These values correspond to one-dimensional electric subbands with $k_y = \pm n\pi/W$, $n = 1,2, ... N_c$; the total number of subbands (or channels) N_c is the largest integer smaller than $k_F W/\pi$ (for a square well lateral confining potential). Each subband carries an equal amount of current, because the loss in forward momentum for large n values is compensated by an increase of the 1D density of states. From the classical correspondence with Eq. (5) it thus follows that $G = (2e^2/h)N_c$. In the actual experiment a local change in electron gas density $n_s = k_F^2/2\pi$ accompanies a change in constriction width. We note that these changes have a similar effect on the number of subbands, and thus on the conductance. In order for the preceding theoretical arguments to apply one needs a channel longer than wide, so that life time broadening of the 1D subbands in the constriction is unimportant. This assumption appears to be validated by our experiment.

The above semi-classical approach can be substantiated by the quantum mechanical Landauer formula [11,19]

$$G = \frac{2e^2}{h} N_c (1 - r) \quad , \tag{6}$$

where r is the fraction of electrons injected in the constriction which scatter back into the wide 2DEG regions. Originally, this type of formula was derived to treat a disordered region connected by means of ideal leads to thermalizing reservoirs, in which all inelastic scattering events are thought to take place exclusively. Apparently, quantum point contacts provide a model system which closely approximates such an idealized wire, in which the absence of a disordered region leads to $r = 0$, and thus to $G = N_c(2e^2/h)$.

We now digress to discuss a more general applicability of the formula $G = N_c(2e^2/h)$ to ballistic transport. First of all, the two-terminal conductance of a 2-DEG in the quantum Hall regime is given by $G = N_L(2e^2/h)$, with N_L the number of occupied Landau levels (still assuming spin degeneracy). This number is given by $N_L = E_F/\hbar\omega_c = k_F l_c/2$, with $l_c = mv_F/eB$ the classical cyclotron radius. This number of Landau levels equals the number of quantum edge states at the Fermi energy [20], which are localized within a distance $2l_c$ from the electron gas boundary. These edge states take the place of the 1D electric subbands which, as explained above, cause the quantized conductance of ballistic point contacts in the absence of a magnetic field. The above point of view has been confirmed very recently by experiments showing that the quantization of the two-terminal conductance of ballistic point contacts is preserved in a magnetic field, the only change being that N_c is replaced by the number of occupied *magneto – electric* subbands [21] in the constricted region [22].

Finally, we remark that under certain conditions *four – terminal* measurements in the ballistic regime can also be understood in terms of the preceding considerations. The four-terminal resistance measured over a constriction is then $R_{4t} = (h/2e^2)(N_1^{-1} - N_2^{-1})$, with N_1 the number of subbands in the constriction and N_2 the number of Landau levels in wide regions near the voltage probes. This formula has been used to describe the suppression of the Sharvin contact resistance by a magnetic field in a four-terminal set-up [23].

4 Coherent electron focussing

In 1965, Sharvin proposed a method to study Fermi surfaces in metals, in which ballistic electrons injected by a point contact are focussed by a longitudinal magnetic field onto a second, collecting, point contact [8]. We have performed [10] an electron focussing experiment in a 2-DEG in the transverse field geometry [24]. The experimental arrangement is shown in Fig. 7. In this geometry the scattering of the injected electrons by the 2-DEG boundary can be studied by comparing the intensities of the focussing peaks associated with multiple reflections, or skipping orbits, of the electrons. The experimental results, shown in Fig. 8, clearly establish that injection of ballistic electrons is realized in this experiment. The large number of classical focussing peaks observed (at 4 K) proves that the scattering at the 2-DEG boundaries indeed is highly specular, as we have assumed thoughout this paper.

At low temperatures interesting finestructure is apparent in the focussing spectra. We believe that this is caused by interference of electrons on different skipping orbits from injector to collector. The experiment thus reflects the spatial coherence of the injected ballistic electrons. We are currently studying a quantitative model for this

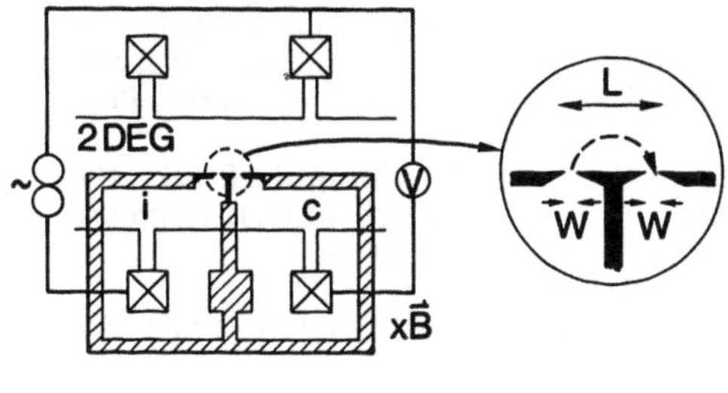

Fig. 7 Layout of the split-gate (shaded area) double point contact structure for transverse electron focussing experiments in a 2-DEG. Crossed squares are ohmic contacts. The gate separates injector (i) and collector (c) areas from the bulk 2-DEG. The dashed line indicates an electron trajectory in a magnetic field.

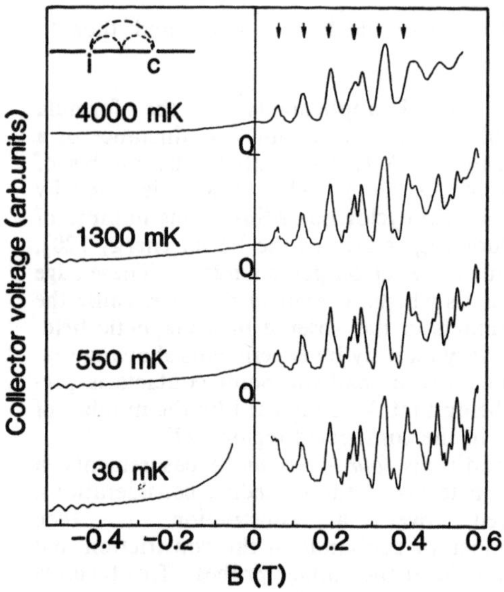

Fig. 8 Transverse electron focussing spectra showing classical, equidistant, focussing peaks at 4 K and fine structure at low temperatures. Calculated positions of the classical focussing peaks are indicated by arrows. The inset illustrates typical trajectories, corresponding to the first and second maxima.

effect. We conclude by noting that the transverse electron focussing geometry of Fig. 7 offers excellent opportunities to study the precise role of the coupling of voltage probes to quantum conductance channels. This is especially so in view of the fact that a point contact may be the least disturbing voltage probe [11,19] which can be realized with the present state of the art.

Valuable contributions to the work described in this paper have been made by M.E.I. Broekaart, L.P. Kouwenhoven, J.M. Lagemaat, P.H.M. van Loosdrecht, J.E. Mooij, J.A. Pals, M.F.H. Schuurmans, C.E. Timmering and J.G. Williamson. The high mobility samples needed for this work were grown by C.T. Foxon and J.J. Harris.

References

1. See the review by P.A. Lee and T.V. Ramakrishnan, Rev. Mod. Phys. **57** (1985) 287.
2. B.L. Al'tshuler, Pis'ma Zh.Eksp. Teor. Fiz. **41** (1985) 530 (JETP Lett., **41** (1985) 648); P.A. Lee, A. D. Stone and H. Fukuyama, Phys. Rev. B. **35** (1987) 1039.
3. C.P. Umbach, P. Santhanam, C. van Haesendonck and R.A. Webb, Appl. Phys. Lett., **50** (1987) 1289;
4. W.J. Skocpol, P.M. Mankiewich, R.E. Howard, L.D. Jackel, D.M. Tennant and A.D. Stone, Phys. Rev. Lett. **58** (1987) 2347.
5. G. Timp, A.M. Chang, P. Mankiewich, R. Behringer, J.E. Cunningham, T.Y. Chang and R.E. Howard, Phys. Rev. Lett. **59** (1987) 732; M.L. Roukes, A. Scherer, S.J. Allen Jr., H.G. Craighead, R.M. Ruthen, E.D. Beebe and J.P. Harbison, Phys. Rev. Lett. **59** (1987) 3011.
6. H. van Houten, C.W.J. Beenakker, B.J. van Wees and J.E. Mooij, Surf. Sci. **196,** (1988) 144; H. van Houten, C.W.J. Beenakker, M.E.I. Broekaart, M.G.J. Heijmans, B.J. van Wees, J.E. Mooij and J.P. André, Acta Electronica, in press.
7. C.W.J. Beenakker and H. van Houten, Phys. Rev. B., in press.
8. Y. V. Sharvin, Zh. Eksp. Teor. Fiz. **48** (1965) 984 (Sov. Phys.-JETP **21** (1965) 655).

9. B.J. van Wees, H. van Houten, C.W.J. Beenakker, J.G. Williamson, L.P. Kouwenhoven, D. van der Marel and C.T. Foxon, Phys. Rev. Lett. **60** (1988) 848.
10. H. van Houten, B.J. van Wees, J.E. Mooij, C.W.J. Beenakker, J.G. Williamson, C.T. Foxon, Europhys. Lett., in press.
11. R. Landauer, I.B.M. J. Res. Dev. **1** (1957) 223; Z. Phys. B **68** (1987) 217.
12. S. Chakravarty and A. Schmid, Physics Reports **140** (1986) 193.
13. B.L. Al'tshuler and A.G. Aronov, Pis'ma Zh. Eskp. Teor. Fiz. **33** (1981) 515 (JETP Lett. **33** (1981) 499).
14. V.K. Dugaev and D.E. Khmel'nitskii, Zh. Eksp. Teor. Fiz. **86** (1984) 1784. (Sov. Phys. JETP **59** (1984) 1038);
15. See also R. Landauer and M. Büttiker, Phys. Rev. B. **36** (1987) 6255.
16. This phenomenon is also encountered in thin superconducting films, P.G. de Gennes and M. Tinkham, Physics (New York), **1** (1964) 107;
17. T.J. Thornton, M. Pepper, H. Ahmed, D. Andrews and G.J. Davies, Phys. Rev. Letters, **56** (1986) 1198; H.Z. Zheng, H.P. Wei and D.C. Tsui, Phys. Rev. B **34** (1986) 5635.
18. Similar experimental results were obtained by D.A. Wharam, T.J. Thornton, R. Newbury, M. Pepper, H. Ahmed, J.E.F. Frost, D.G. Hasko, D.C. Peacock, D.A. Ritchie, G.A.C. Jones, J. Phys. C, to be published.
19. Sharvin contact resistances have been mentioned in the context of theoretical discussions of idealized mesoscopic systems in M. Büttiker, Phys. Rev. B. **33** (1986) 3020; and Phys. Rev. B, **35** (1987) 4123; Y. Imry in *Directions in Condensed Matter Physics*, Vol. 1, ed. G. Grinstein and G. Mazenko, World Scientific, Singapore (1986) 102; See also B. van de Leemput, University of Nijmegen, The Netherlands, Master Thesis.
20. B.I. Halperin, Phys. Rev. B **25** (1982) 2185; P. Streda, J. Kucera and A.H. MacDonald, Phys. Rev. Lett. **59** (1987) 1973.
21. K.-F. Berggren, T.J. Thornton, D.J. Newson and M. Pepper, Phys. Rev. Lett. **57** (1986) 1769; H. van Houten, B.J. van Wees, J.E. Mooij, G. Roos and K.-F. Berggren, Superlattices and Microstructures, **3**, (1987) 497.
22. B.J. van Wees, L.P. Kouwenhoven, H. van Houten, C.W.J. Beenakker, J.E. Mooij, C.T. Foxon and J.J. Harris, subm. to Phys. Rev. Lett.
23. H. van Houten, C.W.J. Beenakker, P.H.M. van Loosdrecht, T.J. Thornton, H. Ahmed, M. Pepper, C.T. Foxon and J.J. Harris, subm. to Phys. Rev. B.
24. V.S. Tsoi, ZhETF Pis. Red. **19** (1974) 114 (JETP Lett. **19** (1974) 70).

Transport Properties of Narrow, Variable Width Channels in the 2DEG of a GaAs:AlGaAs Heterojunction

T.J. Thornton, C.J.B. Ford, D.A. Wharam, R. Newbury, M. Pepper, H. Ahmed, J.E.F. Frost, D.C. Peacock[1]*, D.A. Ritchie, and G.A.C. Jones*

Cavendish Laboratory, University of Cambridge, Cambridge, UK
[1]Also at the GEC Hirst Research Centre, Wembley, Middlesex, UK

1. Introduction

At low temperatures electron transport in narrow channels is markedly different to that expected from the classical Boltzmann conductance. These differences arise because of the wavelike nature of the electrons resulting in quantum interference between coherent electrons and the formation of one dimensional subbands. To observe electron interference the channel must have dimensions of the order of the phase breaking length L_ϕ. In addition, the one dimensional density of states can manifest itself under certain situations if the width of the channel is not much larger than the Fermi wavelength λ_F. To explore this regime we have made use of the high mobility two dimensional electron gas (2DEG) formed at the the interface of a modulation doped GaAs:AlGaAs heterojunction. Schottky barrier gates deposited on the surface of the heterojunction are used to define Hall bar geometries, ring structures and narrow constrictions by electrostatic confinement.

2. Electrostatic Confinement of a 2DEG

The simplest form of split gate heterojunction FET is shown schematically in Fig. 1. These devices have the advantage of producing high mobility channels with continuously variable widths in the range $0.1 \leq W \leq 1\mu m$ (see for example THORNTON et

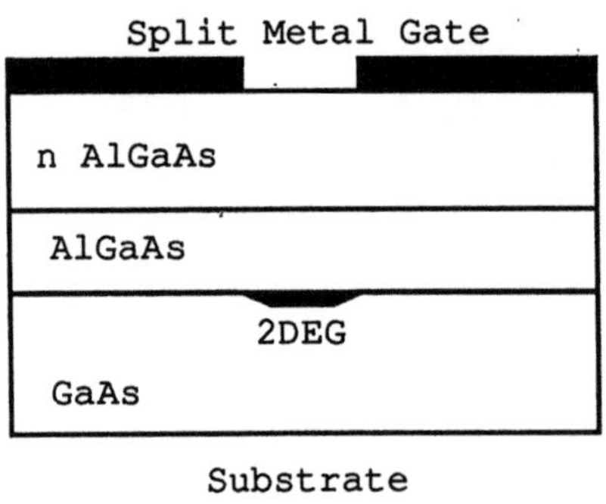

Fig. 1: A schematic cross sectional diagram of a split gate GaAs:AlGaAs heterojunction FET used to define a narrow channel in a 2D electron gas

Springer Series in Solid-State Sciences Vol. 83: **Physics and Technology of Submicron Structures**
Editors: H. Heinrich • G. Bauer • F. Kuchar

al [1], BERGGREN et al [2]). The gap between the gates is typically less than 1μm and to achieve this resolution the gates are patterned in PMMA resist by exposure to an electron beam [1]. After the resist is developed the gate metal is deposited by lift-off. The 2DEG underneath the gates can be removed by a reverse bias leaving only a narrow channel. As the reverse bias is increased the width of the channel is reduced. Electrostatic confinement has also been used to study the transition from 3D to 2D transport in a GaAs MESFET [3].

For the case of ring structures the 2DEG has to be removed from the centre of the loop and this necessitates the use of a dielectric layer [4] as shown in Fig.2. The gate is biased so that it only depletes where it is in contact with the heterojunction surface and not where it is raised above the surface by the dielectric. The PMMA resist in which the ring is drawn is a convenient dielectric because no further processing steps are required after the lithography. A dielectric layer is also used in the multiply connected Hall geometries because this reduces the number of gate contacts required.

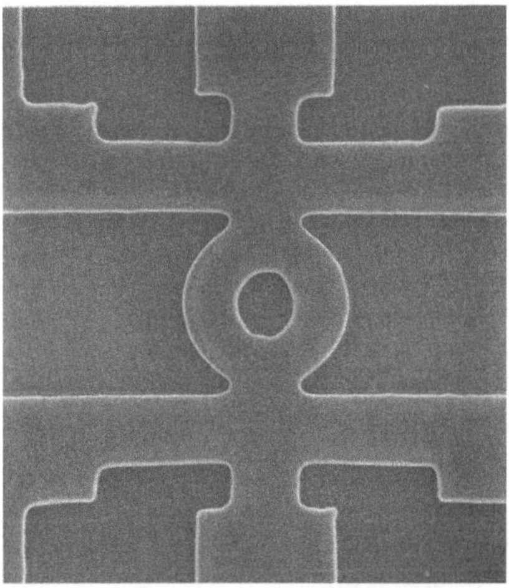

Fig. 2:

SEM micrograph of a resist loop prior to metallisation. A cross-sectional diagram is shown below

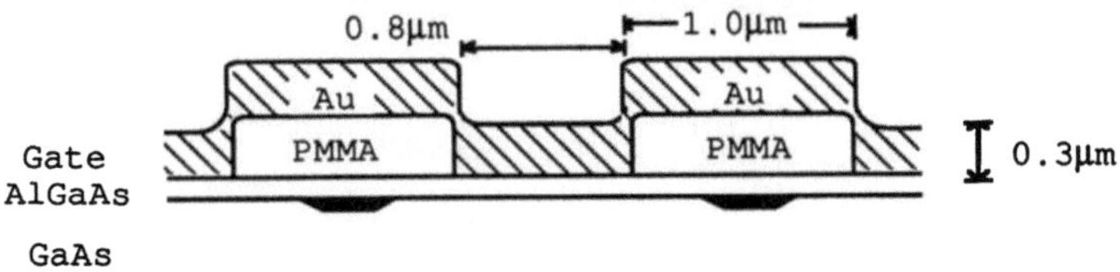

3. The Aharonov-Bohm Effect in Heterojunction Rings

Aharonov-Bohm oscillations have been observed in normal metal rings [5,6,7,8] and recently with larger amplitude in heterojunction rings [9,10]. If the diameter of the ring is smaller than a phase breaking length an electron wavepacket which splits in two as it passes round each arm of the loop will maintain phase coherence until it recombines on leaving the loop. In the presence of a perpendicular magnetic field the flux contained inside the loop produces a phase difference between the wavefunctions which have traversed opposite arms of the loop. If the magnetic field is increased the electron interference modulates the device resistance leading to oscillations which are periodic with period h/e if the waves travel only half way round the loop before recombining.

An Aharonov-Bohm loop similar to that shown in Fig. 2 was made on a heterojunction with a mobility ~$10^6 cm^2 V^{-1} s^{-1}$ at 4.2K in the dark. The sample was mounted in the mixing chamber of a dilution refrigerator and the four terminal resistance was measured using phase-sensitive detection. Currents of the order of 10^{-9}A produced no observable electron heating at temperatures T~100mK. Typical Aharonov-Bohm oscillations are shown in Fig. 3 for a number of different gate voltages. The oscillations are superimposed upon reproducible, aperiodic fluctuations [11,12] and are as much as 5% of the total resistance.

The Fourier power spectra of the oscillations are shown in Fig.4. The peak due to the h/e oscillations gets narrower as the gate bias is increased indicating that the channel width is being reduced.

The oscillations persist up to fields of approximately 2 Tesla superimposed on Shubnikov de Haas oscillations. The period of the latter deviates from the 1/B behaviour expected for Landau quantisation [13] and this is shown in Fig. 5. The deviation occurs at low fields because of the presence of one dimensional subbands. A perpendicular magnetic field raises the energy of the subbands and the magnetoresistance oscillates as

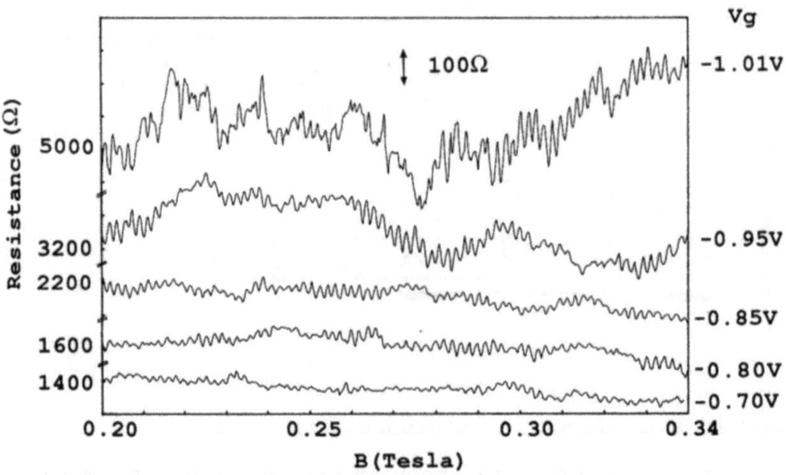

Fig. 3: Aharonov-Bohm oscillations for a number of gate voltages at a temperature T < 100mK

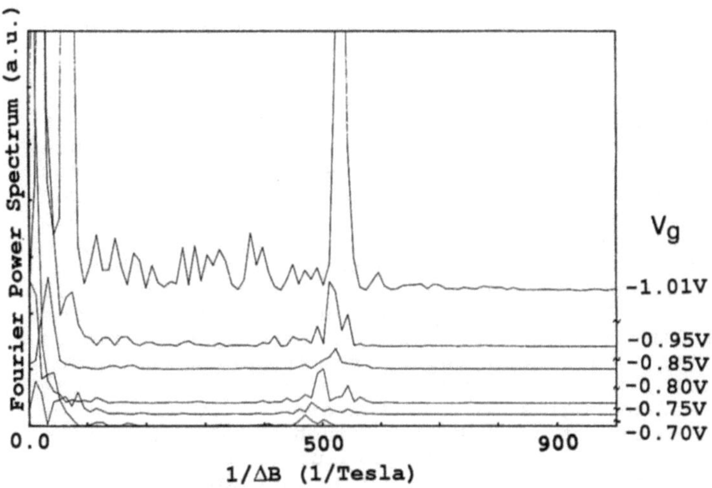

Fig. 4: The Fourier power spectrum of the Aharonov-Bohm oscillations shown in Fig. 3

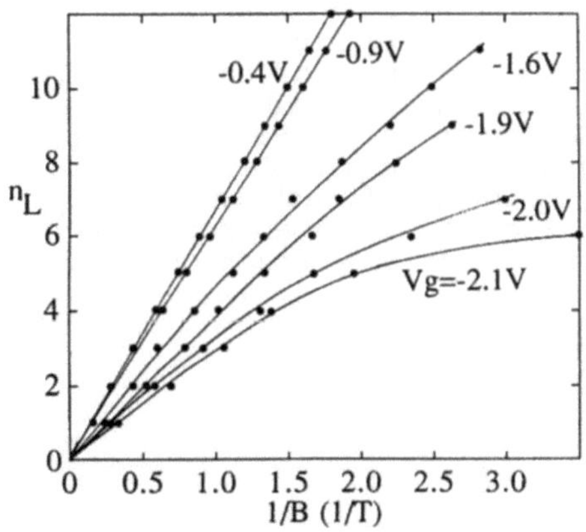

Fig. 5: A fan diagram showing the positions of the minima in resistance plotted against inverse magnetic field, for a range of gate voltages. The solid lines are only a guide to the eye.Taken from [15]

the hybrid levels pass through the Fermi energy. Using a theory developed by BERGGREN et al [2,14] it is possible to take the results in Fig. 5 and use them to estimate the channel width. This has been done by FORD et al [15] who found an approximately linear variation of width with gate voltage. A similar dependence has been seen by ZHENG et al [16].

4. The Hall Effect in Narrow Channels

As well as ring structures we have used electrostatic confinement to define Hall bar geometries with widths W < 0.9 μm and pairs of voltage probes separated by 3μm. From an analysis of the magnetoresistance oscillations we estimate that for the range $-3 \le Vg \le -0.9$ volts the channel width varies with gate voltage as $W(\mu m) = 0.38(3.6 + V_g)$. By extrapolation to zero width this result predicts a pinch-off voltage of -3.6 volts which is larger than the experimental result of -3.4 volts. This is probably because in the narrowest channels ($V_g < -3.0$ volts) the principal effect of changing the gate voltage is to reduce the carrier concentration rather than the width. Figure 6 shows the Hall voltage as a function of gate potential for three different perpendicular magnetic fields. For fields B > 0.1 Tesla the Hall voltage increases as the reverse bias to the gate is increased. This is because of a reduction in the number of carriers in the channel as the width is reduced. High field measurements show that this reduction in the carrier concentration is linear with gate voltage. For magnetic fields less than a certain threshold value the situation is quite different. Initially the Hall voltage remains constant as the width is reduced, then begins to fall and eventually fluctuates about an average value of zero before pinching-off at a gate voltage of -3.4 volts. At higher temperatures (T ~ 1.2K) the fluctuations are not so pronounced and we attribute them to quantum interference between phase coherent electrons.

A similar vanishing of the Hall voltage has been observed by ROUKES at al. [17]. They observed anomalous low field Hall voltages for a number of devices of different widths but were unable to vary the width in the same device. In Fig. 7 we show how the low field anomalies develop as the width of the channel is progressively reduced. Even for quite wide channels (Vg=-1.0V) there is a significant deviation in the Hall voltage. Reducing the width enhances the deviation and the Hall voltage vanishes below a certain threshold field. The threshold field B_t is plotted against gate voltage in Fig. 8 and an approximately linear variation is found. Both the carrier concentration n and

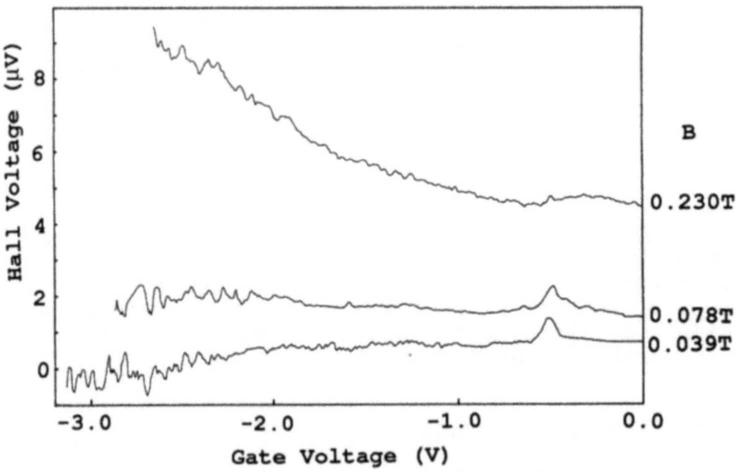

Fig. 6: The Hall voltage as a function of gate voltage for three different values of magnetic field. The narrow channel is defined for Vg < -0.5 volts. T < 100mK and $I_{xx} = 10^{-8}$A

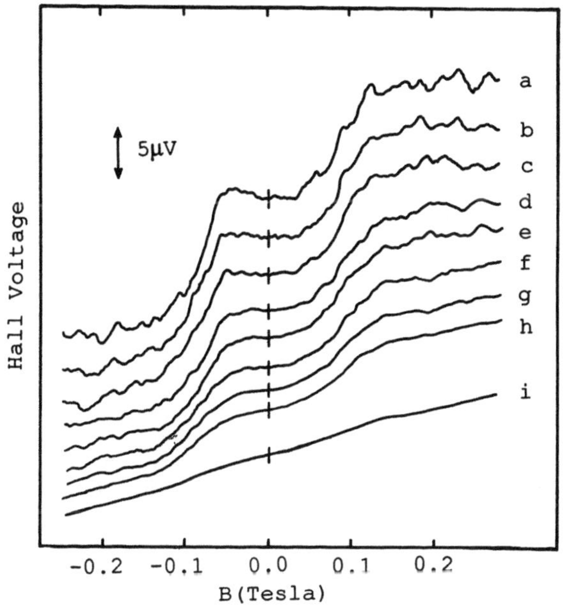

Fig. 7:

The Hall voltage as a function of magnetic field for a number of gate voltages. The current was 10^{-8} A at a temperature of 1.2K. The curves have been offset for clarity

V_g (volts)	
(a)	-2.80
(b)	-2.75
(c)	-2.70
(d)	-2.60
(e)	-2.55
(f)	-2.35
(g)	-2.18
(h)	-2.01
(i)	-1.00

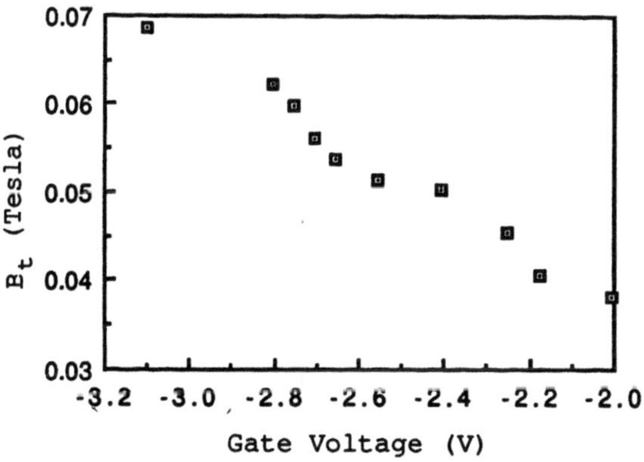

Fig.8: The threshold field B_t for different gate voltages

the width W decrease linearly with gate voltage and for this reason we cannot determine how B_t varies with W and n. However, the linear relation $B_t \sim |V_g|$ is a significantly weaker dependence than that expected from a number of recent theories [17,18].

5. The Quantised Resistance of a Narrow Constriction in a 2DEG

The devices used for the Hall measurements had channel lengths of ~ 3μm between voltage probes. It is possible to achieve much shorter channels by using a single gate of length 0.4μm with a gap of 0.5μm. When electrons are depleted from under the gate the large 2D regions on either side are linked by a narrow constriction. As the constriction is "squeezed" by the gate potential its resistance dominates over that in the wide regions. Our samples were made from heterojunctions with mobilities ranging from 2.5 x 10^5 to 10^6 $cm^2V^{-1}s^{-1}$ and carrier concentrations in the range 2 to 5 x $10^{15}m^{-2}$. This means that the elastic length was always in excess of 2 μm which is considerably longer than the length of the constriction so that electrons pass through ballistically. In addition the constrictions are narrow enough that the density of states will have structure associated with one dimensional subbands.

Figure 9 shows the plateaus in the low temperature resistance of a split gate FET as the width of the constriction is reduced below 0.5μm. Once the background resistance from the large area of 2DEG between the constriction and the voltage probes (typically 100Ω) is subtracted the resistance of the plateaus is found to be given by $R=h/2ie^2$. WHARAM et al [20] have shown that the quantisation is a result of ballistic conduction through one dimensional subbands and they derived the quantised resistance following an approach used by SHARVIN [21]. An equivalent result has been obtained by VAN WEES et al. [22]. Here we derive the result in terms of the flux of electrons through the channel.

For an applied voltage V the number of electrons with energy eV is given by LN(E)eV where N(E) is the density of electron states per unit length and L is the length of the constriction. The number of electrons passing through the constriction in unit

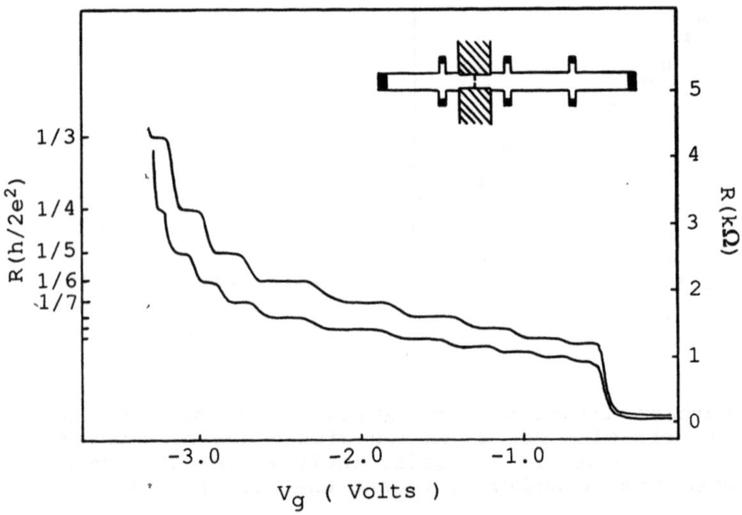

Fig 9: The channel resistance of a narrow constriction at T < 100mK is plotted as a function of gate voltage. The curves show results for different carrier concentrations obtained by shining light on the sample. The inset shows a schematic diagram of the split gate FET used to define the constriction. The split gate itself is 0.5μm wide and 0.4μm long.

time is $J = LN(E)eVv/L$ where v is the electron velocity in the subband. The current is therefore

$$I = N(E)e^2Vv. \tag{1}$$

If the density of states is split into one-dimensional subbands the contribution to the density of states from the nth subband is given by

$$N_n(E) = \frac{g_s}{h}(m/2E)^{1/2} = g_s/hv. \tag{2}$$

Here g_s is the spin degeneracy. Substituting (2) into (1) gives the conductance due to each subband as

$$\frac{1}{R_n} = \frac{I}{V} = \frac{g_s}{h}e^2. \tag{3}$$

Since there is no scattering between subbands they can be treated independently and the total sample resistance is given by

$$\frac{1}{R} = \sum_1^i \frac{1}{R_n}, \tag{4}$$

where the summation is over all i occupied subbands. Equation (4) contains no material parameters and for the spin degenerate case $g_s = 2$ the device resistance is given by $R = h/2ie^2$ in agreement with the experimental result. Provided there is no

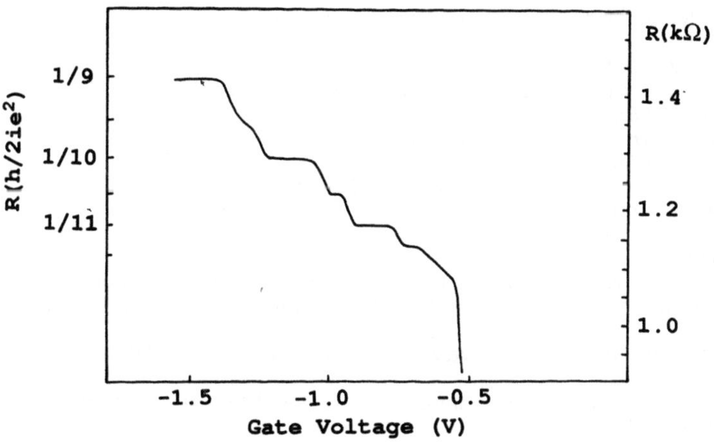

Fig. 10: The channel resistance in a parallel magnetic field, B = 13.6 Tesla, is plotted against gate voltage at T ~ 0.1K. The magnetic field removes the spin degeneracy resulting in additional quantised plateaus at $R = h/21e^2$, $h/23e^2$ and an incipient plateau at $h/19e^2$

scattering between subbands the resistance of the constriction is independent of its length and is determined solely by the number of occupied subbands.

In the presence of a large magnetic field in the plane of the 2DEG but perpendicular to the current the spin degeneracy will be lifted. Now $g_s = 1$ and extra plateaus form with resistance $R = h/2(i+1/2)e^2$. The extra quantisation can be seen in Fig. 10 in which additional plateaus occur in between those originally present at zero field. Experiments of this type will be useful in determining the exact form of the confining potential developed between the gates.

6. Summary

The results described in this article illustrate how Schottky barrier gates can be used to confine a high mobility 2DEG in a GaAs:AlGaAs heterojunction. Devices made in this way have been used to study a number of effects resulting from one dimensional transport. These types of devices are particularly useful because the width of the channels is continuously variable over a large range.

7. References

1. T.J. Thornton, M. Pepper, H. Ahmed, D. Andrews, G.J. Davies: Phys. Rev. Lett. 56, 1198 (1986)
2. K-F. Berggren, T.J. Thornton, D.J. Newson, M. Pepper: Phys. Rev. Lett. 57, 1769 (1986)
3. D.J. Newson, M. Pepper, T.J. Thornton: Philosophical Magazine B, 56, 775 (1988)
4. C.J.B. Ford, H. Ahmed: Proc. of Microcircuit Engineering,1987
5. D.Y. Sharvin, Y.V. Sharvin: JETP Lett., 34 272, 1981
6. B. Pannetier, J. Chaussy, R. Rammal, P. Gandit: Phys. Rev. Lett. 53, 718 (1984)
7. R.A. Webb, S. Washburn, C. Umbach, R.A. Laibowitz: Phys. Rev. Lett. 54, 2696, (1985)
8. S. Washburn, R.A. Webb: Adv.in Phys. 35, 375, (1986)
9. G. Timp, A.M. Chang, J.E. Cunningham, T.Y. Chang, P.Mankiewich, R. Behringer, R.E. Howard: Phys. Rev. Lett. 58, 2814, (1987)
10. C.J.B. Ford, T.J. Thornton, R. Newbury, H. Ahmed, M.Pepper, D. Andrews, G.J. Davies: Superlattices and Microstructures in print
11. P.A. Lee, A.D. Stone, H. Fukuyama: Phys. Rev. B 35, 1039 (1987)
12. T.J. Thornton, M. Pepper, H. Ahmed, G.J. Davies, D. Andrews: Phys. Rev. B 36, 4514, (1987)
13. I.D. Landau, E.M. Lifshitz: In Quantum Mechanics 2nd ed. (Pergamon Press, Oxford, UK. 1975)
14. K-F. Berggren, G. Roos, H. van Houten: to be published.
15. C.J.B. Ford, T.J. Thornton, R. Newbury, M. Pepper, H. Ahmed, C.T. Foxon, J.J. Harris, C. Roberts: J. Phys. C: Solid State Phys. to be published

16. H.Z. Zheng, H.P. Wei, D.C. Tsui, G. Weimann: Phys. Rev. B, 34, 5635 (1986)
17 M.L. Roukes, A. Scherer, S.J. Allen, H.G. Craighead, R.M. Ruthen, E.D. Beebe, J.P. Harbison: Phys. Rev. Lett. 59, 3011 (1987)
18. C.W.J. Beenakker, H. van Houten: to be published
19. V. Srivastava: to be published
20. D.A. Wharam, T.J. Thornton, R. Newbury, M. Pepper, H. Ahmed, J.E.F. Frost, D.G. Hasko, D.C. Peacock, D.A. Ritchie, G.A.C. Jones: J. Phys. C: Solid State Phys.(1988) (20 Mar) to appear
21. Y.V. Sharvin: JETP Lett. 1, 152 (1965)
22. B.J. van Wees, H. van Houten, C.W.J. Beenakker, J.G. Williamson, L.P. Kouwenhoven, D.van der Marel, C.T. Foxon: to be published

Hopping Transport in One-Dimensional Semiconductor Systems

P.N. Butcher[1] *and J.A. McInnes*[2]

[1]University of Warwick, Coventry CV4 7AL, UK
[2]University of Strathclyde, Glasgow, G1 1XH, Scotland, UK

Abstract. The channel conductance of a one-dimensional MOSFET shows reproducible irregular structure as a function of gate voltage when the device is operated near threshold at low temperatures. The voltage scale on which the structure occurs is in the order of k_BT/e. The conductance variations amount to many orders of magnitude when T is taken down into the millikelvin range. These features suggest that the observed structure is due to incoherent hopping. We calculate the conductivity due to this process for a one-dimensional system both numerically and (in an average sense) using percolation theory. The two calculations are in excellent numerical agreement and reproduce the general features of the experimental data.

1. Introduction

There have been a number of investigations of channel conductance in one-dimensional MOSFETS in recent years which are believed to involve incoherent hopping between localised states [1-4]. More data on similar systems will be reported at this meeting. In this paper we outline the theoretical understanding of this conduction process and briefly describe its relationship to hopping in a 2D system which has also been studied in MOSFETS [5,6].

The 1D model is described in Section 2 in which we formulate the rate equations for the occupancies of the localised states. When these equations are linearised in the presence of a weak electric field the result is a set of Kirchhoff equations for a conductance network. The conductivity of the system is the conductivity of this network. We may calculate this in two ways. Firstly, by simulating the network on a computer and executing a direct numerical solution of the equations [7]. We describe results obtained in this way in Section 3. The second way to proceed is to recognise that the average conductivity takes the form $\sigma=\sigma_p \exp(-s_p)$ where approximate analytical expressions may be written for the percolation exponent s_p [8,9,10] and the prefactor σ_p [10]. We take this approach in Section 4. We are able to test the analytical expressions for s_p against numerical calculations of this quantity made by Serota et al [8]. Excellent agreement is obtained with the adjustment of one parameter. We may also test the formula $\sigma = \sigma_p \exp(-s_p)$ against average results obtained from the numerical calculations reported in Section 3. The agreement obtained is good considering the simplicity of the model. Finally, in conclusion, we describe the 1D → 2D cross-over effects when the 1D chain of hopping sites is widened into a broad strip as described by Xie and Das Sarma [12].

Springer Series in Solid-State Sciences Vol. 83: **Physics and Technology of Submicron Structures**
Editors: H. Heinrich · G. Bauer · F. Kuchar

2. The Model

We consider a collection of sites strung out on a straight line of length L. The position of the ith site (i = 1,2,...N) we denote by x_i and the energy of an electron occupying that site we write as ε_i. It is supposed that x_i is equally likely to be anywhere on the line and that ε_i is distributed with a probability density $p(\varepsilon_i)$ which is the same for all sites.

The dynamics of the system is described by a transition rate R_{ij} for an electron occupying site i to jump to an empty site j. (To keep the analysis simple we suppose that the Coulomb repulsion is large enough to guarantee that only one electron can occupy each site.) Then, with f_i denoting the occupation probability of site i, we have the rate equations

$$\frac{df_i}{dt} = \Sigma\ [f_j(1-f_i)\ R_{ji} - f_i(1-f_j)R_{ij}]\ . \tag{1}$$

In thermal equilibrium we have

$$f_i = \overset{o}{f_i} = [1 + \exp\ \{(\varepsilon_i-\varepsilon_F)\}/k_BT]^{-1} \tag{2}$$

where the effective Fermi level ε_F includes a term $k_BT\ell n2$ to allow for spin degeneracy. The thermal equilibrium rates satisfy the detailed balance relation

$$\overset{o}{R}_{ij}/\overset{o}{R}_{ji} = \exp\ [(\varepsilon_i-\varepsilon_j)/k_BT]. \tag{3}$$

Our concern is with the case when a weak electric field E is applied so that an external potential eEx_i appears at site i. To linearise (1) we write

$$f_i = \overset{o}{f_i} + \overset{1}{f_i} \tag{4}$$

and replace (3) by

$$R_{ij}/R_{ji} = \exp\ [(\varepsilon_i + eEx_i - \varepsilon_j - eEx_j)/k_BT] \tag{5}$$

$$\simeq \frac{\overset{o}{R}_{ij}}{\overset{o}{R}_{ji}}\ [\ 1+ \frac{eE(x_i-x_j)}{k_BT}\ ]\ . \tag{6}$$

Then we find that the linearised equations may be put in the transparent form

$$C_i \frac{d}{dt}\ [V_i + Ex_i] = \sum_j g_{ij}\ (V_j-V_i) \tag{7}$$

where

$$C_i = \frac{e^2}{k_BT}\ \overset{o}{f_i}\ (1-\overset{o}{f_i})\ , \tag{8}$$

$$g_{ij} = \frac{e^2}{k_BT}\ \overset{o}{f_i}\ (1-\overset{o}{f_j})\overset{o}{R}_{ij}\ , \tag{9}$$

and we have written

$$V_i = -[Ex_i + (k_BT/e)\ f_i^1/f_i^o(1-f_i^o)] \ . \qquad (10)$$

We recognise (7) as Kirchhoff's equations for a network in which site i has a voltage V_i and is connected to ground through a capacitance C_i and a voltage generator Ex_i and is connected to site j through a conductance g_{ij}. The conductance matrix g_{ij} is symmetrical because of detailed balance.

Equations (7) become particularly simple in the dc case. They reduce to Kirchhoff's equations for a conductance network. The normalisation of the voltage in eq. (10) has been chosen so that the conductivity of the hopping system is identical to that of this conductance network [11].

The behaviour of g_{ij} is dominated by the energy exponentials in the Fermi functions and the Bose factor (which describes the availability of phonons) in R_{ij}. Moreover, because the sites are spatially separated, there is an overlap factor in R_{ij} which describes how fast this function falls off with increasing spatial seperation between the sites. A form for g_{ij} introduced by Ambegoakar et al. [13] has been found to provide an adequate description of most hopping systems:

$$g_{ij} = g_o e^{-s_{ij}} \qquad (11)$$

where

$$s_{ij} = 2\alpha|x_i - x_j| + q\ (\varepsilon_i, \varepsilon_j)/k_BT \qquad (12)$$

with

$$q\ (\varepsilon_i,\ \varepsilon_j) = \tfrac{1}{2}\ [\,|\varepsilon_i - \varepsilon_F| + |\varepsilon_j - \varepsilon_F| + |\varepsilon_i - \varepsilon_j|\,] \ . \qquad (13)$$

In (11) g_o is a characteristic conductance which is determined by the strength of the electron-phonon interaction and α^{-1} is a localisation length. We use this equation throughout the following analysis.

3. Numerical Solutions of Kirchhoff's Equations

In the dc limit eq. (7) reduces to

$$\sum_j g_{ij}\ (V_j - V_i) = 0 \ . \qquad (14)$$

We have solved these equations for 1000 sites subject to the boundary condition that $V_i=0$ at one end of the line and V_i = EL at the other. The evaluation of the conductance is then trivial [7]. We show typical results as a function of the Fermi level position in Figs. 1 and 2 for α^{-1} = 1000Å with the sites distributed over a line of length 20,000Å. The site energies were selected randomly from an interval W=1meV. In the Figures $\sigma^*=\sigma/\alpha L^2 g_o$ is a reduced conductance and $T^*=k_BT/W$ is a reduced temperature which takes the values 1,2,4 and 8×10^{-3} (i.e. T ≈ 10, 20, 40, and 80 mK) as the structure in the curves diminishes. The straight lines on the graphs have slopes $(k_BT)^{-1}$ and serve to indicate the origin of the structure in the energy dependence of the transition rates. As ε_F varies, different conductances dominate the overall conductivity and, because of their exponential variation, produce the large fluctuations shown at low temperature. Raising the temperature smooths out the structure as is illustrated by the curve with T=80mK. The general behaviour shown in these curves is similar to that observed in references 1 to 3. Lee [14] has

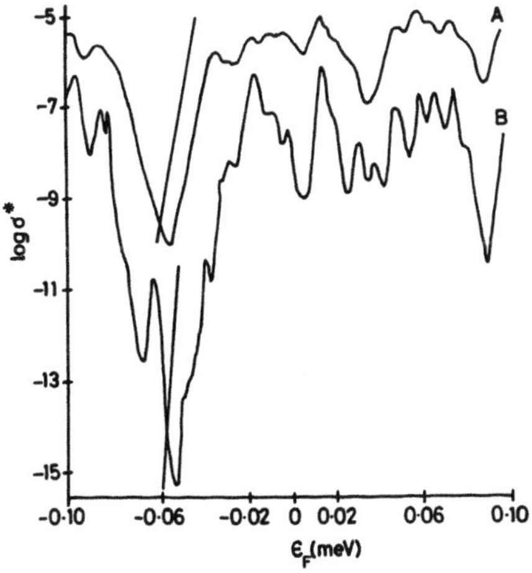

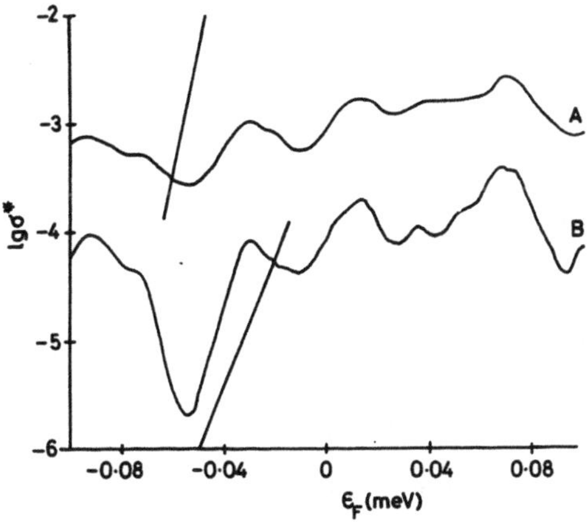

Figure 1 Log σ^* against chemical potential for (a)$T^* = 0.002$ and (b) 0.001.

Figure 2 Log σ^* against chemical potential for (a)$T^* = 0.008$ and (b) 0.004.

calculated similar curves by using a percolation argument. He supposes that σ is dominated by the conductance such that removing any larger conductance will break the chain. The results are similar but the graphs consist of straight line sections including flat-topped parts which are not observed.

4. Percolation Theory

The application of percolation theory to the 1D hopping problem began with the work of Kurkijarvi [15]. His ideas were extended by Serota et al [8]. Butcher and McInnes [9] presented a modified theory in which the exponent s_p in the average conductivity exhibits correct scaling behaviour. In this Section we give a more rigorous derivation of the equation for s_p [10].

To determine s_p in the average conductance $\sigma=\sigma_p\exp(-s_p)$ we remove from the network all conductances g_{ij} for which s_{ij} in eq. (11) is greater than a chosen value s_o. Then we reduce s_o until the network just breaks apart. s_p is identified with the critical value of s_o at which this happens. To estimate s_p when s_{ij} is given by eq. (12) we first remove from the network all sites which are more than $s_o k_B T$ away from the Fermi level because, for these sites, $q/k_BT > s_o$ and they cannot be connected to any other sites however close. The spatial density of the remaining sites is $n=2\rho_F k_B T s_o$ where ρ_F is the density of states at ε_F and they are distributed randomly in space over the length L of the line. We now remove the remaining conductances for which $s_{ij} > s_o$ and seek the value of s_o for which a break finally occurs. We suppose, with Serota, Kalia and Lee [8] that the break will occur at the pair of nearest neighbour sites on the line for which $|x_i-x_j|$ is a maximum. Serota, Kalia and Lee [8] estimate the system averaged value of this quantity to be

$$d_{max} = \frac{1}{n} \ln (nL) . \quad (15)$$

We have checked this relation numerically and it is accurate to a few percent.

We see from these considerations that the percolation threshold is determined by

$$s_p = 2\alpha\, d_{max} + \overline{q}/k_BT \qquad (16)$$

where $\overline{q}$ is an appropriate average of $\overline{q}$. Serota, Kalia and Lee put $\overline{q}=0$ which is certainly an underestimate. However, when $\overline{q}$ is set equal to 0 in eq. (16) and d_{max} is substituted from eq. (15) with $n = 2\rho_F k_B T s_p$ the result is an equation for s_p which is easily solved. In [9] Butcher and McInnes also put $\beta\overline{q}=0$ but improve the analytical and numerical solution of the resulting equation. However, the final formula for s_p is in poor agreement with the numerical percolation data of Serota et al [8]. Here we estimate an upper bound for $\overline{q}/k_BT$ which leads to the consistent introduction of a scaling parameter K with an upper bound of 1.73. Excellent agreement with the numerical percolation data is obtained when K=1.69.

The energies ε_i and ε_j of the two sites i and j at which the break occurs specify a point $(\varepsilon_i,\varepsilon_j)$ lying in the region of the $(\varepsilon_i,\varepsilon_j)$ plane for which $\overline{q}/k_BT < s_p$. If we assume that the point is equally likely to lie anywhere in this region, then an elementary integration yields $\overline{q}/k_BT = 2s_p/3$ so that eq. (16) becomes

$$s_p = K^2 2\alpha d_{max} \qquad (17)$$

with $K=\sqrt{3}$. However, we have certainly overestimated $\overline{q}/k_BT$ and we may expect the correct value of K to be less than $\sqrt{3}$.

When d_{max} is substituted in the eq. (17) from eq. (15) with $n=2k_BT\rho_F s_p$ we may easily manipulate the resulting equation for s_p into the useful form

$$s_p = Kx\, H(z) \qquad (18)$$

where, as in [15], $x=(T_0/T)^{\frac{1}{2}}$ with $k_BT_0=\alpha/\rho_F$ and H(z) is determined by the equation

$$H(z) = [\ln z + \ln H(z)]^{\frac{1}{2}}, \qquad (19)$$

which is readily solved by iteration. In these equations

$$z = Ky/x \qquad (20)$$

where $y=2\alpha L$.

The full curve in Fig. 3 shows H(z) as a function of $\sqrt{z}$. The triangular points were obtained from the numerical percolation data of Serota et al.[8] by writing

$$H(z) = \frac{\langle\langle \ln\rho\rangle\rangle_{med}}{K\sqrt{20}} \qquad (21)$$

and $z = K2\alpha L/\sqrt{20}$. (22)

These relations are appropriate because, in the calculations of Serota et al., $x = \sqrt{20}$ and $\langle\langle \ln\rho\rangle\rangle_{med}=s_p$. Consequently eqs. (21) and (22) follow immediately from (17) and (20). In placing the points on Fig. 3 K has been adjusted to 1.69 to give a least squares best fit.

A general formula for the prefactor σ_p in the average conductivity is given in [11] for a 3D system. It is obtained by summing the contributions to σ from all the conductances with values in the neighbourhood of the

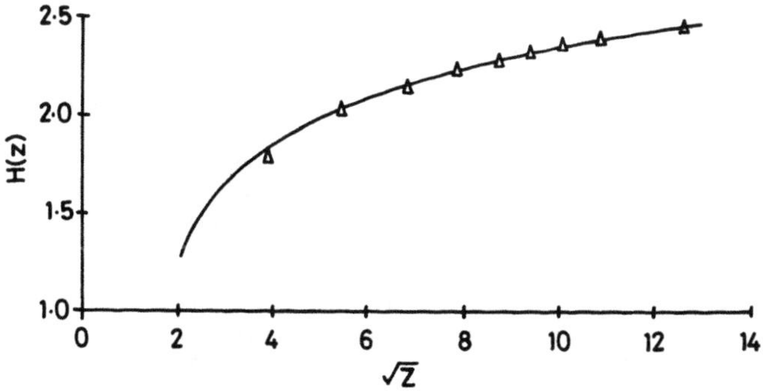

Figure 3
A comparison of H(z) in eq.(19) with values derived from numerical values of s_p [8].

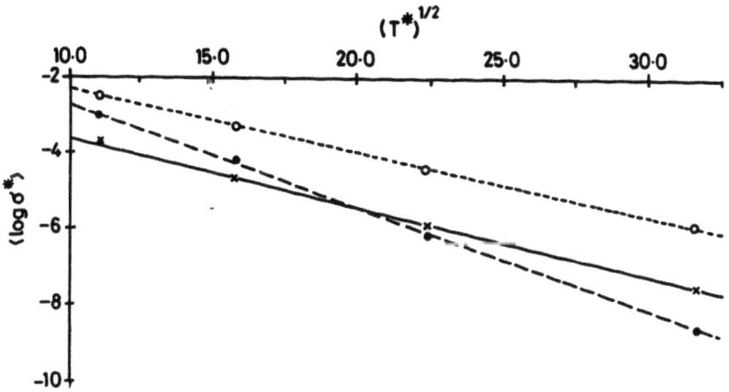

Figure 4 Plots of $\langle \log\sigma^* \rangle$ against $(T^*)^{\frac{1}{2}}$. From Kirchhoff's equations: • . From percolation formulae including the prefactor: × Contribution from the percolation exponent: o.

critical conductance $g_o \exp(-s_p)$. The argument is easily modified to handle a 1D system. Thus we find that

$$\sigma_p = 2g_o\rho_F^2 \int_0^{s_p/2\alpha} x^2 A'(s_p-2\alpha x)dx, \tag{23}$$

where $A'(q) = (k_BT)^2 6q$. Evaluation of the integral in eq. (23) gives

$$\sigma_p=(g_o/8\alpha)[s_p/x]^4. \tag{23}$$

We may use this formula, together with the analytical values of s_p plotted in Fig. 3 to calculate $\sigma=\sigma_p \exp(-s_p)$ which is our theoretical estimate of the average conductivity. In Fig. 4 the analytical results (crosses) are compared with numerical data obtained by McInnes and Butcher in [7] (points) in the form of a plot of $\langle\log(\sigma/\alpha L^2 g_o)\rangle$, where the angular brackets indicate an average over Fermi level, against $T^*=k_BT/W$ with W=1meV. The result is reasonably satisfactory. However, the slope of the numerical line is 50% larger than that of the analytical line.

5. Conclusion

The open circles in Fig. 4 show the temperature variation of s_p alone taken directly from Fig. 3. We see that it is responsible for most of the temperature variation exhibited by the average conductivity which is little affected by the weak temperature dependence of σ_p shown in eq. (23). To remove the slope discrepancy would seem to require the assumption that the exponent in the hopping conductivity formula is not s_p (since our theory correctly reproduces that) but is closer to 1.5 s_p. This has never been the case in other hopping systems but the 1D case is unusual because the calculated conductivity fluctuates at low temperatures over many orders of magnitude. In solving Kirchhoff's equations we have smoothed these oscillations by averaging over the position of the Fermi level. On the other hand, the theoretical conductivity has always been considered to be an average over an ensemble of systems of length L. These two averages are different and we suggest that this difference may be the source of slope discrepancy exhibited in Fig. 4.

We see that the temperature dependence of s_p takes the form $(T_o/T)^{\frac{1}{2}}$ as would be predicted by a simple scaling argument for a 1D system [9] with T_o weakly dependent on L [8,9,10]. For a 2D system the same type of argument yields $s_p \sim (T_o/T)^{1/3}$ with T_o independent of L. The change of index from 1/2 to 1/3 is well understood. So too is the change in the significance of L in the evaluation of T_o. The ability to bypass difficult hops profoundly influences the 2D behaviour and, conversely, the inability to do so is very significant in 1D systems. Finally, the absolute magnitude of the fluctuations become weaker in a 2D system. These dimensional cross-over effects have been calculated using percolation arguments by Xie and Das Sarma [12].

Acknowledgement

The authors are indebted to Alan Fowler for his interest in this work and to the IBM Thomas J. Watson Research Laboratories for financial support.

References

1. A.B. Fowler, A. Harstein and R.A. Webb, Phys. Rev. Letts. 48, 196 (1982).
2. R.F. Kwasniak, M.A. Kasiner, J. McIngalis and P.A. Lee, Phys. Rev. Lett. 552, 224 (1984).
3. R.A. Webb, A. Harstein, J.J. Wainer and A.B. Fowler, Phys. Rev. Letts. 54, 196 (1985).
4. J.J. Wainer, A.B. Fowler and R.A. Webb, Proceedings of the Seventh Conference on the Electronic Properties of 2D Systems (to be published in Surface Science).
5. G. Timp, A.B. Fowler, A. Harstein and P.N. Butcher, Phys. Rev. B. 33, 1499 (1986).
6. G. Timp, A.B. Fowler, A. Hartein and P.N. Butcher, Phys. Rev. B. 34, 8771 (1986).
7. J.A. McInnes and P.N. Butcher, J. Phys. C: Solid St. Phys. 18, L921 (1985).
8. R.A. Serota, R.K. Kalia and P.A. Lee, Phys. Rev. B. 33, 8441.
9. P.N. Butcher and J.A. McInnes, Proceedings of the Seventh Annual Conference on the Electronic Properties of 2D Systems (to be published in Surface Science).
10. P.N. Butcher and J.A. McInnes, to be published in J. Phys. C: Solid St. Phys. (1988).
11. P.N. Butcher, K.J. Hayden and J.A. McInnes, Phil. Mag. 36, 19 (1977).
12. X.C. Xie and S. Das Sarma, Proceedings of the Seventh Conference on the Electronic Properties of 2D Systems (to be published in Surface Science).
13. V. Ambegaokar, B.I. Halperin and J.S. Langer, Phys. Rev. B 4, 2612, (1971).
14. P.A. Lee, Phys. Rev. Letts. 53 20 42 (1984).
15 J. Kurkijarvi, Phys. Rev. B. 8, 822 (1973).

Part V

Submicron Devices: Physics and Applications

Transport Physics in Ultra-Submicron Devices

D.K. Ferry

Center for Solid State Electronics Research,
Arizona State University, Tempe, AZ 85287, USA

Abstract. Technology is rapidly pushing to the fabrication of ULSI "chips", in which individual devices have critical dimensions (e.g. gate lengths) well below 1 μm. Within the next decade or so, we will see the push to reduce these dimensions even further; to well below 0.1-0.2 μm. At this scale, many new and interesting effects, which are currently treated as only high order corrections, will begin to dominate the device performance. Several of these effects are tunneling, surface superlattices, ballistic transport, and quantum interference effects. In this paper, we will treat a few such effects, leaving to other papers in this volume the treatment of further effects. Finally, we treat the need to develop a conceptual non-equilibrium quantum transport theory to handle these problems in realistic device structures.

1. Introduction

Since the introduction of integrated circuits in the late 1950's, the number of individual transistors that can be placed upon a single circuit has approximately doubled every three years. Today, even university design laboratories for the teaching of students can access chip foundries which produce 1.2 μm design rule circuits. Compared with this, many commercial companies are experimenting with the production of chips with critical dimensions of 0.1 μm, and university laboratories have produced individual devices with gate lengths much smaller than this [1,2]. The creation of devices whose spatially important scales may be only a few tens of nanometers opens the door to the study of many new and important physical effects, some of which have been described earlier [3]. Indeed, it can rightfully be said that it will be impossible to understand fully the operation of these devices without a full understanding of these newly appearing physical effects.

What is happening in the reduction of individual feature sizes of a transistor, used as the basic building block for ULSI, is that the critical length (e.g. the gate length or a depletion length) is becoming so small that it is approaching the *coherence length* of the electrons that provide the operation. Over the past several years, it has become evident that this latter length is not the wavelength of the electron itself, but the *inelastic mean free path*, or the length over which the energy coherence is maintained by the electron. With modern modulation doping techniques in heterojunction device structures, this latter length can be more than 1 μm at low temperatures. The consequence is that such small devices must now be treated as quantum mechanical objects, and many phenomena become important that have never been treated in the normal classical and semi-classical treatments of semiconductor devices. While this has served to invigorate studies of quantum

Springer Series in Solid-State Sciences Vol. 83: **Physics and Technology of Submicron Structures**
Editors: H. Heinrich · G. Bauer · F. Kuchar

behavior in device structures, we are limited in that many of these quantum phenomena are only poorly understood at best.

In this paper, a number of quantum effects that are important in devices are reviewed. No comprehensive treatment of each, nor a comprehensive treatment of all effects, is intended. Rather, the selection is governed by those effects which have been shown to already occur in devices, and the fact that others are treated in greater depth by other papers in this volume. Rather, here we try to establish the connection between a few such effects. Nearly all semiconductor devices operate on the principle of hindering the transport of carriers from the source (or emitter) to the drain (or collector) by the presence of a potential barrier, which is modulated by the gate (or base potenial). As the size of devices has been reduced, so-called second-order effects have introduced unintended modifications to this barrier through parasitic effects such as drain-induced barrier lowering [4] (see Fig. 1a). In the ultra-submicron regime, we must begin to consider that many carriers will actually tunnel through the barrier, and encounter strong potential variations (Fig. 1b), further changing the basic operation of the device. In addition, the active channel length can be much less than 0.1 μm after the barriers have been surpassed. Carriers then have the possibility of transiting this region *ballistically* ; e.g. without scattering, or perhaps suffering just a few elastic collisions. Then we can expect to see quantum effects and quantum resonances in this ballistic transport, such as that seen in transport through thin oxides in MOS devices [5,6].

We treat first the problem of tunneling through the case of the resonant tunneling diode. The ideas of lateral surface superlattices, which produce electron localization in arrays of quantum boxes, and the interbox interactions are covered. In each of these two areas, the role of *ballistic* transport arises, and we then give coverage of this area. We mention only briefly the extension of the ballistic ideas to

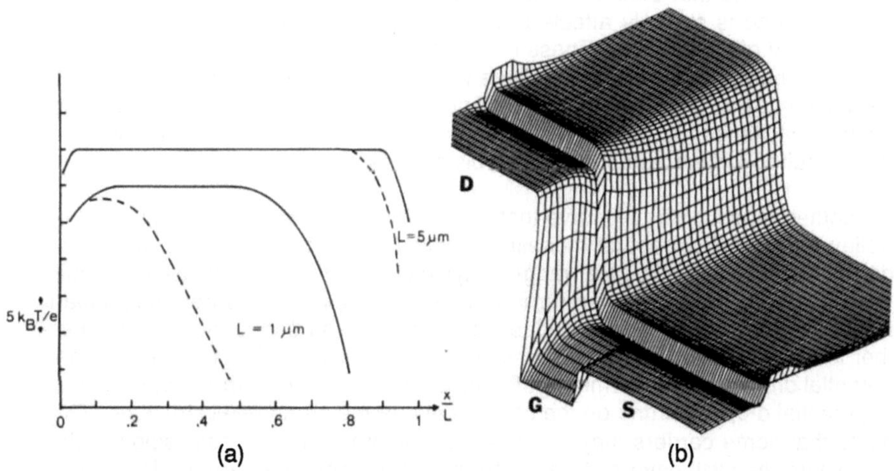

(a) (b)

Fig. 1 The barrier and rapidly varying potential in the channel region of a MOSFET. In (a), the effect of the drain potential on barrier lowering is illustrated, while in (b) we illustrate the strongly varying potential.

the appearance of coherent quantum interference phenomena in low dimensional systems, as there are several excellent papers on this topic included in this volume. Finally, we discuss a major problem that is limiting the full understanding of the quantum effects in real devices, and that is that *there is no adequate theory of quantum transport in inhomogeneous, far-from-equilibrium systems with complicated scattering processes.*

2. Tunneling Devices

While tunneling is included in nearly all introductory texts on quantum mechanics, the full behavior of an actual tunneling device is not nearly understood. The current device which is being studied for its microwave applications, and for the insight into the physics it provides, is the resonant tunneling diode [7,8]. With modern molecular beam epitaxy, the GaAs/AlGaAs system of materials allows one to grow thin barrier regions of AlGaAs and to incorporate a thin quantum well of GaAs between the barriers. It has become apparent that the negative differential conductance (NDC) that exists in the current-voltage curve exists to quite high frequencies [8]. Recent experiments with the resonant tunneling diode often have observed a bistability in the NDC regime [8-10]. It has been postulated that this bistability is the result of charging/discharging the bound states in the well, and/or the localized states just outside the well, although many claim that the bistability is circuit induced. The source of the bistability is controversial and still is being argued. In Fig. 2a, the i-v curve for a typical structure is shown. The bistability usually gives rise to two distinct hysteresis loops as is indicated here. In Fig. 2b, the calculated i-v curve for the first fully quantum mechanical, self-consistent calculation of the i-v curves is shown [11] (the theoretical basis for these curves is discussed below). These latter results clearly indicate that there is one narrow region of hysteresis that extends throughout the NDC region. The NDC slope is quite shallow and is strongly affected by both the doping in the cladding regions and the mobility of the carriers. These results suggest that the *observed* i-v curves depend upon both the intrinsic bistability and the external circuit that produces the measured curves. This is further supported by the fact that there is a large switching capacitance associated with the tunnel barrier that can resonate with the external circuit to induce further observed bistabilities.

Another aspect of the i-v curve that is not fully understood is the magnitude of the valley current. Conceptual modeling predicts peak-to-valley ratios of several hundred, while the experimental range is typically 3:1 for good devices at 300K and perhaps 10:1 at low temperatures (low peak-to-valley ratios generally mean high valley currents due to extraneous transport such as sequential tunneling or another leakage mechanism, as discussed below). In our model [11], we find that the potential drop is non-uniformly distributed across the device structure, with most of the potential drop occurring on the cathode end of the barrier-well structure. This suggests that some carriers may travel ballistically from the cathode region to the barrier, where they will have sufficient energy to overcome the barrier. This would then explain the high valley current as ballistic injection over the barrier. If this is indeed the case, then an undoped region near the barrier would cause more potential to be dropped near the barriers, thus reducing the ballistic contribution to

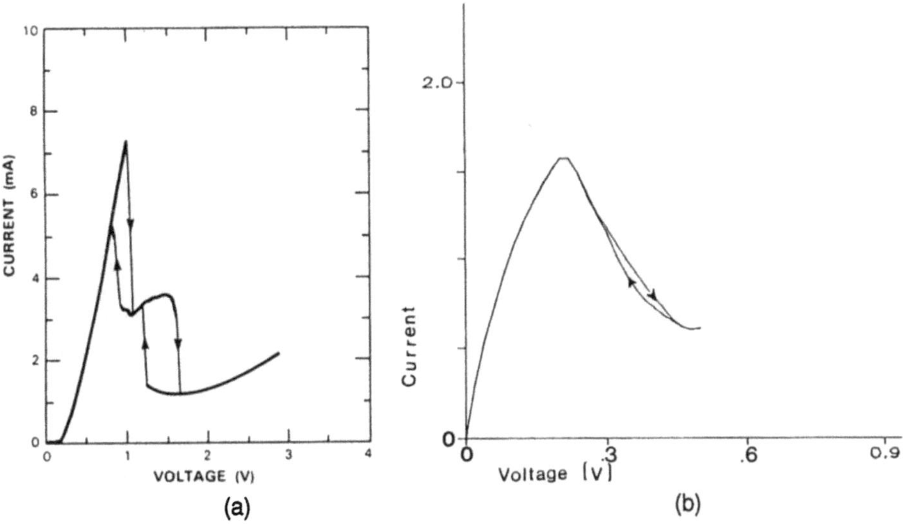

Fig. 2 (a) An experimental i-v curve for a resonant tunneling diode that shows bistability. (b) The theoretical curve, calculated by Wigner function techniques, for a similar diode, which shows the intrinsic bistability.

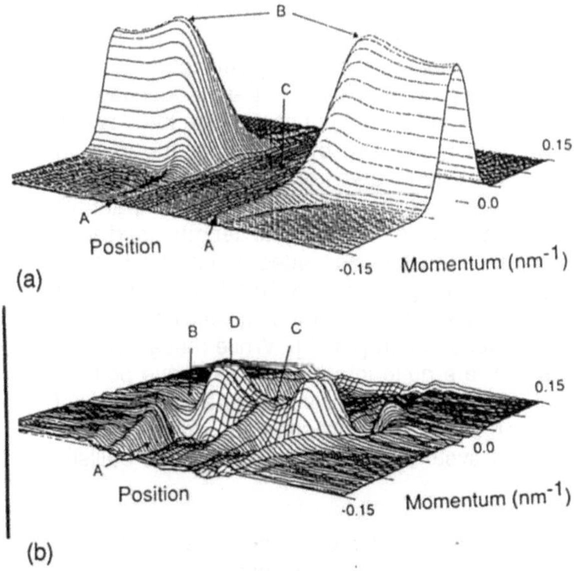

Fig. 3 (a) The equilibrium distribution function, and (b) one with slightly forward bias, from which the equilibrium one has been subtracted.

the valley current. Indeed, experimental devices in which improved peak-to-valley ratios are observed usually include these undoped regions adjacent to the barrier layers.

At low currents in tunneling barriers, the conductance is often much higher than expected for a thermal Fermi-Dirac distribution function [12,13]. One possible cause of this "zero-bias anomaly" is shown in Fig. 3. Here, the equilibrium Wigner distribution is shown in Fig. 3a. The shoulders of the distribution (A) are decidedly non-thermal in nature [14]. There is also a depletion near k=0 at the barrier (B). We find [15] that the initial current comes from carriers in the shoulders of the distribution, which either tunnel through the barriers with high momentum or pass over the barriers. At higher bias levels, the shoulders are depleted, and the depletion at k=0 is filled in by the reservoirs, so that current is carried by carriers tunneling from a thermal distribution (Fig. 3b).

3. Ballistic Transport

Ballistic transport is basically the motion of carriers over a distance for which they do not lose energy by collisions, either with phonons or with other electrons. It appears in most devices in two guises, either as the onset of velocity overshoot in devices in which electrons can be accelerated past their stationary velocity, or in decay of the energy of carriers injected into a semiconductor region with a velocity well above its stationary value (we saw this above as a source of carriers contributing to the valley current). In either case, we recognize that the critical distance over which the ballistic transport exists is the energy relaxation length, which is the same as the inelastic mean free path mentioned in the introduction. In the former guise, it has been sought for many years, but not seen since the devices investigated were too large. This was finally realized by studies of the second guise, in which actual ballistic diodes were fabricated for this purpose [16,17]. From these studies, it was readily apparent that the inelastic mean free path for electrons in GaAs was only 50-60 nm at 300 K. If the gate length used in a transistor to study this effect were longer than this, the effect would be heavily masked by charge redistribution (regions with high velocity would have low carrier densities and conversely). Only recently have devices with sufficiently short gate lengths been fabricated to observe this phenomenon [18,19]. While these devices, and the ballistic diodes mentioned above, are quite interesting, they have pointed out that if we are to study the decay of ballistic electrons, we require a fuller study of the transient details of the electron-phonon interaction to explain their characteristics, yet they are not the best system in which to probe such ballistic transport.

In addition to the special devices mentioned above, carriers passing through (or over) the barriers of Fig. 1 can travel ballistically in the channel. In general, we are interested in the manner in which these carriers lose energy through inelastic processes. There are few ways in which to study the time-dependence of inelastic processes, but one such way is the decay of laser-excited electron-hole plasmas. Lasers now exist with pulse lengths as short as 8 fs, and these lasers have been used to create electron-hole plasmas in semiconductors. The importance of this is that time measurements of the plasma decay can now be carried out with a temporal resolution on this same scale. It is thus possible to probe the detailed dynamics of the electron-phonon interaction on the sub-0.1 ps scale. Needless to say, the results remain somewhat controversial [20-22] both in the interpretation of experimental results and in theoretical understanding. Here, two particular aspects will be considered as they will highlight where significant questions and interesting phenomena occur.

In Fig. 4a, we show measurements from the Rochester group [23], in which an applied electric field is used to accelerate the laser generated carriers. In one curve, we note that the carriers are actually decelerated (to negative velocities) just after the pulse. The understanding of this effect came only upon the conclusion of detailed ensemble Monte Carlo (EMC) calculations by the ASU group to model the experimental situation. Carriers are created in the Γ-valley with energies near the L-valley minimum by the laser excitation. The field then accelerates those electrons with positively directed momentum, while it decelerates those with negative momentum. Those that gain energy (in the positive momentum space) are scattered to the L-valleys, leaving a distribution in the Γ-valley *which is preferentially centered in the negative momentum space.* Only when these carriers have been accelerated through zero does the current begin to rise. This is the classic Jones-Rees effect, normally encountered in the return of carriers to the central valley after inter-valley scattering, and emphasizes the need for fully coupling experimental and theoretical efforts to understand the physics on this short time scale. The applications to devices are clear. Electrons tunneling through (or injected over) the barriers of Fig. 1 may be randomized in momentum, particularly if an inelastic process accompanies the injection. Their response to applied fields is typical to that described here, and this Jones-Rees effect can retard the switching behavior of real devices.

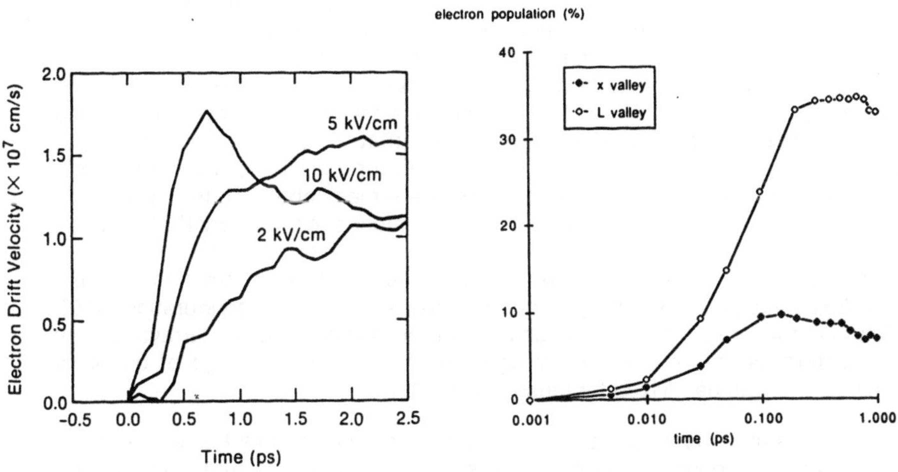

(a) (b)

Fig. 4 (a) Experimental measurements and theoretical calculations of the "Jones-Rees" effect, discussed in the text. (b) Populations of carriers in the satellite valleys after a 20 fs laser pulse, showing the delay in population build-up.

In Fig. 4b, we show another feature of EMC calculations. Here, we plot the populations of carriers in the L- and X-minima after excitation by a 20 fs laser pulse with a photon energy of 2 eV. We note that the carriers do not peak in the satellite valleys until some 0.3-0.5 ps after the laser pulse (such effects are also seen experimentally). The impact is that the carriers are sitting in the central valley localized in the excitation phase volume, and not being scattered to lower energy states by intra-valley LO phonon scattering, which has a considerably longer scattering time. By varying the excitation intensity, we can now study the role of e.g. carrier-carrier scattering as a primary relaxation process. All of these calculations use a semi-classical format, with the scattering processes being evaluated by standard Fermi golden rule techniques. Yet, we are dealing with time scales here that are near the reciprocal of the phonon frequency. A proper treatment of these effects on this time scale requires a more fully quantum mechanical calculation, including the temporal build-up of individual scattering processes, as well as the quantum interference of the individual processes. The importance to devices is obvious: if we are building devices with gate lengths shorter than the inelastic mean free path (even at 300 K), then we must begin to better understand how inelasticity occurs for the carriers.

4. Lateral Superlattices

In very dense arrays of next generation ULSI chips, the devices will appear as if they are a superlattice of structures imposed upon the basic silicon material. We are already facing near-neighbor interactions which change the basic behavior of the architecture through the failure of the partition principle [24]. While such behavior is quite detrimental to the standard mode of operation of these chips, it also opens the door to creation of new, massively parallel methods of operation through the use of cooperative modes of the lateral surface superlattice [25]. While it is easy to think of using the cooperative effects in such an array, it is more difficult to actually carry out this task. One reason is that there is no useful theory for achieving this task. There have been many attempts to apply *neural network* paradigms to this task [26,27], but this is not the best way. The various VLSI chips are characterized by an information dimension that is less than 2 [28], which determines basically the interconnect density and the chip pin-out requirements. On the other hand, neural architectures are three dimensional. Thus, if we are to control the cooperative effects in ULSI arrays, we must seek a theory that is based upon a low information dimension, such as iterated arrays or cellular automata [29]. Unfortunately, only in a very few cases do there exist standard procedures to map a desired functional process, or algorithm, into these hardware configurations [30]. The future of this area, that of highly concurrent processing in device arrays, is very open, as there is little current study going on to control the range of quantum interactions that can be utilized for these effects [31].

Consequently, many groups are now studying coupling mechanisms in such arrays through the ideas of quantum boxes in a variety of structures such as we have suggested [32,33]. The incorporation of such a lateral superlattice into a conventional device structure is discussed elsewhere in this volume [34]. The idea of these quantum boxes is that the electrons can be localized in the boxes by potential barriers (several one-dimensional approaches are described elsewhere in this volume). In Fig. 5, we illustrate one growth approach that has been proposed and a gate structure for a gated method of producing the lateral superlattice. In these cases, the superlattice is developed when the carriers lying in the localized quantum boxes can sense the periodicity and interact with each other.

In condensed matter physics, the inelastic mean free path, or phase-breaking length, arises in yet another context in regard to the field of localization. Here, the localization length describes the extent of the phase coherence of an electron in a random network, where the randomization can be provided either by the use of random site energies, such as in a random alloy on a crystalline lattice, or in randomization of lattice and bond lengths, such as in a truly disordered material, *or in our quantum boxes.* The localization phenomena appear differently in various dimensional structures, but it is the behavior responsible for an important phenomenon -- universal conductance fluctuations, which are discussed in great detail in several other papers in this volume. In systems which have constrained lateral dimensionality, and in which the length of the major conduction path is smaller than the important phase-breaking length, the conductance exhibits fluctuations, with regard to variations of the chemical potential or a magnetic field, of the order of e^2/h. The fluctuations themselves are external characterizations of internal quantum interference of the phase variations of a few possible

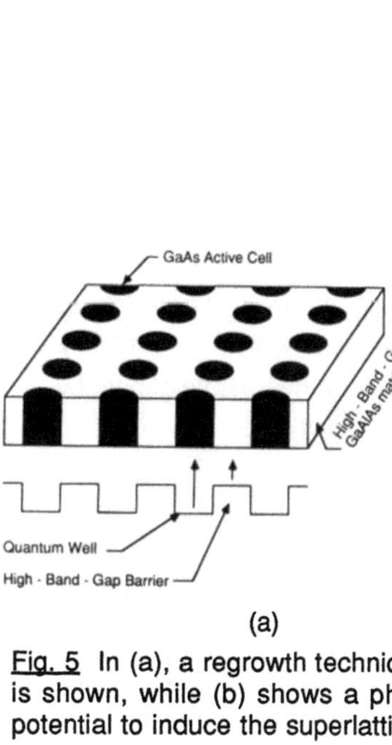

(a)

(b)

Fig. 5 In (a), a regrowth technique for embedding GaAs boxes in a GaAlAs matrix is shown, while (b) shows a photomicrograph of a gate metallization for using a potential to induce the superlattice.

conductance paths, which in turn relate to the various "channels" for conduction. As the chemical potential is varied, or as an external magnetic field is varied (which also shifts individual site energies), various channels are switched on or off, changing the phase interference effects within the sample. Indeed, this phase interference is very much related to a comparable interference effect -- the Aharonov-Bohm effect, in which conducting rings, which enclose a magnetic field, show quantum interference effects in the end-to-end transmission provided once again that the ring dimensions are small compared to the phase-breaking length.

It was recognized that the inelastic mean free path is quite long in inversion layers induced in semiconductor structures because of the much lower electron density and resultant low level of carrier-carrier scattering. Narrow silicon inversion layers, induced in MOS structures were studied for the presence of universal conductance oscillations [35,36]. The development of modulation doping in GaAs-GaAlAs heterojunctions has produced inversion layers on the GaAs side of the heterojunction band discontinuity. Because of the smaller mass of the carriers in GaAs, and the separation of the conduction carriers from the ionized impurities, the mobilities can be exceedingly high in these inversion layers at low temperatures (even to a few times 10^6 cm^2/V-s at 4.2 K). Conducting rings have been fabricated in these modulation doped quantum well structures in order to study the Aharonov-Bohm effect [37,38]. Thin GaAs wires [39] and quantum well structures have been used to study the universal conductance fluctuations [40]. Finally, the high-mobility GaAs structure can support the lateral superlattice discussed above [18,34].

We can study the global transport properties of the quantum boxes in the extended state regime by considering the miniband nature of the lateral superlattice [41]. Then we may hope to see such effects as Bloch oscillations occurring in the transport. On the other hand, if the electrons become fully localized in the quantum boxes, with no long range order, we can try to study the conductance fluctuations mentioned above on a superlattice basis. That is, we can truly construct a realistic Anderson model with our superlattice basis. Localization phenomena can arise either from energy fluctuations on a well defined lattice or by random lattices. Moreover, we should be able to move from one regime to the other by varying the gate voltage.

The Anderson model treats the localized regime in a tight-binding Hamiltonian in which the individual site energies ε_i are assumed to be randomly distributed over a range [-W,W], and the sites are coupled to nearest neighbors by an overlap energy V_{ij}. This leads to the Hamiltonian

$$H = \sum_i \varepsilon_i a_i^+ a_i + \sum_{i,j} V_{ij} a_i^+ a_j ,$$

where a^+ and a are the creation and annihilation operators for an electron at site i or j. The ratio of the range of site energies W to the overlap energy V (W/V) gives a measure of the degree of disorder in the system. A numerical evaluation can be achieved by evaluating the Green's functions in the site representation and using these to evaluate the Kubo formula [42,43]. We chose to illustrate the latter, as it gives a straight-forward method to evaluate the universal conductance fluctuations and the results are expressible in terms of the Landauer multi-channel conductance formula as well. We consider a two-dimensional localized region of volume $\Omega=n_x n_y a^2$, where a is the spacing of the sites and x is the direction of current flow. Then, the conductance is

$$G = \frac{4e^2V^2n_x}{n_y h} Tr\{G''(j'-1,j-1)G''(j,j') + G''(j',j)G''(j-1,j'-1) - G''(j'-1,j)G''(j-1,j') - G''(j',j-1)G''(j,j'-1)\} ,$$

where G" is the imaginary part of the Green's function. To be able to apply this formalism to a stripe of localized region, we bound it with two perfect conductors to act as contacts. We then need to be able to compute the Green's functions anywhere in the sample. The stripe is composed of a finite disordered region and infinite ordered regions as leads on the left and right. We first calculate the Green's functions $G^L(j)$ for the left and G^R for the right semi-infinite perfect lattices. The Green's function at any other point is then built up by using a recursive method which constructs one column at a time from either the left or the right, as

$$G^L(j) = [G^o(j)^{-1} - |V|^2G^L(j-1)]^{-1} \ , \ G^R(j) = [G^o(j)^{-1} - |V|^2G^R(j+1)]^{-1} \ .$$

These then lead to

$$G(j,j) = \{G^o(j)^{-1} - |V|^2[G^L(j-1) + G^R(j+1)]\}^{-1}$$

$$G(j,j') = \left[\prod_{i=j}^{j'} G^L(i)\, V\right] G(j',j') \qquad \text{for } j' > j$$

$$G(j,j') = \left[\prod_{i=j}^{j'} G^R(i)\, V\right] G(j',j') \qquad \text{for } j' < j \ .$$

Results for magnetic field variation are shown in Fig. 6. The apparent periodicity arises from the microscopic loops of area a^2 that exist in this lattice.

The amplitude of the oscillations and fluctuations shown in Fig. 6 are quite sensitive to the level of the Fermi energy, but seem to be independent of the exact details of the model. For example, the same results are obtained if the magnetic field is confined only to the localized region, and not to the contacts. Again, the dominant fluctuation effect arises from the stochastic interference of many electron wave packet trajectories traversing the sample; i.e. the switching of channels in the multi-channel formulation. It is important to note here that the Green's functions fully incorporate all the transverse modes relevant to the multi-channel Landauer formula. While observing the periodic oscillations in magnetoconductance would require some 165 megaGauss, if *a* were 0.25 nm for a regular lattice, only 200 Gauss is required in the structures shown in Fig. 5, which is certainly within the realm possible now. We also note in Fig. 6, that the conductance goes to zero at

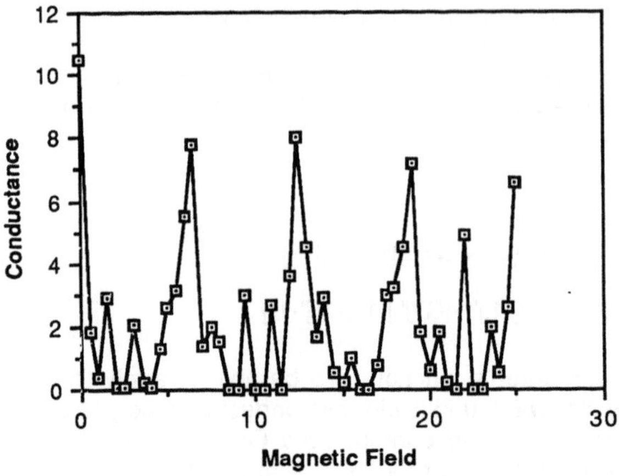

Fig. 6 The conductance, in units of e^2/h, is plotted as a function of the magnetic field, in units of $h/2\pi ea^2$. This plot is for values of W=2V, periodic boundary conditions, and a Fermi Energy E=V.

several values of the magnetic field, which is usually an indication of the quantum Hall effect. Here, we think that the structures are showing behavior similar to that calculated by Hofstadter [44], with a magneto-conductance showing fractal behavior.

5. Quantum Transport

When we are dealing with the transport in semiconductor devices, we are generally treating the response of a very inhomogeneous and far-from-equilibrium structure. In the last section, we treated the quantum transport at T=0 for a localized structure. On the other hand, the study of hot electrons in semiconductor devices is quite old, and is illustrative of the complications that arise due to the high electric fields in such structures. For example, a source-drain potential of 2 V across a 0.1 μm channel length leads to 200 kV/cm average field in the channel, which is a field near to that required for avalanche breakdown. When we now deal with devices that are even smaller, it is apparent that we must begin to study quantum effects in nonlinear, far-from-equilibrium structures that are characterized by time-dependent, inhomogeneous potentials and fields. Moreover, the system is open, with particle, momentum, and energy flow across the boundaries, and with boundary conditions that are ill defined. Thus, it should come as no surprize that this is an area in which quantum transport has made no inroads to date. To be sure, quantum mechanics in quasi-two-dimensional systems has been studied for quite some time, but in this case the quantization is normal to the oxide-semiconductor interface, while the transport is along the interface, and may still be treated by semi-classical means. We are now facing a need to treat the longitudinal carrier flow in these ultra-short quantum devices correctly, as experiments are providing data to show the occurrence of these quantum effects.

In treating the resonant tunneling diode in an earlier section, self-consistent solutions utilizing the Wigner distribution function were alluded to, and these solutions are among the first achieved for far-from-equilibrium, self-consistent quantum structures [11]. Including the potential response to the charge redistribution self-consistently means that we are now solving a dynamic nonlinear system, so that these results are important in this area. Yet, even here the electron-phonon interaction has been approximated and many-body effects ignored.

Generally, there are several approaches to treating transport in far-from-equilibrium systems. These differ mainly by the methods used to simulate the carrier population, and thus the solution vehicle that symbolizes the approach. By these, we mean that one can usually take one of the following representations of the basic wave functions:

- wave function $\psi(\mathbf{r},t)$,
- Green's function $G(\mathbf{r},t;\mathbf{r}',t')$,
- density matrix $\rho(\mathbf{r},\mathbf{r}',t)$,
- Wigner distribution function $F(\mathbf{r},\mathbf{p},t)$.

In fact, each of these differs only in the way in which the wave functions (and the consequent field operators) are incorporated into the basic description. The wave function approach is the most general, and Schrödinger's equation factors into each of the other methods. The Green's function is the appropriate propagator for the solution of this equation, and we return to it below. We treat first the density matrix and its kin, the Wigner distribution function.

The density matrix provides an abstract approach to quantum transport, and it has been used to develop a general discussion of the role that high electric fields play in modifying the electron-phonon interaction [45]. In this approach, which derives from the general derivation of the Prigogine-Resibois equation [46], a retarded quantum kinetic equation for the density matrix is obtained. This quantum kinetic equation can then be cast into a generalized integral equation, which in principle is the only general approach applicable to the far-from-equilibrium structures discussed here. However, the Barker-Ferry equation [47], as this integral equation is now known, has never been solved and few ideas on even where to start have been put forward.

An alternative approach is to use directly the Wigner distribution function, which is a Fourier transform on the off-diagonal elements of the density matrix:

$$F(\mathbf{p},\mathbf{r},t) = \frac{1}{2\pi\hbar}\int dy\, \exp(-i\mathbf{p}\cdot\mathbf{y}/\hbar)\, \rho(\mathbf{r}+\frac{1}{2}\mathbf{y},\mathbf{r}-\frac{1}{2}\mathbf{y})$$

and thus becomes a quantum distribution function of *both momentum and position.* It satisfies normal sum rules, in that the integral over all momentum yields the density distribution (magnitude squared of the wave function), while similarly the integral over all position yields a similar momentum distribution. The solutions shown in Figs. 2,3 were obtained using this approach. The Wigner distribution evolves in time according to the equation (in one dimension)

$$\frac{\partial F}{\partial t} + \frac{p}{m}\frac{\partial F}{\partial x} = \frac{1}{\pi\hbar}\int [V(x+\mu) - V(x-\mu)]\sin(k\mu/\hbar)F(x,p+k,t)d\mu dk \ ,$$

which clearly shows the nonlocal effects of the potential on the distribution. However, the above equation does not produce a unique F, since initial conditions must be specified, and these often contain a great deal of the quantum physics. Such an initial condition is shown in Fig. 3a, which is calculated from the Schrödinger equation with a scattering state basis. More than 64,000 basis states, weighted by a Fermi-Dirac distribution (at 300 K) at the boundaries, were used in this calculation which yields the density matrix itself. This latter is then transformed to the Wigner function as indicated above.

In Fig. 7, the Wigner distribution is shown for a Gaussian pulse impinging upon a tunneling barrier. Several effects are evident here. First, the Wigner distribution is not a positive definite quantity, so great care must be exercised in treating it as a probability function. The negative regions, of course, are related to the uncertainty principle and have a phase space extent whose size is related to the lack of ability to make measurements in such small regions. We also note that, in addition to the incident, transmitted, and reflected Gaussian pulses, there is significant strength in the Wigner function in the neighborhood of zero momentum. This strength is related to the correlations that exist in the problem. For longer times, when the individual pulses have broken away from the barrier, the correlation portion of the Wigner distribution will possess rapid oscillations along the position axis, but the range of the correlations will remain as large as the separation of the various pulses. *These correlations are extremely important and provide the time reversibility of this tunneling system.* If we reverse time, the pulses will recombine into the single initial pulse, provided we do not destroy any of the correlation information. Without this information, the time reversed problem is just the impact of two disjoint pulses. *The use of the Wigner distribution allows a clear indication of the role of inelastic collisions, as these scattering processes destroy the correlation information shown in the figure.* The use of Wigner distribution functions is a powerful method of investigating the non-equilibrium, inhomogeneous system. However, the results shown in Figs. 2,3 are only first approaches, as scattering has been treated in the relaxation time approximation, and no many-body effects have been incorporated. Nevertheless, it is an important

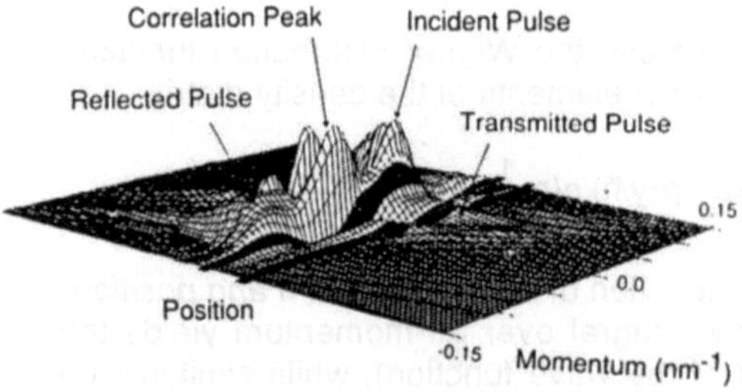

Fig. 7 The Wigner distribution function for a Gaussian wave function impinging upon a tunneling barrier. The various peaks are discussed in the text.

approach that is quite usable. Indeed, it is currently the only approach giving results in the study of transport in far-from-equilibrium, inhomogeneous quantum systems.

The more general (and more familiar) approach has pursued non-equilibrium, finite temperature Green's functions, since the Wigner distribution function is intimately related to the special function $G^<$ [48] (its integral over all ω is essentially F, as we illustrate below). Approaches using the Green's functions must take account of the fact that four independent Green's functions must be determined. These usually are taken as the advanced and retarded functions G_a and G_r, and the two correlation functions $G^>$ and $G^<$. These are of course related, but each of the four must be found to completely determine the time-dependent non-equilibrium solutions. The results are complicated in that the equation of motion for each is strongly coupled to the others. Only in the very special case of slowly varying spatial and temporal variations (which allows the gradient expansion [49] to be used) has the kinetic equation for $G^<$ been shown to converge to the Barker-Ferry equation for the Wigner function [50,51]. This slow progress is not surprising, as it is only recently that the near equilibrium limit of low field transport using Green's functions was shown to converge to the same result obtained by the Kubo formula [52]. We can illustrate part of the problem by considering the relation between $G^<$ and the Wigner distribution F. The "less than" function is

$$G^<(\mathbf{r},t;\mathbf{r}',t') = i\left\langle \Psi^+(\mathbf{r}',t')\Psi(\mathbf{r},t)\right\rangle$$

$$= i\left\langle \Psi^+(\mathbf{R}-\frac{\mathbf{y}}{2},T-\frac{\tau}{2})\Psi(\mathbf{R}+\frac{\mathbf{y}}{2},T+\frac{\tau}{2})\right\rangle .$$

This illustrates that we can relate the two functions as

$$F(\mathbf{R},\mathbf{p},T) = \frac{1}{(2\pi\hbar)^n i}\int d\mathbf{y}\, e^{i\mathbf{p}\cdot\mathbf{y}/\hbar} \lim_{\tau\to 0} G^<(\mathbf{R},\mathbf{y},T,\tau)$$

$$= \frac{1}{(2\pi\hbar)^n i}\int d\mathbf{y}\int \frac{d\omega}{2\pi} e^{i\mathbf{p}\cdot\mathbf{y}/\hbar}\, G^<(\mathbf{R},\mathbf{y},T,\omega) .$$

Buried in this relationship is the fact that $G^<$ is not solely a distribution function. In equilibrium, it is often decomposed into the product of the distribution function and the spectral density A: $G^<=iF(\omega)A(\mathbf{k},\omega)$. One problem is that this decomposition is not known in far-from-equilibrium systems. Thus, it is only in the case of slowly varying systems in space and time that such a decomposition has been achieved through a usable *Ansatz* [50,51]. On the other hand, the basic Green's function approach is sound, and lays a fundamental structure that can be used to pursue the proper solutions. This is a fertile area in which good ideas are needed for solutions to be obtained. Success is necessary if progress is to be maintained in our understanding of the operation of ever smaller semiconductor devices. There exist other approaches, such as the Feynman path integral, but these have great difficulty in applications to time-dependent, open systems that are far-from-equilibrium. The few groups working on this latter problem have had some early success, but whether this will continue is not clear [53,54].

6. Summary

In the preceding sections, I have tried to show that the study of ultra-small semiconductor devices in general involves a great diversity of interesting quantum transport problems: non-local transport due to the non-local effects of potentials, tunneling, ballistic transport, complicated multi-valley processes, and strong non-equilibrium quantum effects. At first glance, it may seem that the various topics that have been discussed here are disjoint and only poorly connected through the use of the inelastic mean free path. However, "ballistic" transport effects were found to be one unifying theme and appeared in each section. The connection is actually much stronger in a more basic sense, and indeed the connection continues throughout the papers in this volume. In 1977, Ilya Prigogine received the Nobel Prize in Chemistry for his pioneering work on the transition from dynamics to thermodynamics. In this work, he created a formalism for the manner in which collisions, and the loss of phase information by individual carriers in a many-body system, create a transition from reversible to irreversible and therefore dissipative behavior, primarily by the introduction of ensemble averaging and the loss of full knowledge of the entire system's dynamics. This was just the latest recognition for more than a century of basic work, by a variety of well-known scientists, addressing the foundations of thermodynamics while trying to understand how it could arise from dynamics. While based in a rather comprehensive and extensive theoretical framework, there is as yet little experimental or theoretical evidence for the details of this important transition in real systems such as devices. We are now at a point where one can construct real physical systems on a spatial scale for which ensemble averaging and phase randomization do not occur. In essence, we are interested then in the manner in which phase-breaking, or the inelastic decay of "ballistic" transport, occurs. Thus, we can recognize that the various physical phenomena discussed here are part of the beginnings of an experimental/theoretical body of data that addresses the important transition from dynamics to thermodynamics. From this body of data, and from the resulting *detailed microscopic* theories that explain the data, we can begin to build the full understanding of this transition. Yet, the problem remains quite difficult and these approaches are only first steps. There must be much further work if we are to succeed.

7. Acknowledgements

The work reported here is the result of many collaborations as well as that of many students. In particular, I would like to acknowledge contributions from N. C. Kluksdahl, A. M. Kriman, R. P. Joshi, M.-J. Kann, R. Mezenner, G. Bernstein, W.-P. Liu, and R. O. Grondin (and permission of the Rochester group to use their results prior to publication). In addition, many helpful discussions came from G. J. Iafrate, R. Chamberlin, K. Hess, L. Eaves, T. C. L. G. Sollner, H. L. Grubin, A.-P. Jauho, C. Jacoboni, L. Reggiani, P. Lugli, U. Ravaioli, and M. A. Osman. Much of the work was supported by the Office of Naval Research (U.S.A.).

References

1. G. Bernstein and D. K. Ferry: Superlatt. and Microstruc. **2**, 147 (1986)
2. Y. Jin, D. Mailly, F. Carcenac, B. Etienne, and H. Landis, Microelectr. Engr. **6**, 195 (1987)
3. J. R. Barker and D. K. Ferry: Sol.-State Electronics **23**, 519 (1981); **23**, 531 (1986)
4. R. Troutman: IEEE Trans. Electron Dev. **ED-26**, 461 (1979)
5. G. Lewicky and J. Maserjian, J. Appl. Phys. **46**, 3032 (1975)
6. M. V. Fischetti, D. J. DiMaria, L. Dori, J. Batey, E. Tierney, and J. Stasiak, Phys. Rev. B **35**, 4404 (1987)
7. L. L. Chang, L. Esaki and R. Tsu: Appl. Phys. Letters **24**, 593 (1974)
8. T. C. L. G. Sollner, P. E. Tannenwald, D. D. Peck, and W. D. Goodhue: Appl. Phys. Letters **43**, 588 (1983); **45**, 1319 (1984)
9. V. J. Goldman and D. C. Tsui: Phys. Rev. Letters **59**, 1623 (1987)
10. O. H. Hughes, M. Henini, L. Eaves, F. W. Sheard, and G. A. Toombs: J. Vac. Sci. Technol., in press
11. N. C. Kluksdahl, A. M. Kriman, C. Ringhofer, and D. K. Ferry, submitted for publication
12. L. Y. L. Shen and J. M. Rowell, Phys. Rev. **165**, 566 (1968)
13. D. S. Lacklison, J. Duggan, J. J. Harris, C. T. B. Foxon, D. Hilgon, C. Roberts, and C. W. Hallon, Appl. Phys. Letters **52**, 305 (1988)
14. A. M. Kriman, N. C. Kluksdahl, and D. K. Ferry, Phys. Rev. B **36**, 5953 (1987)
15. N. C. Kluksdahl, A. M. Kriman, and D. K. Ferry, submitted for publication
16. J. R. Hayes, A. F. J. Levi, and W. Wiegmann: Electron. Letters **20**, 851 (1984)
17. M. Heiblum, D. L. Thomas, C.M. Knoedler, and M. I. Nathan: Appl. Phys. Letters **47**, 1105 (1985)
18. G. Bernstein and D. K. Ferry: Superlatt. and Microstruc. **2**, 373 (1986)
19. G. Bernstein and D. K. Ferry: IEEE Trans. Electron Dev., in press
20. Y. Huang and P. Yu: Sol. State Commun. **63**, 109 (1987)
21. R. W. Schoenlein, W. Z. Lin, E. P. Ippen, and J. G. Fujimoto: Appl. Phys. Letters **51**, 1442 (1987)
22. M. J. Rosker, F. W. Wise, and C. L. Tang: Appl. Phys. Letters **49**, 1726 (1986)
23. K. Meyer, M. Pessot, G. Mourou, R. O. Grondin, and S. Chamoun: Phys. Rev. Letters, in press
24. D. K. Ferry: in *Quantum Transport in Semiconductors*, Ed. by H. L. Grubin, C. Jacoboni, and D. K. Ferry (Plenum Press, New York, in press)
25. J. R. Barker and D. K. Ferry: in *Proc. Int. Conf. on Cybernetics and Society*, (IEEE Press, New York, 1979) p. 762
26. J. J. Hopfield: Proc. Nat. Acad. Sci. (USA) **79**, 2554 (1982)
27. G. Graf, L. Jackel, R. Howard, B. Howard, B. Straughn, J. Denker, W. Hubbard, D. Tennant, and D. Schwarz: AIP Conf. Proc. **151**, 182 (1986)
28. D. K. Ferry: IEEE Circuits and Devices Mag. **1**(4), 39 (1985)
29. D. K. Ferry, L. Akers, and R. O. Grondin: in *Integrated Circuits in the 0.5 to 0.05 Micron Dimensional Range,* Ed. by R. K. Watts (John Wiley, New York, in press)
30. K. Preston and M. Duff: *Modern Cellular Automata* (Plenum Press, New York, 1984)
31. R. T. Bate, private communication
32. G. J. Iafrate, R. K. Reich, and D. K. Ferry: Surf. Sci. **113**, 485 (1982)

33. Y. Tokura and K. Tsubaki: Appl. Phys. Letters **51**, 1807 (1987)
34. D. K. Ferry, G. Bernstein, and W.-P. Liu, in these proceedings
35. J. D. Linci, D. J. Bishop, M. A. Kastner, and J. Melngailis: Phys. Rev. Letters **55**, 2987 (1985).
36. W. J. Skocpol, P. M. Mankiewich, R. E. Howard, L. D. Jackel, and D. M. Tennant: Phys. Rev. Letters **56**, 2865 (1986).
37. S. Datta, M. R. Melloch, S. Bandyopadhyay, R. Noren, M. Varizi, M. Miller, and R. Reifenberger: Phys. Rev. Lett. **55**, 2344 (1985).
38. K. Ishibashi, Y. Takagaki, K. Gamo, S. Namba, s. Ishida, K. Murase, Y. Aoyagi, and M. Kawabe: Sol. State Commun. **64**, 573 (1987).
39. G. P. Whittington, P. C. Main, L. Eaves, R. P. Taylor, S. Thoms, S. P. Beaumont, C. D. W. Wilkinson, C. R. Stanley, and J. Frost: Superlattices and Microstructures **2**, 381 (1986).
40. K. Ishibashi, K. Nagata, K. Gamo, S. Nambu, S. Ishida, K. Murase, Y. Aoyagi, and M. Kawabe: Sol. State Commun. **61**, 385 (1987).
41. R. K. Reich, R. O. Grondin, and D. K. Ferry, Phys. Rev. B **27**, 3483 (1983)
42. D. J. Thouless and S. Kirkpatrick, J. Phys. C **14**, 235 (1981)
43. P. A. Lee and D. Fisher, Phys. Rev. Letters **47**, 882 (1981)
44. D. R. Hofstadter, Phys. Rev. B **14**, 2239 (1976)
45. J. R. Barker, J. Phys. C **6**, 2663 (1973)
46. H. J. Kreuzer, *Nonequilibrium Thermodynamics and Its Statistical Foundations* (Clarendon Press, Oxford, 1981) pp. 312-321
47. J. R. Barker and D. K. Ferry: Phys. Rev. Letters **42**, 1779 (1979)
48. J. Rammar and H. Smith: Rev. Mod. Phys. **58**, 323 (1986)
49. L. P. Kadanoff and G. Baym: *Quantum Statistical Mechanics* (Benjamin/Cummings, Reading, MA, 1962)
50. P. Lipavsky, V. Spicka, and B. Velicky: Phys. Rev. B **34**, 6933 (1986)
51. F. S. Khan, J. H. Davies, and J. W. Wilkins: Phys. Rev. B **36**, 2578 (1987)
52. G. D. Mahan: Phys. Reports **110**, 321 (1984)
53. M. V. Fischetti and D. J. DiMaria, Phys. Rev. Letters **55**, 2475 (1985)
54. B. Mason and K. Hess, private communication; M. A. Littlejohn, private communication

Physics of One-Dimensional MOSFETs

A. Hartstein

IBM Research Division, T.J. Watson Research Center,
P.O. Box 218, Yorktown Heights, NY 10598, USA

1. Introduction

During the past several years many studies have been made on the conductance of quasi-one-dimensional MOSFET systems [1-13]. Related physics has also been explored in GaAs structures [14-16]. One of the key reasons that one would like to use either the MOSFET system or the heterojunction system for these studies is that the Fermi energy is controllable in these systems by the appropriate application of voltage biases or by charging effects. This gives the experimenter a very important parameter to vary in order to understand the proper description of the physics.

In this chapter I will first discuss the nature of the disordered potential in these devices and its effect on the electron wave functions. I will then describe how the situation is affected by the extremely small sample size, the so called mesoscopic regime. One type of MOSFET sample, which has been used to explore these types of effects, will be described. I will then discuss in some detail the study of hopping conduction in the strong localization regime in these samples and the transition to weak localization. The subject of universal conductance fluctuations in the weak localization regime of these samples will be discussed in another chapter in this volume.

2. Theoretical Model

Figure 1 shows a model potential for the inversion layer. The potential energy is assumed to have a random component due to fluctuations present at the interface between the silicon and the silicon dioxide. This potential may arise from the combined effects of roughness at the interface, charge trapping at or near the interface, random doping in the bulk and disorder in the atomic arrangements of the atoms near the interface. A naive picture of localization is indicated by the Fermi energy, E'_F in Fig. 1. Here the potential surface is partially filled by electrons, leaving some regions with barriers to electron penetration. This clearly gives rise to localized electron states.

However, even if the Fermi energy is well above the peaks of the potential fluctuations, as indicated by E_F, localized electron states can still arise. In this case one considers the effects of coherent backscattering on the nature of the electron wave functions. This type of interference effect has the effect of trying to keep an electron from diffusing through the system.

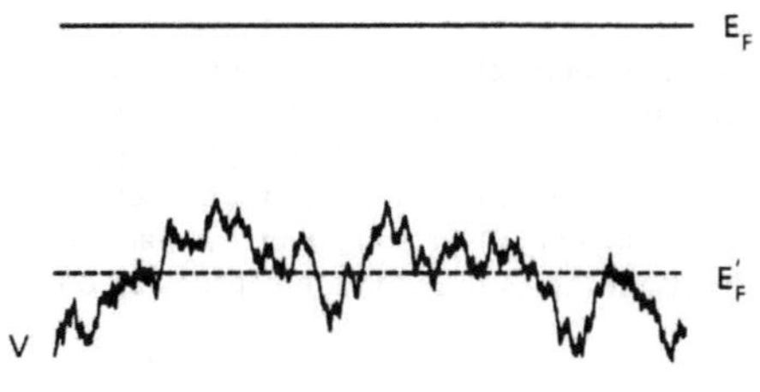

Fig. 1. Model potential for an inversion layer or an accumulation layer. Two possible Fermi energies are indicated

Springer Series in Solid-State Sciences Vol. 83: **Physics and Technology of Submicron Structures**
Editors: H. Heinrich · G. Bauer · F. Kuchar

In effect it has an enhanced probability of remaining near its starting position. This effect always leads to localized states in one and two dimensions, but can lead to either localized or extended states in three dimensions depending on the details. The wave functions can be described as

$$\Psi = \Psi_0(x)e^{-\alpha x}, \tag{1}$$

where $1/\alpha$ is the localization length of the wave function. The term $\Psi_0(x)$ contains all of the local structure in the wave function. In this picture the decay length α is given by [17]

$$\alpha \sim \frac{1}{\xi} e^{-k_F \xi} \tag{2}$$

for a 2D electron gas. Here ξ is a typical correlation length along the interface for the disorder potential, and k_F is the Fermi wave vector. A similar expression obtains in 1D.

The physical phenomena which result from this wave function depend on the localization of the wave function relative to the effective sample size. The effective sample size, L_{Eff}, is the length over which electronic wave functions can remain coherent in the sample. This is generally either the sample length, L, or the inelastic diffusion length, L_{in}, whichever is smaller. If $1/\alpha$ is larger than L_{Eff}, then the physics is described by the weak localization theory. In this regime the physics is dominated by quantum interference effects. If the system is large enough some of these effects are averaged away. The observed negative magnetoresistance, which is the destruction of the localization effect due to an applied magnetic field, is not affected in this way. However, other interference effects, such as the Aharonov-Bohm effect and the related universal conductance fluctuations, are washed out in large systems.

The other limit, $1/\alpha < L_{Eff}$, comprises the strong localization regime. In this regime the counterpart of the weak localization conductance is the direct transfer of charge from one side of the sample to the other via a single wave function. This can be described as a resonant tunneling process [19]. In fact this is the conduction mechanism which dominates at low enough temperatures. At finite temperatures it is also possible to have a hopping conduction mechanism. In this mechanism the electrons transfer from one localized state to another via a phonon assisted tunneling process. This process dominates in most of the experiments which have been reported.

When the size of the sample is small enough, the normal averaged behavior of these conduction mechanisms must be modified. The hopping conduction, which can normally be described as variable range hopping, moves into a range where only 1 or 2 hops dominate the process [9,18]. This does not allow for the normal averaging process to take place. Universal conductance fluctuations [10-12] are another example of a phenomenon which needs the small sample size to survive. It is a consequence of the inapplicability of ensemble averaging in these mesoscopic samples. Another example of a specific phenomenon allowed by the small sample size is the observation of time dependent effects caused by the trapping of individual charges near the inversion layer channel [13].

The other type of small sample size effects are perhaps more obvious. They include the change in effective dimensionality of the conduction process by a simple size effect. This can occur in both the weak localization regime [7] and in the strong localization regime [1]. When sizes get small enough, the wave functions can become quantized for motion in that particular direction. This is indeed the effect that makes inversion layers 2D to start with. A narrow enough sample will lead to 1D quantized behavior. In effect this is the ultimate expression of interference effects in the sample.

3. Experiments

Figure 2 shows idealized top and cross section views of our pinched accumulation layer samples. The substrate is 10 ohm-cm n-type <100> silicon. Two n^+ diffusions are used as the source and drain contacts to the accumulation layer. Two lateral p^+ diffusions are used to control the width of the accumulation layer. The electrons at the surface are in effect pinched into a narrow region because of the electric field applied from these control electrodes as well as the field arising from the built-in potential between the p^+ regions and the n-type substrate. The separation between the control electrodes is about 1 μm, and the length of the narrow conducting region is about 10 μm. The gate oxide is 30 nm thick, and the gate is made of aluminum.

In the actual device, the dimensions are not nearly as ideal as indicated in Fig. 2. The sharp corners are all rounded, and the distance between the control electrodes is not uniform. In fact we have no way of ascertaining the effective width variations in this type of device. Actually, we do not even know the width of the conducting channel, except from our analysis of the experimental data. The effect of an applied gate voltage is two-fold in samples of this geometry. A positive gate voltage will induce electrons at the surface, thereby changing the carrier density and Fermi energy. Moreover, it will tend to broaden the width of the accumulation layer channel. A negative voltage applied to the control electrodes will pinch the channel to narrower widths, but will also tend to change the threshold voltage and thereby reduce the accumulation layer carrier density.

These devices can only be measured at low temperatures, in our case $T < 10$ K. At higher temperatures the substrate is conducting and results in a parallel conduction path between the source and drain contacts. The devices were measured using an AC modulation of the source-drain voltage. The resulting current was synchronously detected using a current sensitive preamplifier feeding into a lock-in amplifier. To avoid hot electron effects the source-drain field was generally kept below 2 mV/cm. Measurements were made in a dilution refrigerator where good data was obtained at temperatures as low as 36 mK. Great care was taken to eliminate extraneous noise signals.

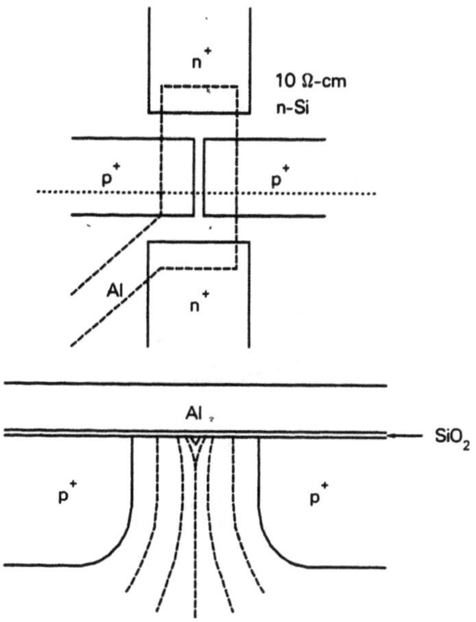

Fig. 2. The upper part shows an idealized top view of a sample. The two n^+ regions are the source and drain. The p^+ regions are the control electrodes. The width between the controls is 1-2 μm and the length of the controls is 14 μm. The lower part shows a cross section along the dotted line. The diffusions are about 1 μm deep, and the oxide thickness is 30 nm. Potential lines are sketched for a positive gate voltage, indicating the narrow conducting channel

The mass of data needed to fully explore the nature of the observed structure in the conductance curves necessitated the use of an elaborate data acquisition system. In order to obtain these data it was necessary to build a noise-free voltage ramp and to carefully interface the equipment to the data acquisition system. The experiment and the computer system were decoupled by use of active filters on all interconnecting lines. These measures were necessary to avoid noise heating of the electrons, particularly at the lowest gate voltages.

Typical data obtained from these samples are shown in Fig. 3. The figure shows the measured conductance as a function of gate voltage for three temperatures. The most notable feature of the data is the large structure observed in the conductance as a function of gate voltage. The structure is reproducible in a given sample, but is sample dependent. This structure was not observed to change over a period of more than six months and numerous temperature cyclings up to 300 K. The relative amplitudes of the peaks can be seen to decrease at higher gate voltages. The widths of the peaks are observed to increase with increasing gate voltage. As the temperature is raised, the peaks broaden and the smaller ones gradually disappear.

Figure 4 shows the temperature dependence of the conductance of these devices at several values of the gate voltage. At gate voltages above 10 V. the curves are linear in $T^{-1/2}$, a result indicative of weak localization transport in a 1D regime. At lower gate voltages the curves are non-linear and as we shall see, the transport is best described by a hopping process. The transition between these regimes is gradual.

The essential physics of 1D transport is contained in Figs. 3 and 4. The low gate voltage, low temperature portion of the figures represent the strong localization regime. The high gate voltage, high temperature regime is best described by weak localization. A smooth transition between these regimes is observed. The curves also contain regions of both 1D and 2D behavior. The details of the temperature dependence of the conductance in the different regimes contains much of this information. The reproducible structure is caused by the extremely small

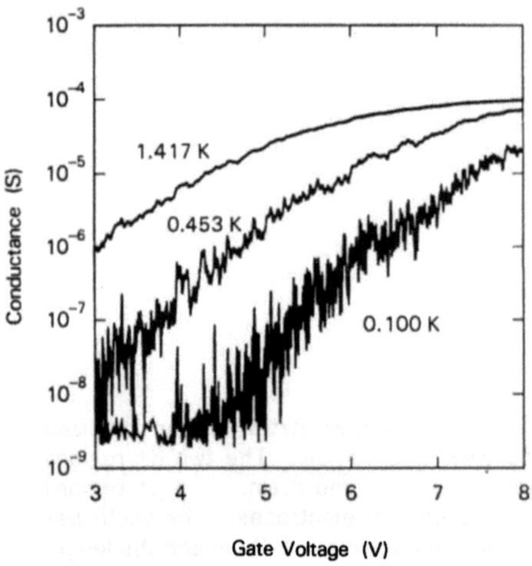

Fig. 3. Conductance is shown as a function of gate voltage for three temperatures. The structure in the figure is real and not noise. In fact the resolution of the figure is insufficient to resolve the sharpest structures

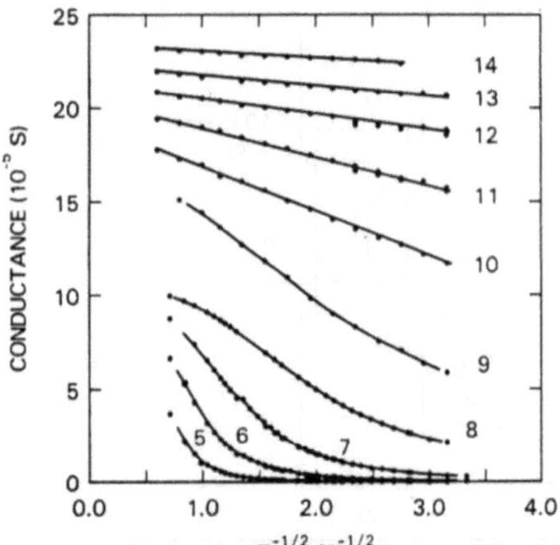

Fig. 4. Temperature dependence of the conductance for several values of the gate voltage.

sample size. In the strong localization regime it is caused by the individual statistics of only 1 or 2 hopping events controlling the conductance of the sample. In the weak localization regime it is caused by universal conductance fluctuations.

4. Variable Range Hopping

The average behavior of these data in the low gate voltage regime can be described by a variable range hopping process. The description of the large peaks in conductance will be treated later. Figure 5 shows the temperature dependence of the low gate voltage conductance, this time plotted on a semi-log scale. These data can be fit to the general form

$$G = G_0 \exp - (T_0/T)^n, \tag{3}$$

where T_0 and n are parameters of the variable range hopping process. We find that for a gate voltage less than 5.8 V., the data are best fit with an exponent $n = 1/2$. For larger gate voltage the best fit is for $n = 1/3$.

This behavior is indicative of variable range hopping. In the low gate voltage regime it is variable range hopping in 1D. In the high gate voltage regime it is variable range hopping in 2D. Variable range hopping is a process in which electrons undergo phonon assisted tunneling to neighboring sites at varying distances. The average hopping distance depends on various parameters, including the temperature. The competition between hopping to near sites and hopping to sites of near energy gives this process its unique temperature dependence. The expressions for this process are given by

$$G = G_0 \exp - (T_0/T)^{1/2} \tag{4}$$

and

$$T_0 = \frac{4\alpha}{N_1 k_B} \tag{5}$$

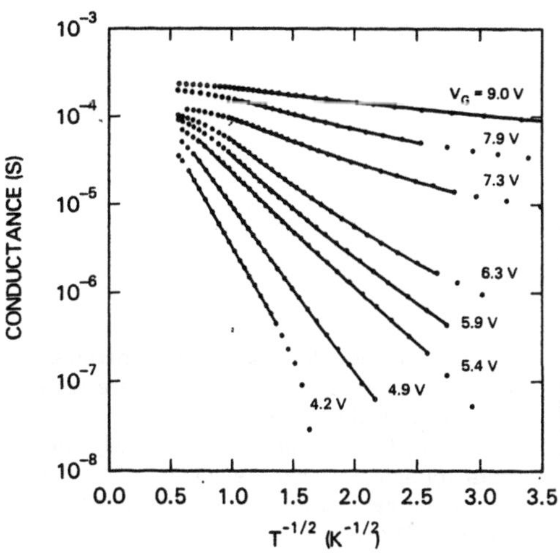

Fig. 5. Temperature dependence of the conductance is shown for several gate voltages. The data were obtained with source, substrate and control electrodes grounded. The solid lines are best least squares fits

for the 1D case. The equivalent expressions derived for the 2D case are

$$G = G_0 \exp - (T_0/T)^{1/3} \tag{6}$$

and

$$T_0 = \frac{27\alpha^2}{\pi N_2 k_B}. \tag{7}$$

In these expressions N_1 and N_2 are the 1D and 2D densities of states, and k_B is Boltzmann's constant.

As the gate voltage is increased the width of the conducting channel increases and a transition occurs between 1D variable range hopping and 2D variable range hopping when the width of the channel is comparable to the average hopping distance. The theory allows us to estimate the effective width of the channel at the transition as w = 30 nm [1]. The main question becomes why is the transition as sharp as observed. Perhaps this is an indication of the rate at which the channel width changes as a function of gate voltage.

One can take the analysis of the variable range hopping data one step further. The values of the parameter T_0 as a function of gate voltage can be obtained from the data of Fig. 5. To reduce experimental fluctuations, we assumed values for the exponents of either 1/2 or 1/3 according to the gate voltage range, and then used a least squared fitting procedure to obtain T_0. Figure 6 shows averaged [8] values of T_0 plotted as a function of gate voltage for both the 1D and 2D regimes.

Equations (5) and (7) give the expected behavior of T_0 in the two regimes. Figure 6 clearly indicates that T_0 is nearly an exponential function of gate voltage. If we assume that N_2 is constant, which has been demonstrated in wide 2D samples at the same carrier densities, we are forced to the conclusion that α must vary exponentially with gate voltage. This is just the type of behavior which is predicted in Eq. (2). Moreover, the relative slopes in the 1D and 2D regions vary by a factor of 2, as required by Eqs. (5) and (7). This indicates that the underlying localization mechanism is the interference process envisioned in the weak localization theory, even though the phenomena we are discussing are strong localization phenomena.

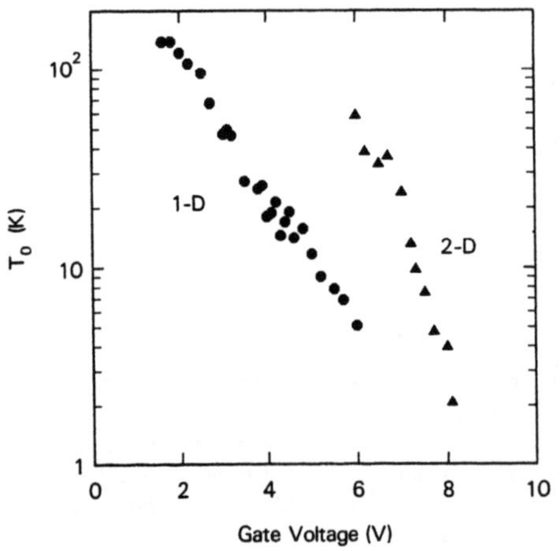

Fig. 6. The variable range hopping parameter, T_0, is plotted as a function of gate voltage for both the 1D and 2D regimes. It is surprising that T_0 varies exponentially with gate voltage, and that the slopes of this dependence vary by a factor of 2 between the 1D and 2D regimes

We can carry this analysis one step further by determining α from (7) and the data in the 2D regime. If we then apply the criterion that a transition from strong localization to weak localization will occur when the localization length, $1/\alpha$, is equal to the effective sample length, the transition is predicted to occur at a gate voltage of 10 V. We have assumed that the effective sample length is the inelastic diffusion length, 0.3 μm, measured in these samples [11]. This is remarkably close to the observed transition as obtained from the temperature dependent data of Fig. 4. Extrapolated data from the strong localization regime correctly predict the onset of weak localization.

5. The Conductance Peaks

We are now in a position to gain some insight into just how small these samples are. The width of the channel was estimated to be 30 nm, and the length is about 10 μm. This gives an area of 0.3 μm^2, and assuming a 2D density of states, it gives an estimate of 500 electrons in the channel. If we consider that only those electron states within a few k_BT of the Fermi energy contribute to the conductance, the number drops to less than 50. This led us to suggest [1] that the explanation for these peaks could be found in the statistical fluctuations in the density states due to the extremely small sample size. LEE [18] has expanded on this suggestion and considered the role of statistical fluctuations in small hopping systems in which only one or two hops dominate the conductance because of percolation effects. AZBEL and co-workers [19-21] have considered the possibility that the structure arises from resonant tunneling.

In order to best examine the nature of the peaks we have looked extensively at a few peaks which show only minimal overlap with other peaks. One such peak is shown in Fig. 7. The largest of the peaks shown in this narrow region of gate voltage is quite well isolated over a reasonable temperature range. Significant overlap sets in above 120 mK. The sides of the peaks are reasonably linear on this semilog plot; and when the gate voltage scale is converted to an energy scale, the magnitudes of the slopes are approximately $1/k_BT$. Note also that the peak positions show a small but systematic shift with temperature.

Figure 8 displays the temperature dependence of the four largest peaks in Fig. 7. Above 200 mK, where the overlap is large, the data are fitted well to the variable range hopping form, (3). For most of the 120 peaks which have been studied, the temperature region below 200 mK can be fitted equally well with either a T^{-1} or a $T^{-1/2}$ exponential temperature dependence. The data displayed by the open circles in Fig. 8 illustrate one example of a peak which deviates

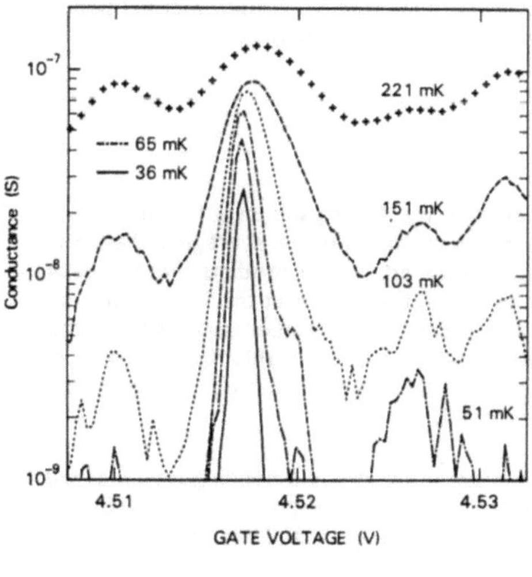

Fig. 7. Conductance as a function of gate voltage on an expanded gate voltage scale for selected temperatures. Only the large peak is displayed for the 65 and 36 mK curves

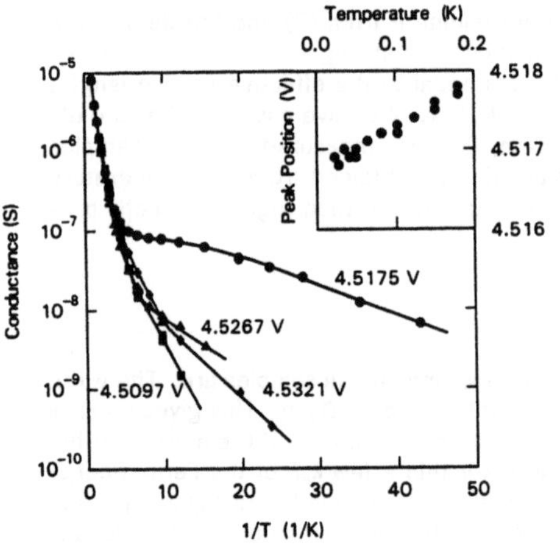

Fig. 8. Temperature dependence of four conductance peaks shown in Fig. 7 at V_G = 4.5097, 4.5175, 4.5267, and 4.5321 V with activation energies 36, 7.2, 13 and 18 μeV, respectively. Solid lines are smooth curves through the data. The inset shows the temperature dependence of the peak position for the peak at $V_G = 4.5175$ V in the temperature regime where the peak is isolated from its neighboring peaks

markedly from this behavior. The data are fitted by $\ln G \propto 1/T$ at low temperatures but flatten off at temperatures where k_BT is approaching the activation energy (7.2 μeV). Five peaks have been observed to behave this way, and 10% of the remaining peaks are better fitted by a $1/T$ dependence than a $T^{-1/2}$ dependence.

The first real theory put forward to explain the existence of the peaked structure was the resonant tunneling theory. The basic idea is that electrons may resonantly tunnel through the sample instead of hopping from site to site. It is known that if an energy level exists for the electrons in the middle of the sample and at the Fermi energy, a resonant transfer can occur. If the situation is symmetrical, the transition probability can be 1, whereas the background tunneling will be exponentially smaller. The width of the tunneling resonance will be exponentially narrow. The $T = 0$ theory [19] demonstrates the existence of large, reproducible, sample-dependent structure in the conductance. The basic idea is that at the special values of the Fermi energy, corresponding to the electron states, constructive interference of the electron wave functions can lead to the extremely high value of the transmission probability.

The more recent finite temperature theory [20, 21] argues that when both thermal population effects and inelastic scattering are taken into account, the temperature dependence of the peaks should be

$$G = G_0 \exp - (T_0/\eta T)^{1/2}, \tag{8}$$

where $1<\eta<4$ and η is a constant for any one peak, with the 4 corresponding to unity transmission probability. At higher temperatures the peaks merge and the average behavior is governed by the variable range hopping formalism. The theory further predicts that the peak positions should not shift with temperature. Another prediction is that at the lowest temperatures the conductance of a peak should increase with decreasing temperature. One of the basic assumptions inherent in the intermediate temperature portion of this theory has been questioned [22]. This refinement of the theory eliminates the intermediate temperature range (8) but leaves the high and low temperature ranges intact.

The second substantive theory which has been proposed involves a hopping model of the conductance in sparse systems where statistical fluctuations are important. LEE [18] has shown that in a 1D hopping system with parameters comparable to those found in our samples, fluctuations not only in the energy of each state but also in the spatial location of the states can indeed produce structure as large as observed experimentally. The percolation problem reduces to only 1 or 2 critical hops.

When a single critical hop dominates the current, the conductance in the linearized approximation with Boltzmann statistics is

$$G = G_{ij} \exp [-2\alpha_{ij}r_{ij} - (1/2k_BT)(|E_i - \mu| + |E_j - \mu| + |E_i - E_j|)], \tag{9}$$

where μ is the chemical potential. This shows that a peak at low temperature will be simply activated, and that off of the peak the slope of the curves should correspond to $1/k_BT$. If a single peak persists to high enough temperature without overlap from neighboring peaks, the Boltzmann approximation is no longer valid. The simple exponentials in (9) are replaced by Fermi and Bose functions. This can lead to the kind of deviations from a simple activated temperature dependence observed in Fig. 8.

When two hops are taken in series, the situation qualitatively changes. If each hop is described by (9), one can form the series combination and solve for the peak position. One obtains

$$E_p = (E_i + E_l)/2 + k_BT(\alpha_{kl}r_{kl} - \alpha_{ij}r_{ij}) + (k_BT/2)\ln(G_{ij}/G_{kl}), \tag{10}$$

where i, j, k and l describe the four sites involved in the hopping process. This type of peak shifts linearly in temperature with a magnitude and sign given by the asymmetry of the hopping sites involved.

The experimental data as represented by Figs. 7 and 8 are in better agreement with this hopping model than with the resonant tunneling model [9]. The temperature dependence of the main peak shown in Fig. 8 points to this model. The inset to Fig. 8 indicates the temperature dependence of the peak position, which is only consistent with the hopping model. In passing we should note that it is possible to have both types of conductance mechanisms acting in parallel so that the experiments do not disprove the resonant tunneling model, they only show that the mechanism has not been directly observed in this system.

As we have discussed earlier, the resonant tunneling mechanism should become dominant at low enough temperature. The temperature scale involved depends on the channel length of the sample. As samples become shorter, resonant tunneling will become important at higher temperatures. This conductance mechanism has been observed in samples of length 0.5 μm. at temperatures below 200 mK [23].

Fig. 7 contains additional structure which has not been discussed. Since we have already discussed the large peaks in terms of individual hopping states and the intervening regions as thermal excitation to these states, it is difficult to understand the origin of this substructure. We do, however, know that the localized wave functions arise from interference effects. Perhaps the substructure arises from these types of effects in a similar way as the universal conductance fluctuations do. Perhaps these interference effects are then manifest in the hopping conduction via the overlap of the wave functions.

6. Summary

In this chapter I have discussed some of the physical phenomena which result from localized wave functions in low dimensional disordered systems. The basic physics arises from quantum mechanical interference effects. These define the localized nature of the wave functions.

Many different phenomena are a consequence of these considerations. I have discussed variable range hopping, resonant tunneling and single hopping processes in some detail. Many other effects are discussed elsewhere in this volume.

7. Acknowledgments

The work which I have discussed was a collaborative effort among several people. I would like to acknowledge the contributions of A. B. Fowler, S. B. Kaplan, J. J. Wainer and R. A. Webb.

8. References

1. A.B. Fowler, A. Hartstein and R.A. Webb: Phys. Rev. Lett. 48, 196 (1982)
2. R.G. Wheeler, K. K. Choi, A. Goel, R. Wisnieff and D. E. Prober: Phys. Rev. Lett. 49, 1674 (1982)
3. W.J. Skocpol, L.D. Jackel, E.L. Hu, R.E. Howard and L.A. Fetter: Phys. Rev. Lett. 49, 951 (1982)
4. R.R. Kwasnick, M.A. Kastner, J. Melngalis and P.A. Lee: Phys. Rev. Lett. 52, 224 (1984)
5. R.G. Wheeler, K.K. Choi and R. Wisnieff: Surf. Sci. 142, 19 (1984)
6. W.J. Skocpol, L.D. Jackel, R.E. Howard, H.G. Craighead, L.A. Fetter, P.M. Mankiwich, P. Grabbe and D.M. Tennant: Surf. Sci. 142, 14 (1984)
7. C.C. Dean and M. Pepper: J. Phys. C 17, L1287 (1982)
8. A. Hartstein, R. A. Webb, A. B. Fowler and J. J. Wainer: Surf. Sci. 142, 1 (1984)
9. R.A. Webb, A. Hartstein, J.J. Wainer and A.B. Fowler: Phys. Rev. Lett. 54, 1577 (1985)
10. J.C. Licini, D.J. Bishop, M.A. Kastner and J. Melngailis: Phys. Rev. Lett. 55, 2987 (1985)
11. S.B. Kaplan and A. Hartstein: Phys. Rev. Lett. 56, 2403 (1986)
12. W.H. Skocpol, P.M. Mankiewich, R.E. Howard, L.D. Jackel, D.M. Tennant and A.D. Stone: Phys. Rev. Lett. 56, 2865 (1986)
13. K.S. Ralls, W.J. Skocpol, L.D. Jackel, R.E. Howard, L.A. Fetter, R.W. Epworth and D.M. Tennant: Phys. Rev. Lett. 52, 228 (1984)
14. P.C. Main, L. Eaves, R.P. Taylor, G.P.Whittington, S. Thoms, S.P. Beaumont and C.D.W. Wilkinson: In The Physics of Semiconductors, ed. Olof Engstrom, 1591 (World Scientific, Singapore 1986)
15. G. Timp, A.M. Chang, P. Mankiewich, R. Behringer, J.E. Cunningham, T.Y. Chang and R.E. Howard: Phys. Rev. Lett. 59, 732 (1987)
16. G. Timp, A.M. Chang, J.E. Cunningham, T.Y. Chang, P. Mankiewich, R. Behringer and R.E. Howard: Phys. Rev. Lett. 58, 2814 (1987)
17. D. Vollhardt and P. Wolfle: Phys. Rev. Lett. 45, 842 (1980)
18. P.A.Lee: Phys. Rev. Lett. 53, 2042 (1984)
19. M. Ya. Azbel and P. Soven: Phys. Rev. B 27, 831 (1983)
20. M. Ya Azbel, A. Hartstein and D.P. DiVicenzo: Phys. Rev. Lett. 52, 1641 (1984)
21. M. Ya Azbel and D.P. Divicenzo: Phys. Rev. B 30, 6877 (1984)
22. A.D. Stone and P.A. Lee: Phys. Rev. Lett. 54, 1196 (1985)
23. A.B. Fowler, G.L. Timp, J.J. Wainer and R.A.Webb: Phys. Rev. Lett. 57, 138 (1986)

MOSFETs Under Electrical Stress – Degradation, Subthreshold Conduction, and Noise in a Submicron Structure

*F. Koch, M. Bollu, and A. Asenov**

Physik-Department, Technische Universität München,
D-8046 Garching, Fed. Rep. of Germany

Abstract. We show that electrical degradation of a micron-sized FET device leads to a submicron structure with interesting transport properties. The conductance gate voltage characteristic of a degraded FET shows evidence for quantum effects in the form of a resonant-tunneling process which causes a peak-to-valley conductivity change by many orders of magnitude. Noise is observed from single trap states created by the electrical stress.

1. Introduction

Small is beautiful when it comes to microelectronic ICs. More to the point, small devices are inherently fast. The FETs that rapidly switch currents in Megabit circuits in order to charge up trench cell capacitors have channel lengths of the order of one µm. Test structures with this µm-dimension are readily available to the experimentalist.

The step into the submicron world is laborious and expensive. The tools for making smaller lateral structures, although they exist, are available to only a few well-heeled laboratories. Several groups were successful in fabricating 1-dimensional (1D) FET-structures /1-4/. 1D means that the width of a 2D-channel has been made comparable to or even smaller than some characteristic length such as the most probable hopping distance /1/, the inelastic scattering length /2,3/ or the elastic mean free path /4/. Channel widths down to 500 Å have been achieved. Fowler et al. /5/ have worked with FETs whose channel lengths were 0.5 µm being not much larger than the localization length.

Examining the operation of a micron-sized transistor in a real circuit one comes to realize that electrical inhomogeneity along the direction of current flow really makes it a submicron structure. With gate- and drain voltages (V_g and V_D respectively) both in the 5 V range there exists a short zone just before the drain contact where the potential drops rapidly (Fig. 1). One speaks of the pinch-off region, with severely altered transport properties and avalanche generation of carriers. In addition, this zone is altered and degraded in a characteristic manner after a period of operation.

Visible evidence for the inhomogeneity of an operating device is the emission of light from the drain region. A short channel transistor examined under the microscope, when operated with $V_D > V_g$, actually lights

*) Permanent address: Institute of Microelectronics, Sofia (Bulgaria)

Springer Series in Solid-State Sciences Vol. 83: **Physics and Technology of Submicron Structures**
Editors: H. Heinrich · G. Bauer · F. Kuchar

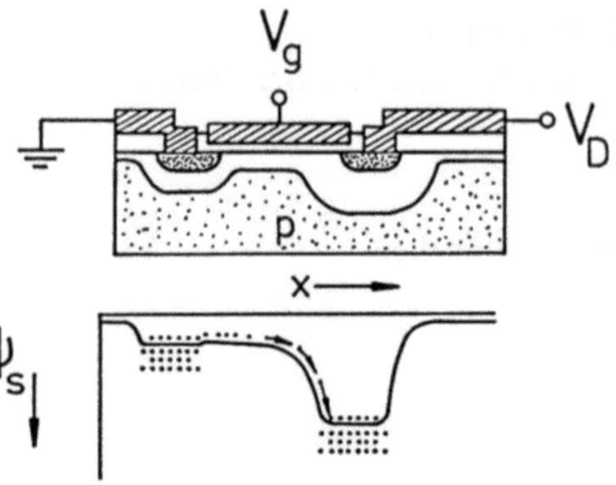

Fig. 1: Schematic cross-sectional view of a FET. For large V_D the potential drops rapidly in a small region near the drain.

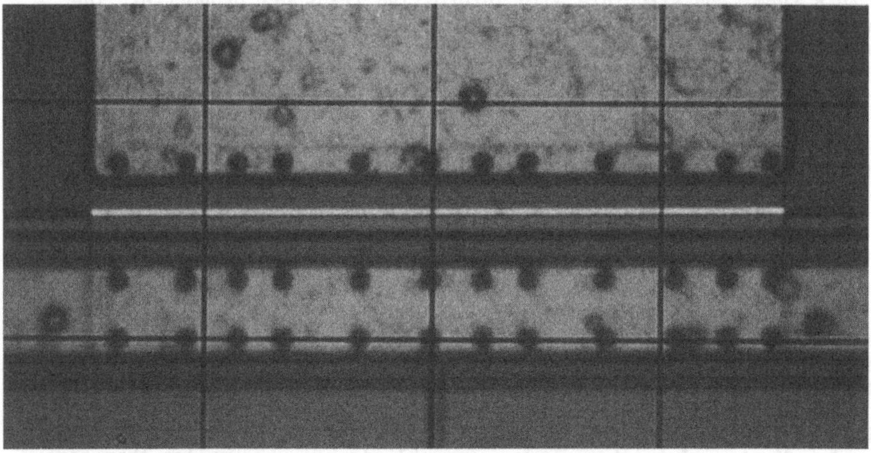

Fig. 2: Emission of visible light (bright strip) from a MOS-FET under electrical stress (after M. Herzog)

up. M. Herzog took the picture shown in Fig. 2. The brightly lit strip identifies the drain side of this n-channel device. A Megabit-DRAM in operation would look like a bill-board flashing for ∿ nsec intervals whenever a current pulse is generated to charge a capacitor cell.

In the following Sec. 2 we show how the electrical inhomogeneity can be exploited to make a very short channel (∿ 0.1 µm) device. Sec. 3 explores the defect states that define the short channel. We make use of 2D-device modelling to characterize electrically the short channel. Sec. 4 examines current transport in the short structure with particular attention to the subthreshold regime at low temperature, and Sec. 5 presents some observations on electronic noise in the small device.

2. Degradation - The Poor Man's Way to a ∿ 0.1 μm Device

Some of the hot carriers generated in the region of the electric field spike will collide with the bounding Si-SiO_2 interface. It appears that holes which are injected into the interface region in the pinch-off zone are responsible for the electrical-stress-induced degradation /6,7/. It turns out that the stress-induced defects represent a net negative charge after being occupied by an electron (acceptor state), thus giving rise to the formation of a potential barrier in an n-channel FET. The spatial distribution of the interface traps N_{it} along the channel reflects the position and width of the field spike that has caused them. The effect tends to saturation at a density which is of the order of several times 10^{12} cm^{-2}. We need not pursue here further the question of the physical origin of the N_{it}. It suffices to note that the degradation effect provides a high density of trap states in a narrow strip of width ∿ 0.1 μm.

We show in Fig. 3 how the point charges in occupied traps provide a potential barrier blocking the onset of conduction in the device until a much higher V_g is applied. It is an inherently fluctuating, rough barrier because of the granular nature of randomly positioned point charges. For each 1 μm of the FET width, the damaged strip has an area of 10^{-9} cm^2 and

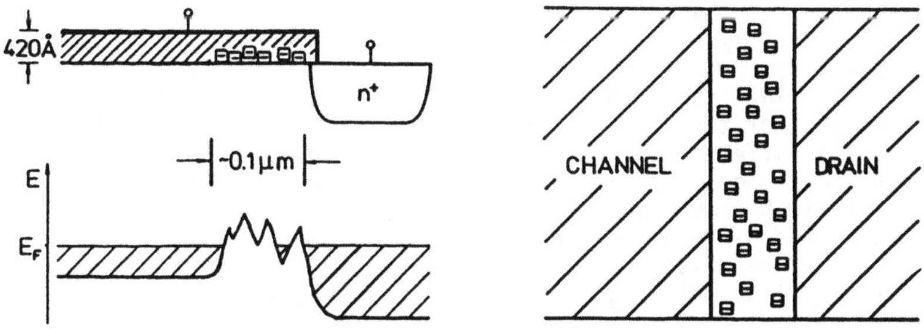

Fig. 3: Occupied interface traps, the "random-potential" barrier, and a top-view of the damaged strip.

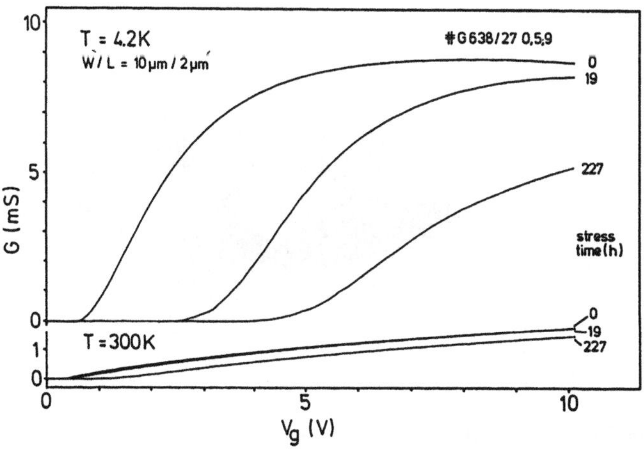

Fig. 4: Conductance vs. gate voltage after 0, 19, and 227 hours of stressing (V_g = 8 V, V_D = 8 V at 300 K) for temperatures 4.2 and 300 K.

contains 10^2 - 10^3 trapped charges. For voltages V_g ranging up to and somewhat above the new threshold voltage V_T after degradation, the channel resistance is dominated by the transport in the barrier strip. We have created a genuine submicron transistor with a different V_T and in series with the undamaged remainder of the channel. The latter is a convenient "Cu-wire" connection to the external measuring circuit.

Electrical measurements, as in Fig. 4, clearly show the expected shift of V_T, in particular for temperatures in the liquid He range. The conductance curve successively shifts to higher V_g as the stress time is increased. Note the tendency to saturation and the decreasing slope dG/dV_g. At room temperature the shift is much less, an effect that is explained in the next section in terms of the T-dependent occupation of the traps.

3. Modelling the Degraded Transistor

In order to understand quantitatively the distribution of potential and currents in the device after the defect states have been created, it is necessary to resort to numerical calculations. The challenge to do so arises when trying to make sense of the very marked T-dependence in Fig. 4. The observations cannot be accounted for in terms of thermal activation across a fixed barrier potential. In Refs. /8,9/ several different kinds of defect states were examined ranging from fixed charge, a distribution that has a constant density for all energies to one that peaks exponentially at the conduction band edge. Only the last of these gives a reasonable account of the T-shift data. The clue to understanding the temperature dependence of the V_T-shift lies in knowing how the band-bending ψ_s depends on V_g for different T. Because of the strong peaking of the N_{it} vs. energy at the band edge, the T-dependent Fermi energy E_F very effectively controls the trap occupation. At high T fewer of the states are filled for a given band bending because E_F lies lower. This thermal scanning of the N_{it} distribution is the central point in Refs /8,9/. It serves to establish the $N_{it}(E)$ by matching calculations to the experiments on subthreshold current.

Fig. 5 is taken from the calculations in Ref. /8/ and serves as an example of the simulations. Curve b gives ψ_s for a given V_g at T = 77 K.

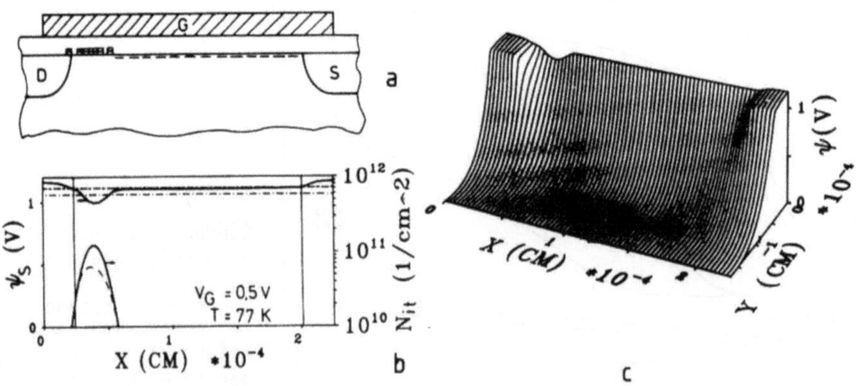

Fig. 5: Results of a modelling calculation for a degraded device.

The assumed spatial distribution of traps is given by the scale on the right side of the figure. Available states are those given by the solid line, filled states by the broken line. The corresponding 2D potential distribution appears in the lower part of the figure. It shows clearly the reduced ψ_S in the region of the defect states. This is the potential barrier blocking transport in the channel. The barrier height changes with ψ_S as more or less of the traps are filled. The modeling of course works with a smoother distribution of charge. It ignores the random placement of point charges.

Given the potential as in Fig. 5c it is a straightforward matter to calculate the currents that flow for small applied V_D ($\sim$ meV). In the subthreshold regime it is mainly the diffusion current in the spatial charge density profile described by the potential. Comparing calculated and measured subthreshold currents has served to obtain the energy distribution of the defect states.

4. Subthreshold Current Structures - The Case for Resonant Tunneling

The physical origin of the potential barrier is the small number of randomly positioned discrete charges. Modelling describes only the average potential, not the random and large fluctuations that result from the granular nature of the charge. Typically there is a single point charge every 300 - 100 Å in the barrier.

In Fig. 6 we show in a logarithmic plot the conductance G in S as measured at 4.2 K after successive stress cycles (V_g = 3 V, V_D = 8 V, T = 65 K). The threshold V_T, which in this plot represents roughly the point where the logarithmic curve changes slope, is shifted by as much as 8 V. The oxide thickness is 420 Å for this device, so that the shift indicates a defect density of up to 4 x 10^{12} cm^{-2}. We note the distinct, randomly peaked structures that evolve with increasing stress. The peak-to-valley ratio of the strongest peaks is as much as 4 orders of magnitude.

There are two conduction mechanisms which can account for this peaked structure. Variable range hopping (VRH) conductance in 1D is expected to be randomly peaked as a function of Fermi energy /10,11/. A 2D generalisation /12/ provides essentially the same result except that the peak amplitude is slightly reduced. The origin of the peaks in the framework of VRH can be understood as follows: In VRH an electron traverses the sample by jumping from one localised state to another. As the transition probability between two states depends exponentially on their energetic and spatial separation, the conductance is approximately given by that pair of states whose transition probability is the smallest (weak link). The peaks and valleys correspond to a sudden jump of the weak link to a new pair of states. The overall conductance in the VRH-model should obey the well-known Mott law, $G \sim \exp(-(T_0/T)^{1/3})$ in 2D. We do not observe an overall increase of G with temperature between 1 and 20 K, which excludes VRH from being the dominant transport mechanism in our sample.

The T-dependence of our conductance peaks is in good agreement with the predictions by Stone and Lee /13/ who introduced inelastic processes in Azbel's resonant tunneling model /14/. Tunneling through a potential barrier is strongly enhanced when the energy of the incoming particle coincides with the energy of some bound state in the barrier. Fig. 7 illustrates this situation in the case of a degraded FET. In the lower part of

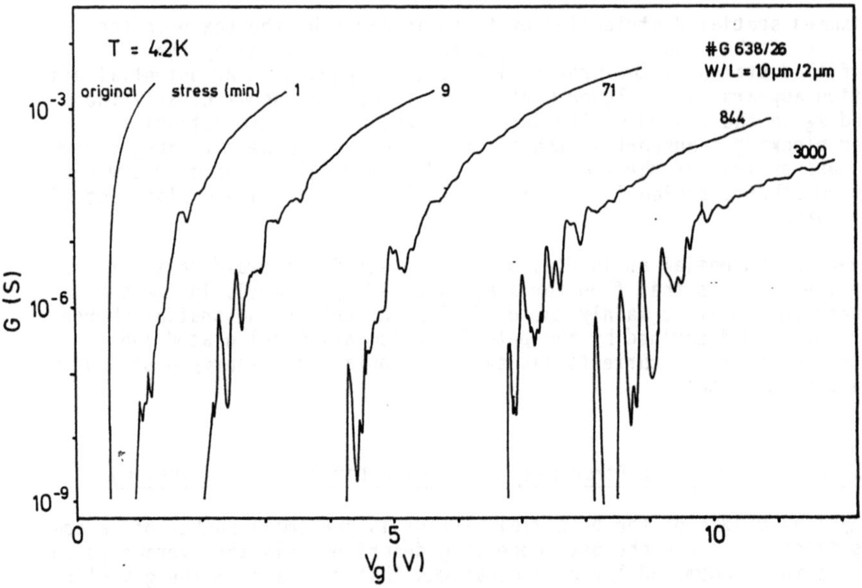

Fig. 6: Subthreshold conductance after various stress cycles (V_g = 3 V, V_D = 8 V, T = 65 K). The device width is 10 μm. Oxide thickness is 420 Å.

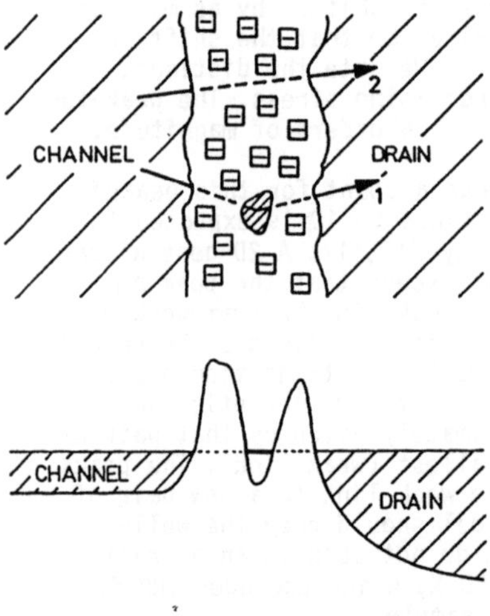

Fig. 7: Tunneling paths across the barrier. The resonant tunneling (path 1) is expected to dominate.

the figure the potential is drawn along the resonant tunneling path (path 1). Tunneling directly through the entire sample (path 2) is exponentially unlikely. It should be emphasized that Azbel's resonant tunneling takes place via one single site. In modern devices resonant states are realized by subbands of a 2D electron system in between an epitaxially grown double barrier structure /15/.

In the case of resonant tunneling the conductance peaks when E_F coincides with the energy of a resonant state. For the case that kT is larger than the linewidth Γ of such a state and E_F is off resonance, the conductance is dominated by activated hopping to the resonance /16/, and this will give linear flanks to the peaks in a log G vs. E_F plot with slope $\pm$ 1/kT. Stone and Lee /13/ predict that the peak maximum should decrease with temperature in such a way as to keep the integrated transmission (area under the peak) constant. If $kT < \Gamma$, the peak shape and amplitude should be independent of temperature.

The relation between gate voltage and Fermi energy is masked by the variation of the barrier potential itself with V_g. From the modelling considerations in the previous section we know that the average barrier height changes and that successively new trap sites are filled with increasing V_g. The rate of change in barrier potential with V_g can also be described in terms of a density of states. From the flanks of the peaks in the case $kT > \Gamma$ we obtain typically 1/10 of the full 2D density of states, which has also been reported by Fowler et al. /5/.

Azbel /14/ pointed out that a resonant state can only have a large transmission amplitude if it is close to the center of the sample (within the localisation length L_0). The peaks in the conductance vs. V_g are expected to be well separated, if there is at most one highly transmitting state within kT at E_F. This means that $D(E_F)L_0WkT \lesssim 1$. $D(E_F)$ denotes the density of states at the Fermi energy and W the width of the sample. Assuming $L_0 \approx 100$ Å this condition is satisfied in our device.

Generally, resonant transmission is expected to dominate VRH conduction below some temperature T_c at which the most probable hopping distance is larger than the sample length L /13/. Therefore T_c strongly increases, if L is reduced. In Ref. /5/ a transition from resonant tunneling to Mott hopping type of conduction is found to be in the 100 mK range for 0.5 µm long MOSFETs. In our deliberately fabricated short channel transistors resonant tunneling can be observed up to 20 K.

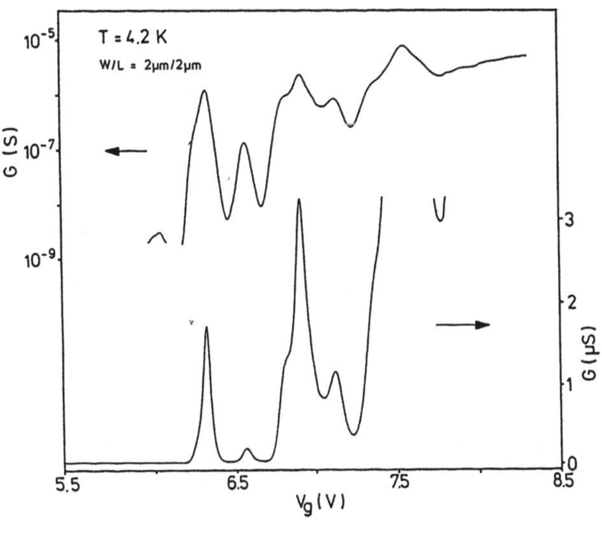

Fig. 8: Subthreshold conductance of a stressed 2 µm device in logarithmic and linear displays. The threshold V_T is ∿ 8 V.

We show in Fig. 8 an additional example of the subthreshold current structure in a totally different device. Both the logarithmic and linear plots are given for a transistor with the threshold V_T shifted to ∿ 8 V. The exact pattern of peaks differs from that of the previous Fig. 6, and yet there are similarities. The relative magnitudes and spacing of the peaks are similar. In several examples we find peaks that appear as a sequence and lead one to speculate. Could these be multiple resonant bound states of a given leakage path? Or does a new peak signify a newly occupied trap state in the vicinity of the given resonant tunneling junction? These questions remain to be answered in future work.

5. Noise. "To be or not to be (Occupied)", that is the Question.

A most natural consequence of the model of shallow interface traps that become filled with the increased band bending is that these traps can have a mind of their own. They can randomly in time absorb an electron and emit it again. The great sensitivity of the channel conductance in the threshold régime to a single added electron means that one can monitor the effect of a single state that changes occupation. Switching signals have by this time been reported by a number of workers /17-20/. The point to be made here is that with the electrical stressing more of such switching traps are created. At finite temperatures they all can fill and empty at their own characteristic pace depending on their activation energy and energy position relative to the Fermi level. By carefully tuning the spectral width of the measuring system, the temperature and position of E_F one can select to observe a chosen few of these. We show in Fig. 9 one example of such switching, in particular how by slowly sweeping V_g the signal comes and goes. Initially there is a rare, once in a while

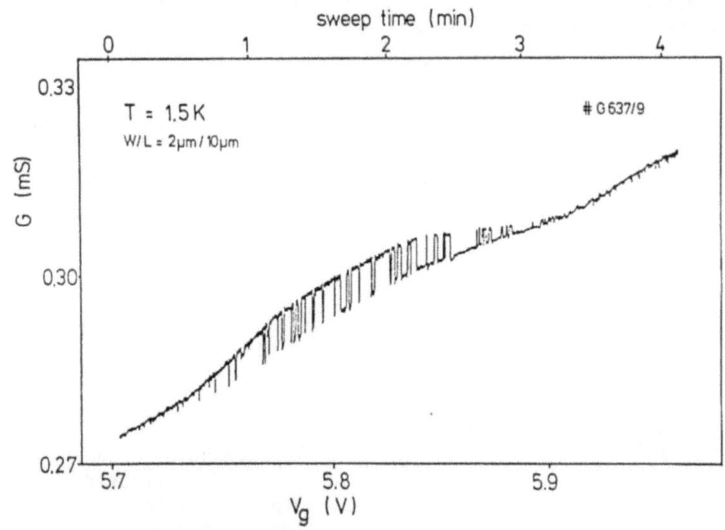

Fig. 9: Example of a trap switching signal varying with the applied V_g. The device is degraded only lightly with V_T-shift of ∿ 2 V.

appearance of the lower conductivity state. This becomes more frequent with rising V_g. The up- and down-states have about equal weight in the middle of the time-V_g sweep. At the upper end of the V_g sweep the down-state stabilizes.

For many of the switching signals we have studied the T-dependence of the characteristic frequency in order to obtain values of the trap energy E_T and the capture cross-section /19,20/. The E_T vary greatly, from a few tens of meV to 100 or more meV. The capture cross-section is found to be temperature activated with no unique value for the activation energy which can account for 1/f-noise observed in larger devices.

6. Acknowledgements and Apologies

We thank the Siemens A.G. for financial and material support in the way of test samples. The many fruitful discussions with W. Weber, W. Hänsch, and O. Jäntsch are gratefully acknowledged. The DAAD provided support for the visit of A. Asenov. M. Bollu received support as a Siemens-Stipendiat. M. Herzog provided the picture in Fig. 2. The motivation to look for light emission came from discussions with an interested colleague (Prof. R. Kaiser). T. Poppe supplied the data in Fig. 8.

The apologies go to the many friends and colleagues whose work we did not cite and discuss explicitly for lack of space and time.

References

1/ R.A. Webb, A. Hartstein, J.J. Wainer, A.B. Fowler, Phys. Rev. Lett. 54, 1577 (1985)
2/ S.B. Kaplan, A. Hartstein, Phys. Rev. Lett. 56, 2403 (1986); W.J. Skocpol, P.M. Mankiewich, R.E. Howard, L.D. Jackel, D.M. Tennant, Phys. Rev. Lett. 56, 2865 (1986); L. Eaves, these proceedings
3/ T.J. Thornton, M. Pepper, H. Ahmed, D. Andrews, G.J. Davies, Phys. Rev. Lett. 56, 1198 (1986)
4/ B.J. van Wees, H. van Houten, C.W.J. Beenakker, J.G. Williamson, Phys. Rev. Lett. 60, 848 (1988); H. van Houten, these proceedings
5/ A.B. Fowler, G.L. Timp, J.J. Wainer, R.A. Webb, Phys. Rev. Lett. 57, 138 (1986)
6/ K.R. Hofmann, Ch. Werner, W. Weber, G. Dorda, IEEE Trans. Electron Devices, vol. ED-32, 691 (1985)
7/ Ye Qiu-Yi, A. Zrenner, F. Koch, C. Zeller, G. Dorda, Appl. Phys. Lett 52(7), 561 (1988)
8/ A. Asenov, M. Bollu, F. Koch, J. Scholz, Appl. Surf. Sci. 30, 319 (1987)
9/ M. Bollu, A. Asenov, F. Koch, to be published
10/ P.A. Lee, Phys. Rev. Lett. 53, 2042 (1984)
11/ J.A. McInnes, P.N. Butcher, J. Phys. C 18, L 921 (1985)
12/ X.C. Xie, S. Das Sarma, Phys. Rev. B 36, 4566 (1987)
13/ A.D. Stone, P.A. Lee, Phys. Rev. Lett. 54, 1196 (1985)
14/ M. Ya. Azbel, Solid State Comm. 45, 527 (1983)
15/ See for example the article by Satoshi Hiyamizu in this book
16/ M.Ya. Azbel, A. Hartstein, D.P. Di Vincenzo, Phys. Rev. Lett. 52, 1641 (1984)
17/ K.S. Ralls, W.J. Skocpol, L.D. Jackel, R.E. Howard, L.A. Fetter, R.W. Epworth, D.M. Tennant, Phys. Rev. Lett. 52, 228 (1984)

18/ M.J. Uren, D.J. Day, M.J. Kirton, Appl. Phys. Lett. 47, 1195 (1985)
19/ M. Bollu, A.J. Madenach, F. Koch, J. Scholz, H. Stoll, in: Proceedings of 9th International Conference on Noise in Physical Systems, ed. C.M. van Vliet, p. 217, World Scientific, Singapore (1987)
20/ M. Bollu, F. Koch, A. Madenach, J. Scholz, Appl. Surf. Sci. 30, 142 (1987)

Physics and Fabrication of Metal-Semiconductor Microstructures

E. Rosencher and P.A. Badoz

Centre National d'Etudes des Télécommunications, B.P:98, Chemin du Vieux Chêne, F-38243 Meylan Cedex, France

1. INTRODUCTION

Thanks to advances in ultra-high vacuum technology, it has recently become possible to realize epitaxial semiconductor/metal/semiconductor (SMS) structures [1, 2] using a Si/$CoSi_2$/Si sandwich. The rather small lattice mismatch (~ 1.2 %) between Si and $CoSi_2$ crystals as well as their similar cubic structures allows the production of monocrystalline Si/$CoSi_2$ and Si/$CoSi_2$/Si heterostructures. These SMS structures open the way to promising ultra-low base resistance devices for millimeter wave applications.

In this paper, we intend to review the main results on the fabrication and physics of those metal-semiconductor microstructures. This latter category adresses mainly two different kinds of structures : the SMS-transistor where the $CoSi_2$ film is intended to be continuous and permeable base transistors (PBT) where discontinuities in the metallic film are intentionally introduced by nanolithography techniques. Though this paper is mainly focused on the electrical properties of these structures, some informations on the morphological quality of these sandwiches are given in Sec. II, which are necessary for the understanding of the transport properties. In Sec. III, parallel transport in the ultra-thin metal films will be addressed while the perpendicular transport through the SMS structure will be developed in Sec. IV. We shall finally describe the current status of the Si/$CoSi_2$/Si permeable base transistor in Sec. V.

2. FABRICATION AND MORPHOLOGY OF Si/$CoSi_2$/Si HETEROSTRUCTURES

Until recently, $CoSi_2$ layers were grown by solid phase epitaxy (SPE) on <111> Si surfaces. Co layers are electron gun evaporated under a pressure less than 5 x 10^{-8} Torr with the sample kept at room temperature and then annealed at 650°C (± 25°C) for 10 minutes in order to obtain the $CoSi_2$ layers. Extensive details may be found in [3].

Grazing angle scanning electron microscopy (SEM) and transmission electron microscopy (TEM) have shown that smooth $CoSi_2$ layers could be grown for a thickness up to 20 nm. Above this limit, the metal layers are rough and discontinuous, leading to a milky appearance of the wafers [3]. Cross-sectional TEM observations show that, for a thickness less than 20 nm, the interface between Si and $CoSi_2$ is rather smooth, with the presence of bowl-shaped $CoSi_2$ intrusions in Si, while the $CoSi_2$/vacuum interface is much smoother (Figure 1) [4]. Plane view TEM photographs of a $CoSi_2$/Si structure are shown in Figure 2a and b. The shape of the Moiré fringes indicates that strain fields are present in the $CoSi_2$ layers and reflects the 1.2 % lattice mismatch between Si and $CoSi_2$. Moreover, no pinholes are observable over a few square micrometers [5].

Springer Series in Solid-State Sciences Vol. 83: **Physics and Technology of Submicron Structures**
Editors: H. Heinrich • G. Bauer • F. Kuchar

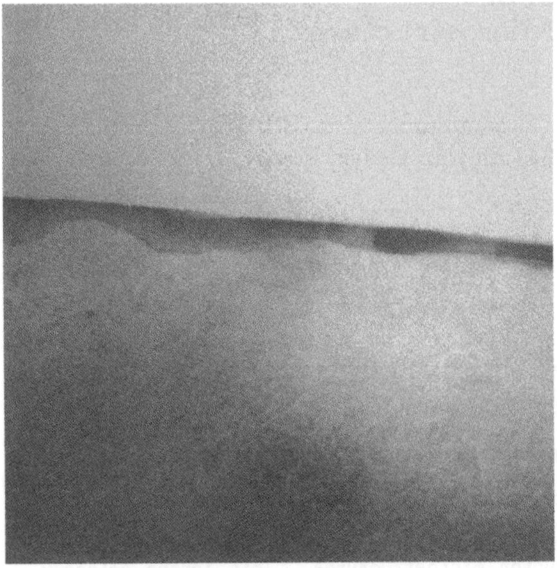

Fig. 1 Cross-sectional TEM image of a 5 nm thick $CoSi_2$ layer on Si substrate. One notes the bowl-shaped $CoSi_2$ intrusion in Si bulk for an otherwise sharp $CoSi_2$/Si interface.

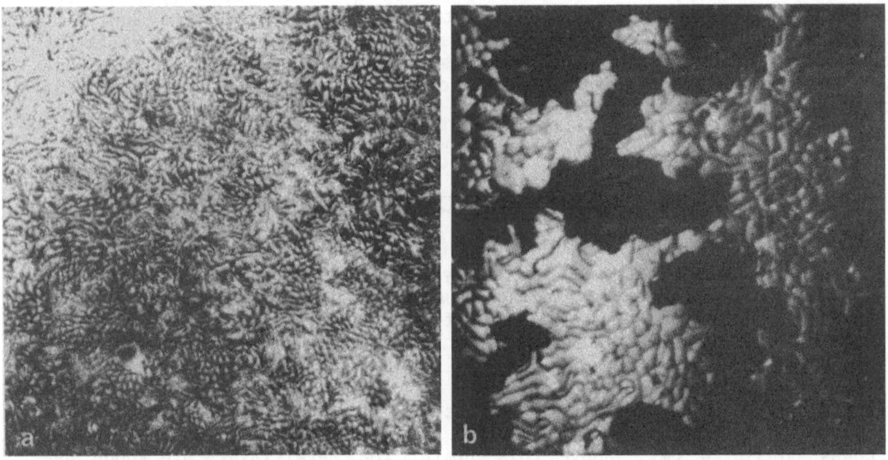

Fig. 2 TEM plane view observations of 5 nm thick $CoSi_2$ layers. (a) The bright field image evidences the Moiré fringes (due to lattice mismatch and strains) and the absence of pinholes. (b) The dark field image using a diffraction spot characteristic of type B grains indicates an equivalent density of both grains.

These pinholes, when they are present, appear as regions where Moiré fringes are absent, as described in [6]. Figure 2b is a plane view TEM observation which stresses the orientation of the $CoSi_2$ crystal relative to the Si one in the <111> direction. It is clear that $CoSi_2$ is a mixture of two types of grains : type A grains where the Co planes are an extension of the Si bulk planes and type B grains where the $CoSi_2$ lattice is rotated 180° around the direction relative to the Si lattice [4].

The realization of monotype (only A or B grains) and pinhole-free $CoSi_2$ layers by SPE has recently been reported by various groups [7]. The growth conditions are rather complex, depending on the $CoSi_2$ layer thickness, the nature of the Co/vacuum interface, etc. Anyway, the physical origin of the pinhole formation during growth or subsequent annealing is still unclear.

For SMS transistors, silicon layers are deposited by MBE on $CoSi_2$ films at 650°C and doped by Sb coevaporation under a pressure in the 10^{-9} Torr range. Figure 3 is a cross-sectional TEM photograph of a $Si/CoSi_2/Si$ heterostructure. It is clear that the roughness of the interfaces has dramatically decreased, the transition between $CoSi_2$ and Si material occurring within one monolayer over long distances. This effect (we call it "planarisation") is still unexplained. Moreover, grazing angle SEM observations indicate that the growth mechanism of Si over $CoSi_2$ is three-dimensional (Figure 4). This results in a high density of sub-grain boundaries, twins, etc. in the epitaxial Si layer. Finally, let us note that high resolution TEM (HRTEM) observations indicate that, for such small $CoSi_2$ film thickness (< 2 nm), the $CoSi_2$ lattice is entirely strained in the $Si/CoSi_2/Si$ sandwich in order to fit the Si bulk lattice [4].

It has to be noted that von Känel and his coworkers [8] have recently reported the growth of $CoSi_2/Si$ heterostructures using coevaporation of Co

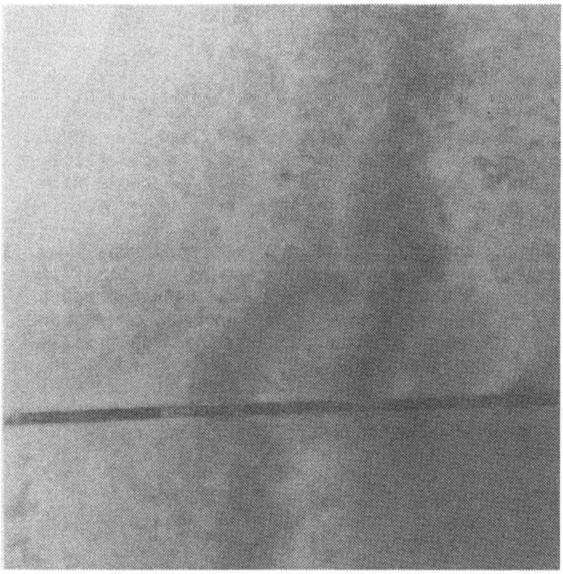

Fig. 3 Cross-sectional TEM image of a $Si/CoSi_2/Si$ heterostructure. Note the planarisation of interfaces relative to Figure 1.

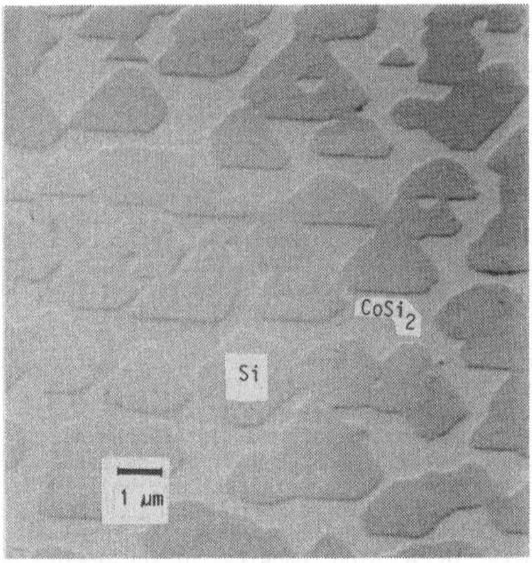

Fig. 4 SEM image of 23 nm thick Si epilayer on top of $CoSi_2$ (8 nm)/Si_{bulk} structure.

and Si upon Si wafers with evaporation rates that have to be exactly stoichiometric. Two-dimensional Si growth on top of $CoSi_2$ has been claimed by these authors for growth temperatures less than 450°C, yielding strained and pinhole-free layers when the film thicknesses are kept below 5 nm for $CoSi_2$ and 2 nm for Si. This technique opens the way to the study of metal-semiconductor superlattices.

3. PARALLEL TRANSPORT IN Si/$CoSi_2$ HETEROSTRUCTURES

Because of their outstanding crystalline quality, $CoSi_2$ films are ideal candidates for the study of electron transport in ultra-thin metallic films. Indeed, other possible systems (like Au on NaCl) are usually unstable with respect to exposure to air and temperature cycles.

Hensel et al. were the first to show that down to very low thickness, i.e. 10 nm, the film resistivity exhibits little dependence on the $CoSi_2$ film thickness [9]. These thicknesses are much less than the bulk transport scattering length of ~ 100 nm as determined by magneto-resistance measurements, so that boundary scattering of the carriers is essentially specular in this system. In order to account for this phenomenon, these authors have used the well-known Fuchs-Sondheimer theory derived from the classical Boltzmann's equations, introducing a specularity parameter p which is the fraction of electrons specularly reflected from the interfaces. The resistivity ρ of a film of thickness d is thus given by

$$\rho = \rho_\infty \left[1 - 3/2k \int_0^1 \frac{(u - u^3)(1 - p)(1 - \exp(-k/u))}{1 - p\exp(-k/u)} du\right]^{-1} \qquad (1)$$

where ρ_∞ is the bulk resistivity and k the ratio of the film thickness d to the mean free path λ_e, i.e. $k = d/\lambda_e$. Figure 5a shows the set of calculated curves in $CoSi_2$ film for specularity parameter p ranging from 0 to 1 (i.e.

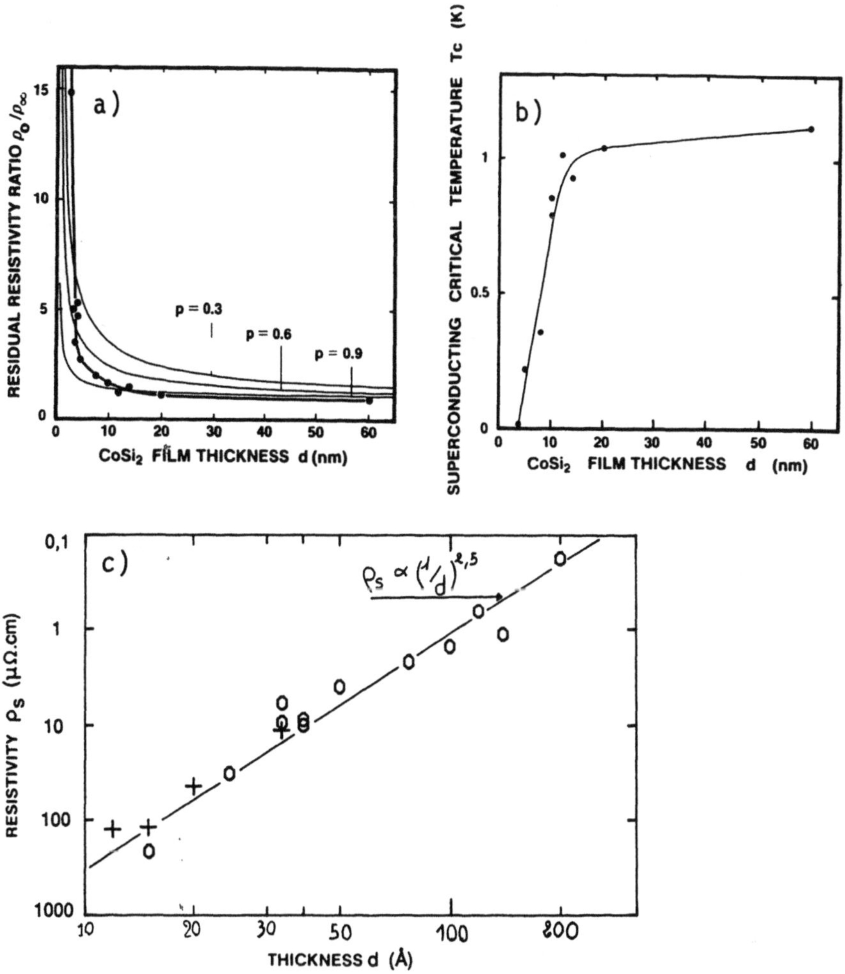

Fig. 5 $CoSi_2$ film resistivity ρ_0 measured at 4 K (a) and superconductivity critical temperature T_c (b) as a function of film thickness d. The variation of ρ_0 as a function of the inverse of the film thickness 1/d follows a power law (solid line) which can be explained using a quantum transport theory (see text) (c).

from purely diffuse scattering to purely specular) compared to experimental data from Badoz et al. [10]. It is clear that for film thicknesses lower than 10 nm, the variation of the film resistivity with thickness is far steeper than expected from Fuchs-Sondheimer theory. Figure 5b shows the values of the superconducting critical temperature T_c of $CoSi_2$ films as a function of thickness. Here again, there is an abrupt drop of T_c in $CoSi_2$ films thinner than 10 nm. It has to be noted that these results are different in nature from the usual experiments in thin metal films near the localisation regime ($R_\square \sim \hbar/e^2 \sim 4000\ \Omega_\square$). In our experiments, the sudden change in T_c and ρ occurs in films with resistances in the 30 $\Omega_\square$ range.

In order to investigate the influence of $CoSi_2$/Si interface roughness, resistivity and Hall measurements in ultra-thin $CoSi_2$ layers obtained by SPE (o) and coevaporation (+) have been compared : the latter -provided by Von Känel and coworkers- are indeed smoother than the former. Figure 5c shows that no influence of interface roughness is observed : this result is indicative of a fundamental origin of the resistivity enhancement for very low thicknesses [11].

Fishman and Calecki have developed a model based on the quantum transport of carriers in those ultra-thin metal films [12]. Indeed, the de Broglie wavelength of electrons in $CoSi_2$ (about 7 Å) is of the same order of magnitude as the investigated metal layer thicknesses. Their model predicts a resistivity versus thickness function given by

$$\rho = \rho_v + \rho_s \tag{2}$$

with

$$\rho_s^{-1} = [\frac{e^2}{4\pi\hbar^6}][\frac{d^5}{(\lambda\Delta)^2}]\frac{6}{N(N+1)(2N+1)}\sum_{\nu=1}^{N}(\frac{k_{\nu F}^2}{\nu^2}) \tag{2'}$$

where ρ_v is the bulk resistivity, Δ is the mean thickness over which transport is forbidden, N is the number of the last occupied 2D subband and $k_{\nu F}$ is the Fermi wave vector of the νth 2D subband. λ is related to the mean

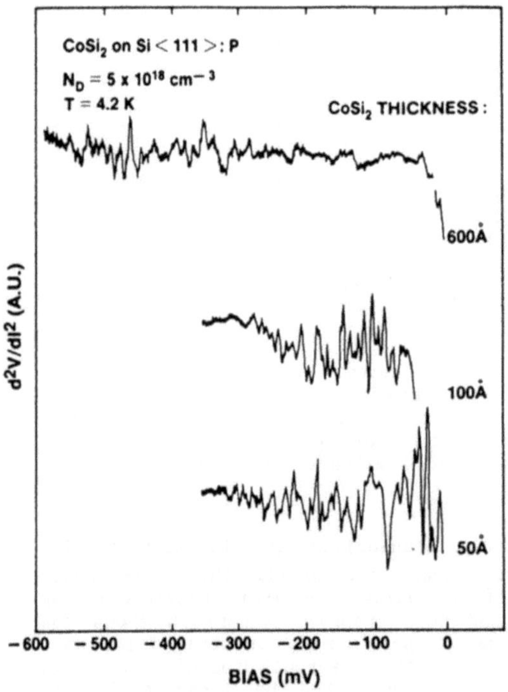

Fig. 6 High bias tunneling spectra of $CoSi_2$/Si tunnel diodes with metal thicknesses of 50, 100 and 600 Å.

distance between potential fluctuations. Figure 5c shows that the agreement between theoretical prediction and experimental data is excellent for $\Delta \sim 4.5$ Å, i.e., the thickness of one monolyer and $\lambda \sim 2$ Å. The low values of λ and Δ, as well as the lack of experimental influence of layer roughnesses, indicate that those fluctuations are likely due to diffusive centers at the $CoSi_2$/Si interface (such as ill-coordinated magnetic Co atoms [4]) rather than thickness fluctuations.

Quantization of the metallic electron gas in ultra-thin $CoSi_2$ films has been studied by tunneling spectroscopy (TS) in degenerate Si/$CoSi_2$ Schottky diodes [13]. Sharp features are observed in the TS spectra whose behaviour as a function of $CoSi_2$ film thickness depends on the energy range. From - 150 meV up to + 150 meV relative to the $CoSi_2$ Fermi level, the peak positions are not dependent on the film thickness ; additional experiments indicate that those features are due to phonon emission by the relaxation of hot electrons in the depletion layer of Si [14]. For higher energies (from - 200 meV to - 600 meV), a set of sharp features is observed, whose position depends in a complex way on the metal film thickness (see Fig. 6). This complex behaviour is likely to be due to the high value of the electron energy at the Fermi level (~ 4.5 eV), leading to a small value of the de Broglie wavelength (~ 7 Å). Small fluctuations in film thickness thus lead to dramatic changes in the energy distribution of quantized electrons [14]. Experiments are currently under progress to investigate the influence of $CoSi_2$ film roughness on tunneling spectra.

4. PERPENDICULAR TRANSPORT IN Si/$CoSi_2$/Si HETEROSTRUCTURES

The first evidence of a transistor effect in a monolithic SMS structure was given by Rosencher et al. in 1984 [15]. Figure 7 shows the energy band diagram of a SMS transistor. This device consists basically of two back to back Schottky diodes. One of these diodes, the emitter, is forward biased, while the other one, the collector, is reverse biased. The carrier transport between the emitter and the collector is the subject of intensive investigations [5, 17-19]. Two mechanisms may indeed be involved, with relative weight strongly dependent on technology.

1. Electrons are emitted via thermionic emission from the forward biased emitter junction in the metal. A fraction of those carriers cross the metal films via ballistic transport and are collected by the reverse biased collector junction [15, 18]. This mechanism is described by the solid lines in Figure 7.

2. Pinholes are present in the metal film, in which silicon channels are imbedded. The electrons are transferred from the emitter to the collector via those semiconducting channels and the current flow is controlled by the barrier lowering in the pinholes (dashed line in Figure 7) [16, 17].

A theoretical model has been developed in [19] in order to evaluate the different weights of mechanisms 1 and 2. The conclusion is that, for usual doping levels, a single 150 nm radius pinhole in the metal base is enough to short circuit the whole ballistic transport in a 20 µm x 20 µm SMS transistor ! It is thus clear that TEM observations, which investigate only a few square micrometers of a device, have no statistical significance in drawing conclusions on the predominance of one mechanism over the other. We have thus developed an electrical measurement, a transconductance technique described in [16] and [19], which allows us to measure the relative weights of mechanisms 1 and 2. This technique, based on the screening of the collector potential by the metallic $CoSi_2$ film when no pinholes are present, ensures that, in "pinhole-free" SMS, electron transport occurs almost entirely through the metal base. Independent evidence of the dominant role of hot electron transfer through the base in SMS-T is given in Figures 8a and 8b

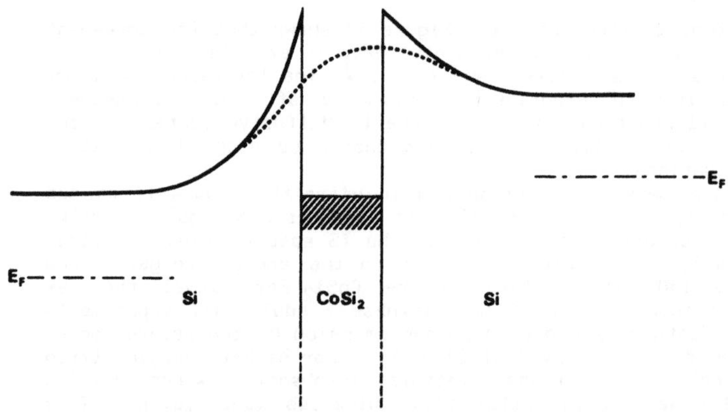

Fig. 7 Schematic energy-band diagram of a SMS transistor in a semiconductor- metal-semiconductor junction (solid line) and in a semiconductor pinhole channel (dashed line).

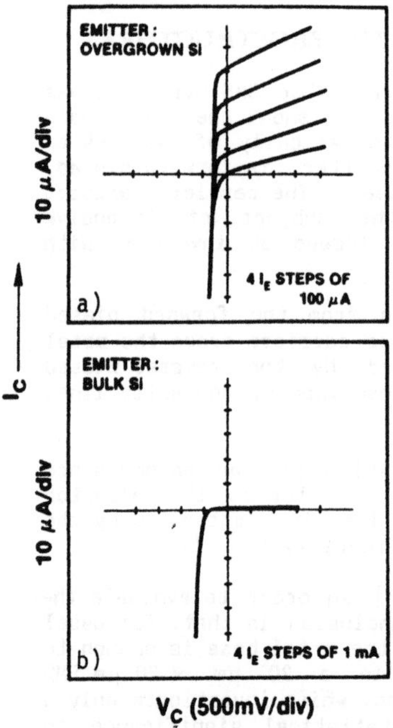

Fig. 8

Fig. 9

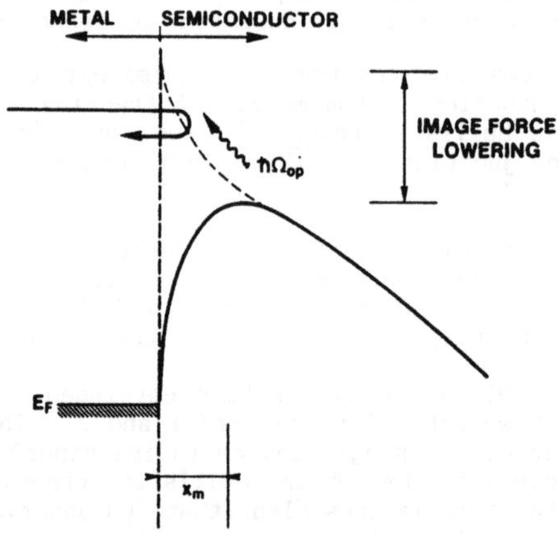

Fig. 8 Common base current-voltage characteristics of a SMS-T with a 7 nm thick base using either the regrown Si (c) or the bulk Si (b) as emitter. The measurements are performed at room temperature.

Fig. 9 Electron potential energy versus distance from the collector metal-semiconductor interface. An example of an electron backscattering event is symbolized.

showing the common-base characteristics of a SMS structure (7 nm base thickness) using either the overgrown (Figure 8a) or bulk (Figure 8b) silicon as the emitter. The transfer ratios α are 15 % and less than 10^{-4}, respectively, for the same values of emitter current I_E (500 μA). Since the pinhole current must be of the same order of magnitude in both directions, the value 15 % is clearly due to the transfer through the metal base. The asymmetry of current gain is consistent with the already reported systematic difference in barrier heights ϕ_{ms} between the $Si_{bulk}/CoSi_2$ ($\phi_{ms} \sim 0.63$ eV) and $Si_{epi}/CoSi_2$ ($\phi_{ms} \sim 0.69$ eV) junctions [20] : the injected electron energy is well above the collector barrier when emitted from the overgrown Si and below when emitted from the bulk Si.

The ballistic transport [21] in "pinhole-free" SMS-T is described by a mean free path $\lambda_B(T)$, so that the emitter-to-collector transfer ratio of electrons is expected to be

$$\alpha = \alpha_o (T) . \exp (- d/\lambda_B (T)) \quad (3)$$

where T is the measurement temperature and α_o is the current gain extrapolated to zero metal base thickness. The departure of α_o from unity is due to collector as well as emitter losses :

$$\alpha_o = \alpha_c \cdot \alpha_q \cdot \alpha_e \quad (4)$$

where α_c is the current gain upper limit associated with scattering in the Si collector, α_q is the quantum mechanical transmission of the base-collector potential barrier and α_e is the emitter efficiency coefficient. The collector scattering contribution is expected to follow

$$\alpha_c = \exp (- x_m/\lambda_{ph}) \quad (5)$$

where x_m is the position of the maximum of the collector barrier potential in the image force approximation and λ_{ph} is the mean free path in the Si collector (see Figure 9).

Figures 10a and b compare the above theory with experiment. Figure 10a shows the transfer ratios α obtained on samples with various values of $CoSi_2$ base thickness, at room temperature and at 77 K. Error bars correspond to the dispersion of values for different devices fabricated on the same wafer. The results clearly show that Eq. (3) is verified, with values of λ_b of 8 ± 1.5 nm at 300 K and 35 ± 5 nm at 77 K. This agreement is strongly in favor of ballistic theory. Moreover, the λ_B values are close to the mean free path deduced from resistivity measurements [22]. This indicates that the same scattering mechanisms control both the electron transport close to the Fermi level and the hot electron relaxation for energies in the 0.7 eV range above E.

Figure 10b shows α versus $V_{BC}^{-1/4}$ curves, taken at different temperatures, where V_{BC} is the base-collector bias. Equation 5 clearly holds since $V_{BC}^{-1/4}$ is directly proportional to x_m [21]. Furthermore, a mean free path is extracted from the slope of the curves and its temperature variation is shown in the inset of Figure 10. The very low values obtained for λ_{ph} (< 2 nm) remain to be understood.

Another problem requiring explanation is the overall low values of α_o (~ 0.3 at RT) corresponding to $\alpha/\alpha_c \sim 0.6$. The answer is most probably related to the quantum nature of the electron. Indeed, the electron energy E_1 in the metal is in the 5 eV range while its value E_2 in the semiconductor is in the 50 meV range, i.e. the Schottky barrier lowering. Consequently, the abrupt change in the electron wavelength leads to a quantum reflection at the metal-semiconductor interface. If the crude model of the abrupt-step barrier is assumed, the quantum transmission coefficient α_q is [22]

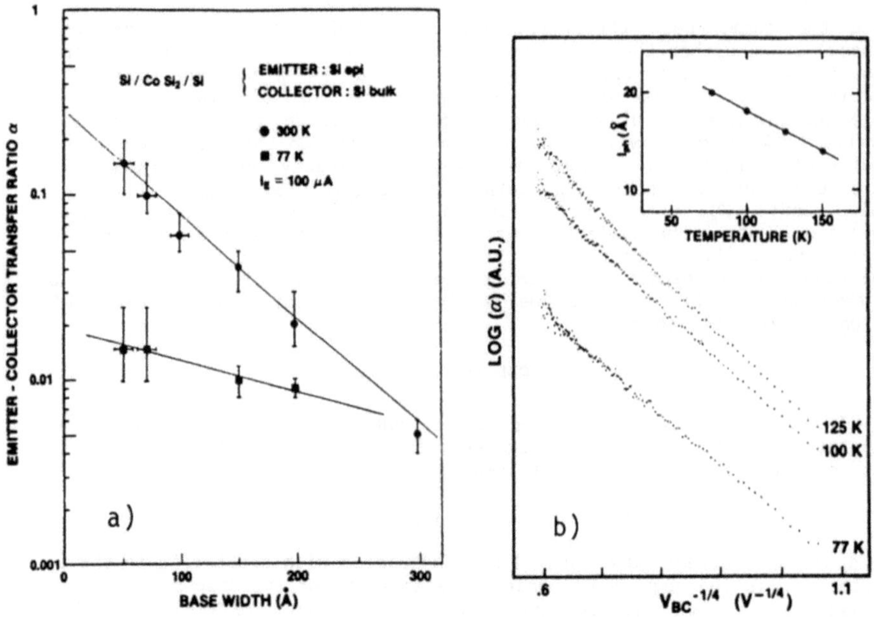

Fig. 10 Transfer ratio α versus $CoSi_2$ base thickness measured at 77 K and 300 K in SMS transistors (a). Transfer ratio α versus base-collector bias at different temperatures. The mean free path λ_{ph} deduced is shown in the inset as a function of temperature (b).

$$\alpha_q = 1 - [(m_2*E_1)^{1/2} - (m_1*E_2)^{1/2}/((m_2*E_1)^{1/2} + (m_1*E_2)^{1/2})]^2 \qquad (6)$$

where m_i* is the effective mass in the ith medium. Taking $m_1* = 1$ in $CoSi_2$ and $m_2* \sim 0.3$ in Si, one obtains $\alpha_q \sim 0.5$, which is in fair agreement with the experimental data of $\alpha_0 \sim 0.3$ taking into account the collector backscattering.

All these results, as well as those obtained by Sze and his coworkers [21], show that the quantum reflection is a severely limiting factor for the device interest of SMS transistors. These results suggest, in order to reduce this reflection, the use of highly asymmetrical SMS structure, for instance by use of two different semiconductors and/or metals.

5. Si/CoSi$_2$/Si PERMEABLE BASE TRANSISTORS

The idea of the permeable base transistor (PBT) is to take advantage of mechanism 2 described in the preceding section. The main advantages of this device are :

> For small enough Si channels, PBTs behave like thermionic devices so that their transconductance is very high [23].

> There are no fundamental limitations such as quantum reflection in SMS transistors.

> It is easily shrinkable with no problems of punchthrough such as in bipolar transistors.

> Compared to bipolar transistors, there is no problem associated with minority carrier storage time and the base access resistance can be extremely low (≦ 1 Ω).

Two ways are possible to imbed semiconductor channels in a metal grid.

1. The first one is to use the natural porosity of a metal layer on the surface of a semiconductor, leading to a natural permeable base transistor (also called a metal grid transistor [24]). Discontinuous $CoSi_2$ films [6, 16] but also W layers deposited during Si CVD growth [24] have been used. However, the high input capacitance, as well as the lack of control in the geometry of metal openings, is not in favour of such a device.

2. The second way is to define the opening by lithographic techniques. Bozler and Alley [25] were the first to realize such a structure. They have used W grids of 320 nm periodicity and CVD deposited GaAs as the semiconductor, though the W/GaAs system is not epitaxial.

The fabrication of $Si/CoSi_2/Si$ permeable base transistors with submicron Si channel dimensions has recently been reported by Glastre et al. [26]. These authors have solved the combined problems of submicrometer lithography and epitaxial growth in the following way.

Since Co has no volatile compound at room temperature with usual etching gases (F, Cl...), $CoSi_2$ lines cannot be etched from a continuous layer by a dry etching technique. Only sputtering may be used, which is detrimental to the Si surface on which Si overgrowth is to take place.

A counter mask technique has thus been used : Si wafers are thermally oxidized in wet ambient at 900°C, leading to a 120 nm thick SiO_2 film. A continuous 30 nm W layer is deposited by sputtering. An electron beam exposure dose in the 200-300 $\mu C/cm^2$ range is then used to expose a 360 nm thick spin coated PMMA resist layer which serves as an etch mask for the underlying W film. After development, W is etched in a SF_6 plasma and acts as an etch mask for the CHF_3 plasma etching of SiO_2 The remaining W film is then dissolved in a ferricyanide solution. Low energy electron diffraction (LEED) and Auger spectroscopy indicate the existence of a highly perturbed Si surface layer about 5 nm thick [27]. A sacrificial oxide of 15 nm thickness is thus grown at 900°C and consequently dissolved by a HF dip in a glove box connected to the molecular beam epitaxy (MBE) machine. After a 900°C thermal flash for 10 s, sharp 7 x 7 structures in the LEED diagrams as well as Auger spectroscopy indicate excellent crystallinity and cleanliness of the Si surface.

A 30 nm thick Co layer is then evaporated on the sample kept at room temperature under a pressure of 3×10^{-9} Torr. The system is annealed at 650°C for 10 min yielding a 90 nm thick $CoSi_2$ film. In situ LEED control shows two different behaviours, depending on whether the LEED spot is focused on the $CoSi_2$ film (sharp 1 x 1 pattern) or on unreacted Co covering SiO_2 layers (no pattern). The unreacted Co is then dissolved in an oxidizing solution.

Figure 11 shows the scanning electron microscopy (SEM) observations of $CoSi_2$ films in submicron lines and in a 5 µm line, respectively. The huge density of pinholes in large area thick $CoSi_2$ films as observed in Figure 11 is well known and has been ascribed by Ishibashi and Furukawa to the smaller surface energy of Si relative to $CoSi_2$, which tends to denude the Si surface [28]. As a matter of fact, weak 7 x 7 structures can be observed in the LEED

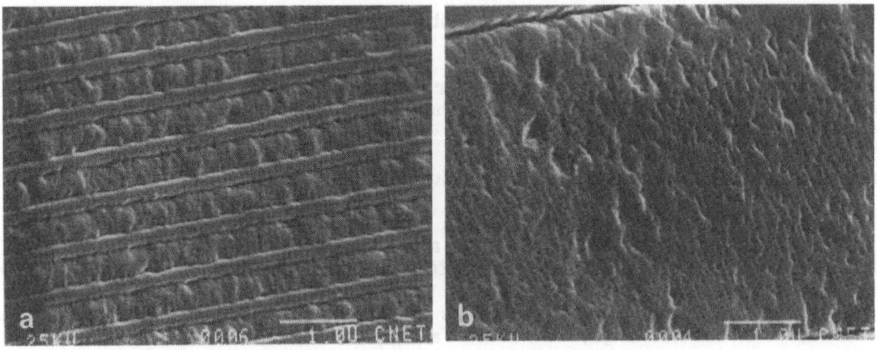

Fig. 11 SEM photographs of a 90 nm $CoSi_2$ reacted at 650°C in 0.4 µm (a) and 5 µm (b) lines, respectively. Note the reduction of pinhole density in the submicron lines and at the $SiO_2/CoSi_2$ border line compared to large $CoSi_2$ areas.

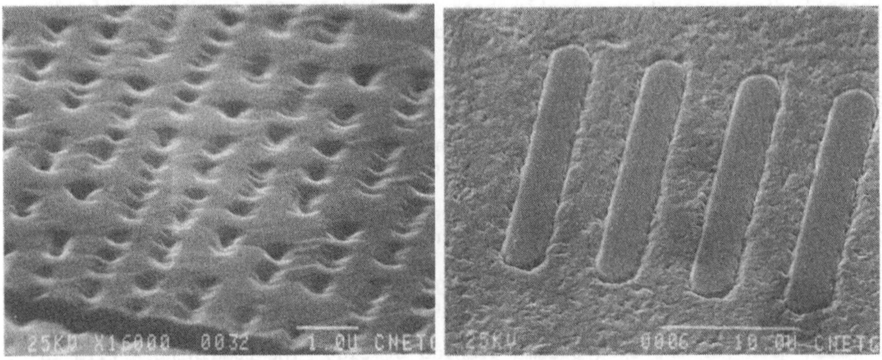

Fig. 12 SEM view of the surface of silicon overgrown on top of 60 nm thick $CoSi_2$ grating of 0.8 µm (a) and 8 µm (b) period, respectively.

patterns on $CoSi_2$, indicative of denuded Si surfaces. It is however clear from Figure 11 that the density of pinholes is dramatically reduced in the submicron $CoSi_2$ lines. The same remark applies to the $SiO_2/CoSi_2$ border line over a distance of 0.8 µm. Following the arguments of [28], it is believed that strains due to the lattice mismatch between both crystals are relieved at the $CoSi_2$ film periphery.

After $CoSi_2$ formation in submicron lines, unreacted Co and SiO_2 lines are chemically dissolved in the glove box. The Si wafer is reloaded in the MBE chamber and annealed at 900°C for 10 sec. 7 x 7 LEED patterns and Auger spectra indicate a perfect Si surface, which is not surprising since this Si surface was protected by a SiO_2 film during the whole experiment. The Si/SiO_2 interface is indeed known to be atomically smooth and defect-free [29]. Si is then evaporated under a pressure of 2×10^{-9} Torr at a deposition rate of 0.18 nm/sec : the first 800 nm are 3×10^{16} cm^{-3} Sb doped and the last 200 nanometers are Sb degenerated to ensure a good ohmic contact. As previously, LEED patterns confirm the good crystallinity of the Si deposited layer. Figures 12a and 12b show SEM photographs of the

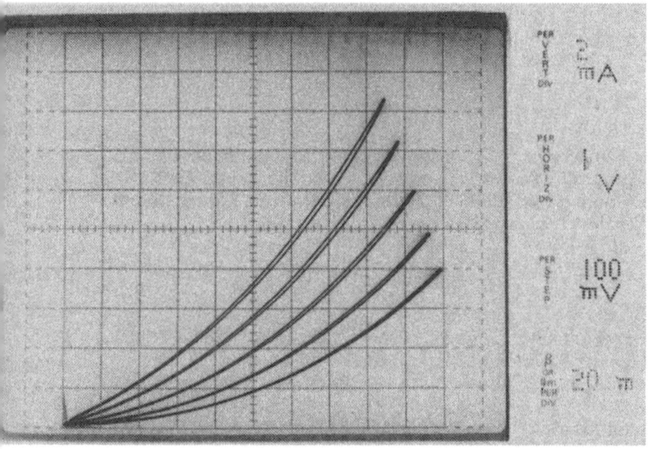

Fig. 13 Common emitter characteristics of a $Si/CoSi_2/Si$ permeable base transistor with a 0.4 µm period $CoSi_2$ grating. The emitter area is 200 µm (emitter-base bias from - 0.4 V to 0.4 V with 0.1 V steps).

resulting Si surface for 0.8 and 8 µm period gratings, respectively. Si on top of the 4 µm lines is crystalline but has grown three-dimensionally and is a mixture of A and B grains. On the other hand, Si deposited on top of submicron lines exhibits a smooth surface with some residual defects above the $CoSi_2$ lines : silicon grains have been seeded in the Si lines and have merged through lateral epitaxial growth on top of the $CoSi_2$ grating. It is not clear, however, whether Si atoms preferentially stick in the Si openings and realize a pure A type lateral epitaxy whatever the orientation of $CoSi_2$ crystallites in the lines or whether $CoSi_2$ submicron lines are single type and Si thus grows two-dimensionally. This question is currently under study using transmission electron microscopy.

Finally, a 150 nm W layer is deposited ex situ followed by a classical 1 µm Al evaporation. The W film serves as a barrier against Al diffusion in silicon. The emitter region (200 µm x 200 µm) is then defined by optical lithography and reactive ion etching. The resulting structure is a mesa structure which is electrically tested using the Si epilayer as the emitter, the $CoSi_2$ film as the base and the rear side as the collector. Figure 13 shows common emitter I-V characteristics of a PBT with 0.4 µm wide Si channels. The measured transconductance is 100 $A/V.cm^2$ for a collector current of 40 A/cm^2 (corresponding to 1 $kW.cm^{-2}$), which is a factor of 5 below theoretical expectation [23].

This discrepancy is due to a residual contamination of the Si surface between the $CoSi_2$ lines, which results in a parasitic junction. Work is in progress to develop a more efficient cleaning procedure of the submicron Si channels.

6. CONCLUSION

The physics and technology of metal-semiconductor microstructures are clearly a rapidly expanding new field of research. The success of new growth techniques such as codeposition opens the way to new studies such as hot electron spectroscopy [30, 31], Fowler-Nordheim emission [32], etc. Finally, the search for epitaxial semiconducting silicides is still of paramount importance : it would offer to the $Si/CoSi_2$ system all the possibilities of the III-V and II-VI families.

ACKNOWLEDGEMENTS

The authors thank F. Arnaud d'Avitaya, Y. Campidelli, J. Chroboczek, S. Delage, G. Glastre, J.C. Pfister, C. Puissant and G. Vincent from CNET, A. Briggs, C. Calecki and G. Fishman from CNRS, J. Henz, H. von Känel and M. Ospelt from ETH Zurich for contributing to this work.

REFERENCES

[1] S. Saitoh, H. Ishiwara, S. Furukawa : Appl. Phys. Lett. 37, 203 (1980).
[2] J.C. Bean, J.M. Poate : Appl. Phys. Lett. 37, 643 (1980).
[3] F. Arnaud d'Avitaya, S. Delage, E. Rosencher, J. Derrien : J. Vac. Sci. Technol. B3, 770 (1985).
[4] C. d'Anterroches, F. Arnaud d'Avitaya : Thin Solid Films 137, 351 (1986).
[5] E. Rosencher, P.A. Badoz, C. d'Anterroches, G. Glastre, G. Vincent, F. Arnaud d'Avitaya : Mat. Res. Soc. Symp., vol. 91, 415 (1987).
[6] R.T. Tung, A.F.J. Levi, J.M. Gibson : Appl. Phys. Lett. 48, 635 (1986).
[7] J.L. Batstone, R.T. Tung, J.M. Phillips, J.M. Gibson : Mat. Res. Soc. Symp. (To be published).
[8] J. Henz, H. von Känel, M. Ospelt, P. Wachter : Surface Science 189/190, 1055 (1987).
[9] J.C. Hensel, R.T. Tung, J.M. Poate, F.C. Unterwald : Phys. Rev. Lett. 54, 1840 (1985).
[10] P.A. Badoz, A. Briggs, E. Rosencher, F. Arnaud d'Avitaya, C. d'Anterroches : Appl. Phys. Lett. 51, 169 (1987).
[11] J.Y. Duboz, P.A. Badoz, E. Rosencher, J. Henz, M. Ospelt, H. von Känel, A. Briggs : (To be published in Appl. Phys. Lett.).
[12] G.Fishman, D. Calecki : (To be published).
[13] E. Rosencher, P.A. Badoz, A. Briggs, Y. Campidelli, F. Arnaud d'Avitaya : In Proceedings of the first international symposium of silicon molecular beam epitaxy, ed. J.C. Bean, The Electrochemical Society, Vol. 85-7, p. 268 (1985).
[14] P.A. Badoz, E. Rosencher, F. Arnaud d'Avitaya : Submitted to Physical Review.
[15] E. Rosencher, S. Delage, Y. Campidelli, F. Arnaud d'Avitaya : Electron. Letters 20, 762 (1984).
[16] E. Rosencher, S. Delage, F. Arnaud d'Avitaya, C. d'Anterroches, K. Belhaddad, J.C. Pfister : Physica B134, 106 (1985).
[17] J.C. Hensel, A.F. Levi, R.T. Tung, J.M. Gibson : Appl. Phys. Lett. 47, 151 (1985) See also Ref. 10.
[18] E. Rosencher, P.A. Badoz, J.C. Pfister, F. Arnaud d'Avitaya, G. Vincent, S. Delage : Appl. Phys. Lett. 49, 271 (1986).
[19] J.C. Pfister, E. Rosencher, K. Belhaddad, A. Poncet : Solid State Electron. 29, 907 (1986).
[20] S. Delage, P.A. Badoz, E. Rosencher, F. Arnaud d'Avitaya : Electronics Lett. 22, 207 (1986).
[21] S.M. Sze : In Physics of Semiconductor Devices (Wiley-Interscience, New York, 1969) Chap. 11.
[22] P.A. Badoz, E. Rosencher, S. Delage, G. Vincent, F. Arnaud d'Avitaya : In Proceedings of the 18th International Conference of Physics of Semiconductors (1986, Stockholm) (To be published).
[23] A. Marty, J. Clarac, J.P. Bailbe, G. Rey : IEEE Proc. 130, 24 (1983).
[24] J. Lindmayer : Proc. IEEE, 1751 (1964).
[25] C.O. Bozler, G.D. Alley : IEEE Trans. Electron Devices 27, 1128 (1980).
[26] G. Glastre, E. Rosencher, F. Arnaud d'Avitaya, C. Puissant, M. Pons, G. Vincent, J.C. Pfister : Appl. Phys. Lett. (in press).

[27] A. Rohatgi, P. Rai-Choudhury, S.J. Fonash, P. Lester, RAmbu singh, P.J. Caplan, E.H. Pointdexter : J. Electrochem. Soc. 133, 408 (1986).
[28] K. Ishibashi, S. Furukawa : IEEE Trans. Electron Devices ED-33, 322 (1986).
[29] C. d'Anterroches : J. Microsc. Spectrosc. Electron. 9, 147 (1984).
[30] O.L Nelson, D.E. Anderson : J. Appl. Phys. 37, 66 (1965).
[31] J.R. Hayes, A.F.J. Levi, W. Wiegmann : Electron. Lett. 2, 851 (1984).
[32] M. Heiblum : Solid State Electron. 24, 343 (1981).

Resonant Tunneling Barrier Structures and Their Applications to the Resonant Tunneling Hot Electron Transistor

S. Hiyamizu, S. Muto, T. Inata, S. Sasa, T. Fujii, K. Imamura, H. Ohnishi, and N. Yokoyama*

Fujitsu Limited, 10–1 Morinosato-Wakamiya, Atsugi 243–01, Japan

1. INTRODUCTION

In 1981, we started to develop a new ultrahigh-speed device with a multilayer structure of III-V compound semiconductors, as one of the future electron devices. The device was required to exhibit new functions based on quantum mechanical effects and to operate faster than any conventional devices. Recently, we developed a resonant tunneling hot electron transistor (RHET) [1] which has a resonant tunneling barrier (RTB) structure as an emitter barrier and exhibits new functions due to the negative differential resistance (NDR) of the RTB structure. In this paper, we describe an RHET - one of the most promising candidates for the future electron devices - and extremely excellent NDR characteristics of device-quality RTB structures grown by MBE which lead to much improved device performance of RHET.

2. GaAs-AlGaAs RESONANT TUNNELING HOT ELECTRON TRANSISTOR

A GaAs-AlGaAs RHET was developed in 1985 [1], based on the study of GaAs-AlGaAs hot electron transistors [2,3]. Figure 1 shows a schematic cross-section of the RHET with an energy band diagram. The emitter, base and collector layers are Si-doped n^+-GaAs ($n = 1 \times 10^{18}$ cm^{-3}). An undoped

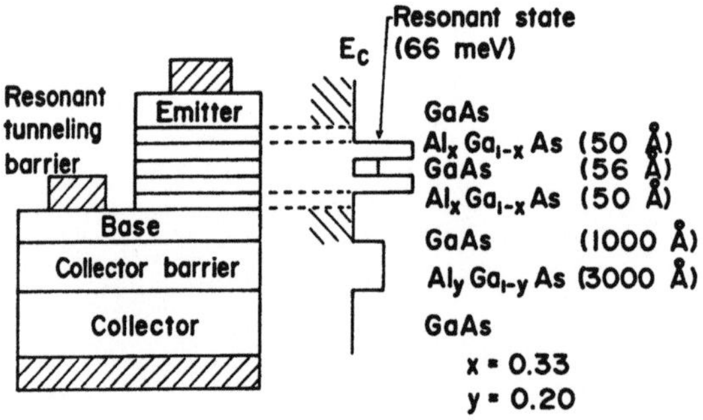

Fig. 1 Schematic cross-section of GaAs-AlGaAs RHET

*) present address: Faculty of Engineering Science, Osaka University 1-1 Machikaneyama, Toyonaka, Osaka 560, Japan

Springer Series in Solid-State Sciences Vol. 83: **Physics and Technology of Submicron Structures**
Editors: H. Heinrich · G. Bauer · F. Kuchar

RTB with a 56 Å -thick GaAs well layer and 50 Å -thick $Al_xGa_{1-x}As$ barrier layers (x = 0.33) is introduced between the emitter and the base layers. The base layer (1000 Å -thick) is isolated from the collector layer by an undoped $Al_yGa_{1-y}As$ collector barrier layer (y = 0.22, 3000 Å -thick).

The operation principle of the RHET is illustrated with schematic band diagrams in common-emitter configuration (emitter earth) in Fig. 2. A single energy level (E_1) is indicated in the quantum well. A constant positive voltage is applied between the emitter and collector. (a) When the base-emitter voltage (V_{BE}) is zero, there is no electron injection and no collector current with a positive collector voltage. (b) When base-emitter voltages of around $2E_1/q$ are applied (q: the electron charge), electrons are injected into the base by resonant tunneling through the RTB structure. Electrons injected into the base are ballistically or nearly ballistically transferred to the collector through the base, resulting in collector current flow. (c) When the base-emitter bias is further increased, the collector current is decreased, because the resonant tunneling current is reduced under the off-resonant condition of the RTB structure.

Figure 3 shows the collector current as a function of the base-emitter voltage in common-emitter configuration. The collector current exhibits a peak due to the resonant tunneling current. It is worth noting that the collector current does not exhibit a negative differential resistance with respect to the collector-emitter voltage, but with respect to base-emitter voltage. This is a fundamentally different trait of RHET from a real-space hot electron transfer device [4], which exhibits a negative differential resistance in drain current with respect to the drain voltage.

By using the negative differential resistance characteristics of RHET, we can make various kinds of electric circuits with only a single RHET and a few resistors such as a frequency multiplier [1], an Exclusive-Nor logic [1], and a flip-flop circuit which can be applied to logic and/or memory circuits [5,6]. These circuits become so much simpler than those of conventional devices because of the new function of RHET, that they can be expected to operate much faster than conventional circuits. These results indicate that RHET has very high potential for applications.

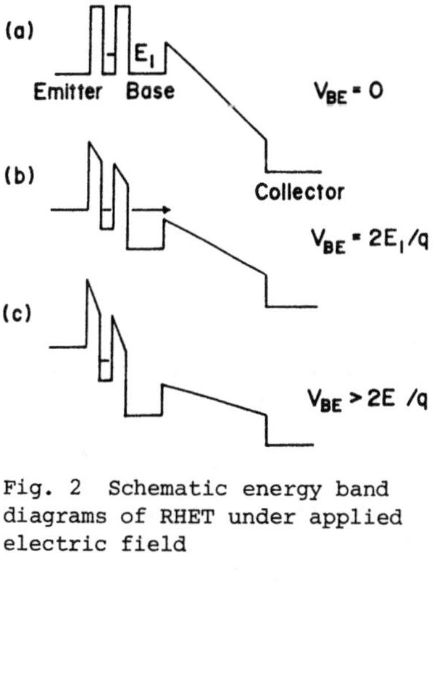

Fig. 2 Schematic energy band diagrams of RHET under applied electric field

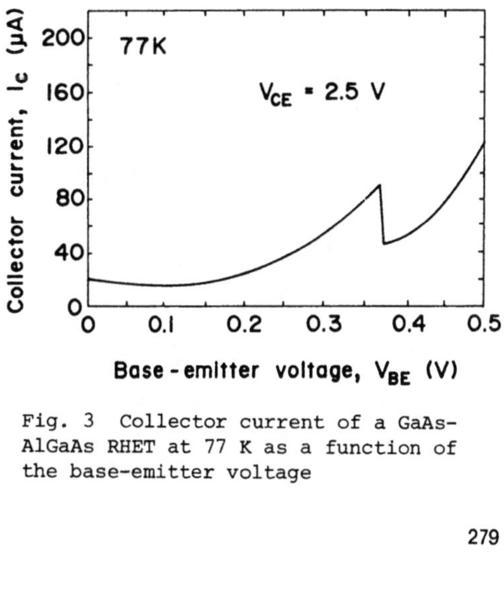

Fig. 3 Collector current of a GaAs-AlGaAs RHET at 77 K as a function of the base-emitter voltage

Recently a GaAs-AlGaAs RHET was improved to exhibit a current gain of 5.1 and a peak-to-valley current ratio (collector current) of 2.6 at 77 K for a 250 Å -thick base layer [7]. This device performance, however, is still not sufficient to allow applications of GaAs-AlGaAs RHETs to practical circuits. The required characteristics are a peak-to-valley current ratio of more than 10 [8] with peak current density of about 1 x 10^5 A/cm^2 [9] for a RTB structure and the current gain of more than 15 [8] for an RHET. We evaluate RTB structures with this criterion in this study.

3. RESONANT TUNNELING BARRIER STRUCTURES

3.1 GaAs-AlGaAs Resonant Tunneling Barrier Structures

In 1973, TSU and ESAKI [10] proposed resonant tunneling barrier (RTB) structures for the first time, and CHANG et al.[11] observed resonant tunneling current through a GaAs-AlGaAs RTB structure in the next year. Since SOLLNOR et al. [12] demonstrated excellent negative differential resistance (NDR) characteristics for improved GaAs-AlGaAs RTB structures grown by well-advanced molecular beam epitaxy (MBE) in 1983, much effort has been made to further improve NDR characteristics of GaAs-AlGaAs RTB structures for device applications by changing various structural parameters [13-20]. Figure 4 shows the structural parameters of a GaAs-AlGaAs RTB, which are the barrier width, well width, barrier height, spacer thickness, and doping concentration in an n-GaAs region.

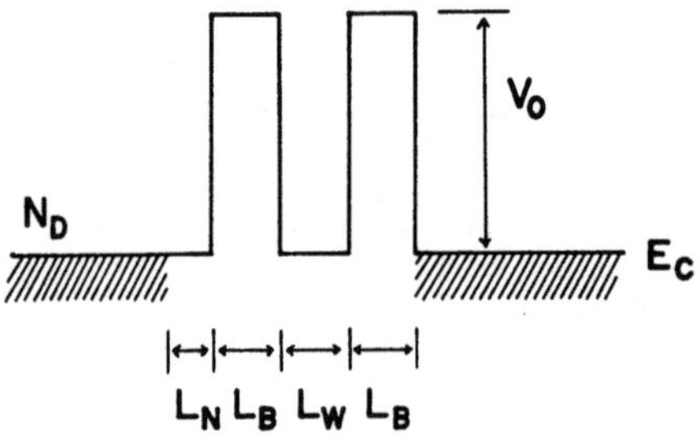

Fig.4 Parameters of a resonant tunneling barrier structure: barrier width (L_B), well width (L_W), barrier height (V_0), spacer thickness (L_N), and doping concentration (N_D)

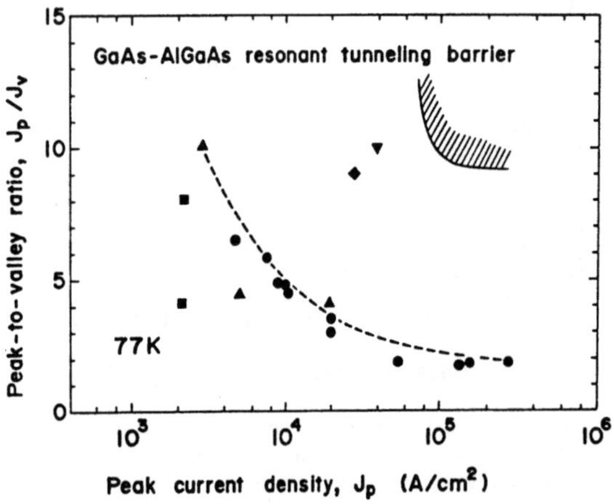

Fig. 5 Peak-to-valley current ratio of GaAs-AlGaAs RTBs at 77 K as a function of peak current density:(▲) after Refs. 13,14, 15; (■) after Ref. 16; (●) after Ref. 19, 20; (◆) after Ref. 17; (▼) after Ref. 18

Figure 5 shows the peak-to-valley current ratio (J_p/J_v) observed at 77 K as a function of peak current density (J_p) for GaAs-AlGaAs RTBs grown by MBE [13-20]. At low peak current density of about 10^3 A/cm^2 we have high peak-to-valley current ratio of around 10, but with increasing peak current density, peak-to-valley current ratio decreases monotonically, and it becomes as low as 2 at J_p = 1 to 2 x 10^5 A/cm^2. Although there are two exceptionally good data (◆, ▼)[17,18], it seems difficult to achieve the required characteristics (indicated by a shaded region in Fig. 5) with GaAs-AlGaAs RTB structures. Hence, we investigated RTB structures of other materials, i.e., InGaAs-InAlAs RTB structures grown on InP substrates.

3.2 InGaAs-InAlAs RTB Structures Lattice-Matched to InP Substrates

In order to increase peak current density by one order of magnitude while keeping peak-to-valley current ratio at around 10, one of the best attempts is to reduce the electron effective mass of a barrier layer, because the tunneling probability through a single potential barrier is given by eq. (1) and can be enhanced significantly with the reduced electron effective mass [21].

$$T \propto \exp\left\{-2d[2m^*(V_0 - E_0)]^{1/2}/\hbar\right\}, \qquad (1)$$

where d is the potential barrier width, V_0 is the potential barrier height, E_0 is the electron energy, m^* is the electron effective mass in the barrier layer, and $2\pi\hbar$ is Plank's constant.

The electron effective mass in a barrier layer for InGaAs-InAlAs RTB structures lattice-matched to InP ($0.075m_0$, where m_0 is electron mass in vacuum) is much smaller than that of GaAs-$Al_xGa_{1-x}As$ RTB structures ($0.123m_0$ for x = 0.67). Calculated peak current density of an InGaAs-InAlAs RTB structure (41 Å -thick barrier layers and a 61.5 Å -thick well layer) becomes about 20 times larger than that of a GaAs-$Al_{0.67}Ga_{0.33}As$ RTB with the equivalent barrier height [21]. Hence, we can expect much improved NDR characteristics in InGaAs-InAlAs RTB structures. In addition, an InGaAs-InAlAs heterostructure lattice-matched to InP has excellent properties for high-speed device application, such as high electron mobility, high electron saturation velocity, large conduction band offset and high maximum doping concentration [22], which are superior to those of a GaAs-AlGaAs heterostructure. This suggests the high application potential of an InGaAs-InAlAs RTB structure.

$In_{0.53}Ga_{0.47}As$-$In_{0.52}Al_{0.48}As$ RTB structures, lattice-matched to InP, were grown on (100)-oriented n^+-InP substrate at 470 °C by MBE [21,23]. Figure 6 shows a schematic cross-section of the $In_{0.53}Ga_{0.47}As$-$In_{0.52}Al_{0.48}As$ RTB structure with its energy band diagram. It consists of a Si-doped n-InGaAs layer (0.2 µm, n = 1 x 10^{18} cm^{-3}), an undoped InGaAs layer (15 Å), an undoped InAlAs barrier layer (L_B), an undoped InGaAs well layer (thickness L_W), an undoped InAlAs barrier layer (L_B), an undoped InGaAs spacer layer (15 Å) and a Si-doped n-InGaAs layer (0.2 µm, n = 1 x 10^{18} cm^{-3}). The top and bottom layers were heavily Si-doped n^+-InGaAs (n = 2 x 10^{19} cm^{-3}) for non-alloyed ohmic contacts (Au). The thickness of the barrier layer, L_B, (thickness of the well layer, L_W) was changed from 41 Å to 32.2 Å (from 44 Å to 61.5 Å). The potential barrier height is 0.53 eV.

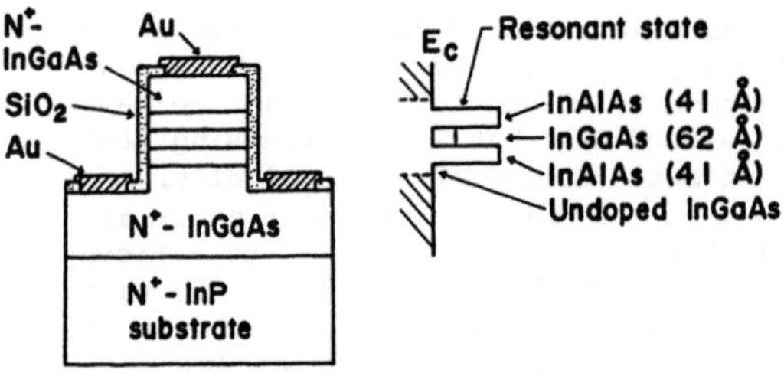

Fig. 6 Schematic cross-section of an $In_{0.53}Ga_{0.47}As$-$In_{0.52}Al_{0.48}As$ RTB structure, lattice-matched to InP

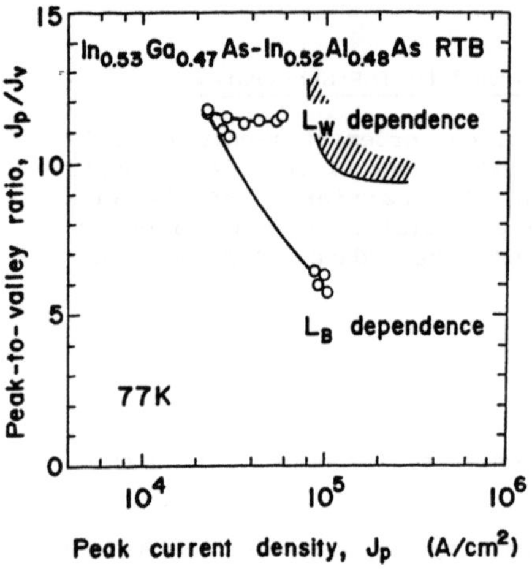

Fig. 7 Peak-to-valley current ratio of InGaAs-InAlAs RTBs at 77 K as a function of peak current density

Peak-to-valley current ratio of InGaAs-InAlAs RTB structures observed at 77 K is shown as a function of peak current density. For L_B-dependence (L_B = 41 Å, 32.2 Å; L_W = 61.5 Å), J_p/J_v decreases with increasing J_p, but for L_W-dependence (L_W = 44 Å, 52.7 Å, 61.5 Å; L_B = 41 Å), J_p/J_v remains to be more than 11, even if J_p increase to 5.5 x 10^4 A/cm^2. The best data we obtained is J_p/J_v = 11.5 with J_p = 5.5 x 10^4 A/cm^2.

The temperature dependence of the peak current density (J_p) and the valley current density (J_v) in the range of 10 K $<$ T $<$ 300 K was also investigated in the InGaAs-InAlAs RTBs [24]. The value of J_p is almost independent of temperature, but J_v starts to increase with increasing temperature from 100 K. At 300 K, J_p/J_v = 5.5 was obtained with J_p = 4.8 x 10^4 A/cm^2 for an InGaAs-InAlAs RTB with L_B = 46.9 Å and L_W = 32.2 Å, which is the best NDR characteristics observed at room temperature for lattice-matched RTB structures.

3.3 Pseudomorphic InGaAs-InAlAs Resonant Tunneling Structures

Further improved NDR characteristics were achieved in pseudomorphic $In_{0.53}Ga_{0.47}As$-$In_{1-x}Al_xAs$ RTB structures [25,26]. In this system, we can increase the barrier height from 0.53 eV to 1.20 eV by increasing AlAs mole fraction, x, of $In_{1-x}Al_xAs$ from 0.52 (x value for lattice-matching to InP) to 1. Hence, we can use much thinner barrier layer in this system, which would result in a high J_p value with a large J_p/J_v ratio.

$In_{0.53}Ga_{0.47}As$-$In_{1-x}Al_xAs$ pseudomorphic RTB structures (x = 0.65, 0.74, 1) were grown on (100)-oriented, n^+-InP substrates at 470 K by MBE. The structure of the RTB diode is basically the same as shown in Fig. 6. An $In_{0.53}Ga_{0.47}As$ well layer is of 15 atomic layers (approximately 44 Å -thick) and an $In_{1-x}Al_xAs$ (x = 0.65, 0.74, 1) barrier layer is of 9 atomic layers (about 26 Å -thick). In this structure only two InAlAs barrier layers are not lattice-matched to an InP substrate (3.7% lattice-mismatch), but their thickness is much thinner than the critical thickness (about 100 Å) for formation of misfit dislocations in strained multilayer structures [27]. Hence, the pseudomorphic InGaAs-InAlAs RTB structures can be expected to be free from misfit dislocations.

Figure 8 shows the peak-to-valley current ratio of $In_{0.53}Ga_{0.47}As$-$In_{1-x}Al_xAs$ pseudomorphic RTB structures at 77 K as a function of the peak current density. As the x value increases, the peak-to-valley current ratio steeply increases and reaches as high as 35 with $J_p = 2.3 \times 10^4$ A/cm^2 at x = 1. This J_p/J_v value is the highest ever achieved for RTB structures and is almost three times higher than the best data obtained for previous GaAs-AlGaAs and InGaAs-InAlAs lattice-matched RTBs (shown in Fig. 5 and Fig. 7).

The current-voltage characteristics of the $In_{0.53}Ga_{0.47}As$-AlAs RTB at 77 K is shown in Fig. 9.

Even at room temperature, the $In_{0.53}Ga_{0.47}As$-AlAs RTB exhibited an excellent peak-to-valley current ratio of 14 with peak current density of 2.3×10^4 A/cm^2 . This J_p/J_v ratio is also more than twice as large as the best value (5.5) reported hitherto.

These results indicate that the stress in $In_{0.53}Ga_{0.47}As$-AlAs RTBs does not affect the NDR characteristics. The stress in AlAs in the direction perpendicular to the interface becomes compressive, so it turns out to reduce the AlAs barrier height (from 1.36 eV to 1.2 eV)and the electron

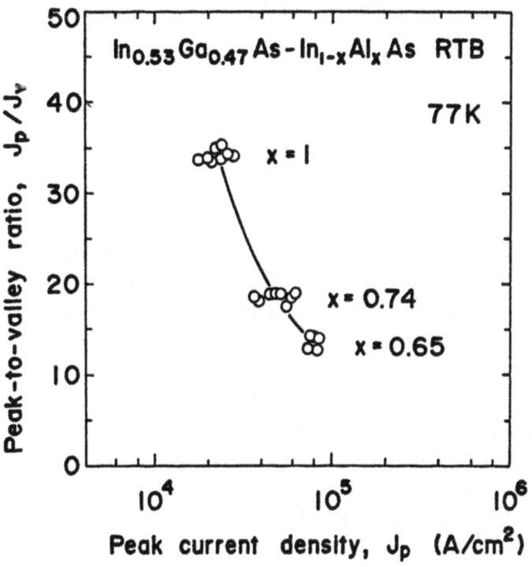

Fig. 8 Peak-to-valley current ratio of $In_{0.53}Ga_{0.47}As$-$In_{1-x}Al_xAs$ pseudomorphic RTBs at 77 K as a function of peak current density

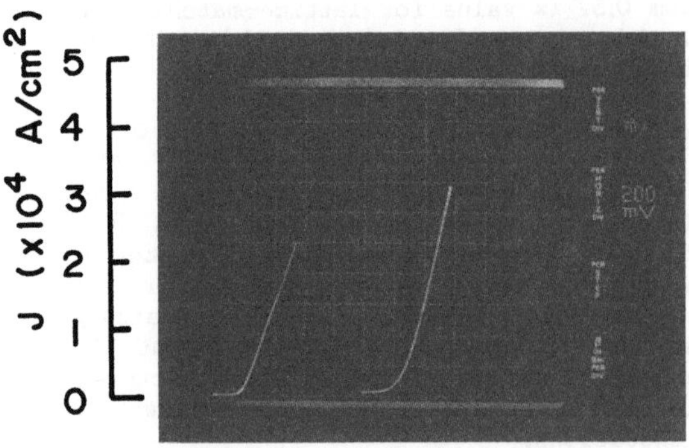

Fig. 9 Current-voltage characteristics at 77 K for the $In_{0.53}Ga_{0.47}As$-AlAs RTB structure

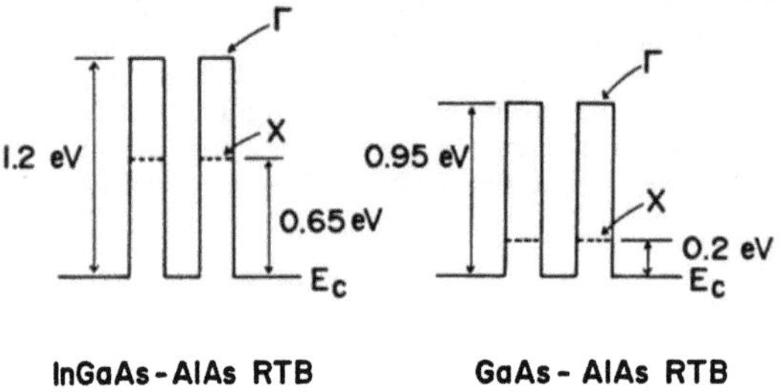

Fig. 10 Energy band diagram of $In_{0.53}Ga_{0.47}As$-AlAs RTB and GaAs-AlAs RTB.

effective mass (from $0.15m_0$ to $0.14m_0$) in the AlAs layer [25,26], which contributes to enhance the resonant tunneling current.

The energy band diagram of an $In_{0.53}Ga_{0.47}As$-AlAs RTB is shown in Fig. 10. The barrier height is 1.2 eV for the Γ-band and 0.65 eV for the X-band, which is more favorable than those of a GaAs-AlAs RTB (0.95 eV for the Γ-band and 0.2 eV for the X-band) and a $In_{0.53}Ga_{0.47}As$-$In_{0.52}Al_{0.48}As$ RTB (0.53 eV for the Γ-band and 0.55 eV for the L-band). This may partly lead to more excellent NDR characteristics in the $In_{0.53}Ga_{0.47}As$-AlAs RTB.

3.4 Application to RHET

Using an $In_{0.53}Ga_{0.47}As$-$In_{0.52}Al_{0.48}As$ RTB (L_W = 38 Å, L_B = 44 Å), $In_{0.53}Ga_{0.47}As$-$In_{0.53}(Ga_{0.5}Al_{0.5})_{0.47}As$ RHET was developed to improve the current gain and peak-to-valley current ratio in 1987 [28]. Figure 11 shows a schematic cross section of the InGaAs-In(GaAl)As RHET. A quarternary In(GaAl)As layer (2000 Å) was used for the collector barrier. The base layer was a 250 Å thick n-$In_{0.53}Ga_{0.47}As$. The whole structure of this RHET is lattice-matched to an InP substrate. The peak-to-valley current ratio of as high as 19.3 was achieved at 77 K, which is about eight times higher than that of GaAs-AlGaAs RHETs. The maximum current gain observed was 25 at 77 K. This is also about five times larger than that of GaAs-AlGaAs RHETs. These results almost meet our criterion required for RHET.

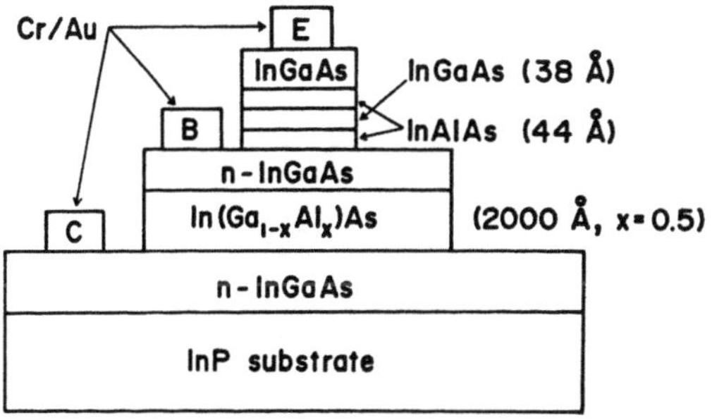

Fig. 11 Schematic cross-section of InGaAs-In(GaAl)As RHET, lattice-matched to InP

4. SUMMARY

We developed device-quality $In_{0.53}Ga_{0.47}As-In_{0.52}Al_{0.48}As$ RTBs and pseudomorphic $In_{0.53}Ga_{0.47}As-In_{1-x}Al_xAs$ RTBs (x = 0.65, 0.74, 1) by MBE. We achieved a peak-to-valley current ratio (J_p/J_v) of 35 with peak current density (J_p) of 2.3 x 10^4 A/cm^2 at 77 K and J_p/J_v = 14 with J_p = 2.3 x 10^4 A/cm^2 at room temperature in the pseudomorphic $In_{0.53}Ga_{0.47}As$-AlAs RTB, which is the best negative differential resistance (NDR) characteristics ever reported for any RTB structures. This will bring much improved device performance of RHET as well as new transistors with a RTB structure such as a resonant tunneling bipoler transistor (RBT) [29,30]. In addition, we applied an InGaAs-InAlAs RTB structure lattice-matched to InP to an RHET and confirmed the apparently improved device performances (the peak-to-valley current ratio of 19.2 and a current gain of 25 at 77 K), which almost satisfy our criterion for practical use of this new functional device.

Acknowledgements

The development of RHETs was performed under the management of the R & D Association for Future Electron Devices as a part of the R & D Project of Basic Technology for Future Industries, sponsored by the Agency of Industrial Science and Technology (MITI), Japan.

References

1. N. Yokoyama, K. Imamura, S. Muto, S. Hiyamizu, H. Nishi: Jpn. J. Appl. Phys. 24, L853 (1985)
2. N. Yokoyama, K. Imamura, H. Ohnishi, H. Nishi, S. Muto, K. Kondo, S. Hiyamizu: Jpn. J. Appl. Phys. 23, L311 (1984)
3. N. Yokoyama, K. Imamura, H. Ohnishi, H. Nishi, S. Muto, K. Kondo, S. Hiyamizu: IEDM Technical Digest (1984) p.532
4. A. Kastalsky, S. Luryi: IEEE Electron Device Lett. EDL-4, 334 (1983)
5. N. Yokoyama, K. Imamura: Electron. Lett. 22, 1228 (1986)
6. N.Yokoyama: Extended Abstracts of 18th Conf. Solid State Devices and Materials, Tokyo, 1986, p.347
7. T. Mori, H. Ohnishi, K. Imamura, S. Muto, N. Yokoyama: Appl. Phys. Lett. 49, 1779 (1986)
8. This is required to obtain enough noise margin for logic circuits with RHETs

9. This high peak current density is necessary to achieve a very short charging time (about 1 [ps]) of a capacitor between the base and the emitter of an RHET
10. R. Tsu, L. Esaki: Appl. Phys. Lett. 22, 562 (1973)
11. L.L. Chang, L. Easki, R. Tsu: Appl. Phys. Lett. 24, 593 (1974)
12. T.C.L.G. Sollnor, W.D. Goodhue, P.E. Tannenwald, C.D. Parker, D.D. Peck: Appl. Phys. Lett. 43, 588 (1983)
13. M. Tsuchiya, H. Sakaki, J. Yoshino: Jpn. J. Appl. Phys. 24, L466 (1985)
14. M. Tsuchiya, H. Sakaki: Appl. Phys. Lett. 49, 88 (1986)
15. M. Tsuchiya, H. Sakaki: Jpn. J. Appl. Phys. 25, L185 (1986)
16. T.J. Shewchuk, P.C. Chapin, P.D. Coleman, W. Kopp, R. Fischer, H. Morkoc: Appl. Phys. Lett. 46, 508 (1985)
17. H. Morkoc, J. Chen, K. Rebby, S. Luryi, T. Henderson: Appl. Phys. Lett. 49, 70 (1986)
18. W.D. Goodhue, T.C.L.G. Sollnor, H.Q. Le, E.R. Brown, B.A. Vojak: Appl. Phys. Lett. 49, 1086 (1986)
19. S. Muto, T. Inata, H. Ohnishi, N. Yokoyama, S. Hiyamizu: Jpn. J. Appl. Phys. 25, L577 (1986)
20. S. Muto, N. Yokoyama, S. Hiyamizu: Proc. High Speed Electronis, Stockholm, 1986 (Springer-Verlag, Berlin, Heidelberg, 1987) p.72
21. T. Inata, S. Muto, Y. Nakata, T. Fujii, H. Ohnishi, S. Hiyamizu: Jpn. J. Appl. Phys. 25, L983 (1986)
22. S. Hiyamizu, T. Fujii, S. Muto, T. Inata, Y. Sugiyama, S. Sasa: J. Crystal Growth 81, 349 (1987)
23. S. Muto, T. Inata, Y. Sugiyama, Y. Nakata, T. Fujii, S. Hiyamizu: Jpn. J. Appl. Phys. 26, L220 (1987)
24. Y. Sugiyama, T. Inata, S. Muto, Y. Nakata, S. Hiyamizu: Appl. Phys. Lett. 52, 314 (1988)
25. T. Inata, S. Muto, Y. Nakata, S. Sasa, T. Fujii, S. Hiyamizu: Jpn. J. Appl. Phys. 26, L1332 (1987)
26. T. Inata, S. Muto, S. Sasa, T. Fujii, S. Hiyamizu: Extended Abstracts 19th Conf. Solid State Devices and Materials, Tokyo, 1987, p.359
27. J.W. Mathews, A.E. Blakeslee: J. Crystal Growth 27, 118 (1974)
28. K. Imamura, S. Muto, H. Ohnishi, T. Fujii, N. Yokoyama: Program of 45th Annual Device Research Conf., Santa Barbara, VIA-3
29. F. Capasso et al: IEEE Electron Device Lett. EDL-7, 573 (1986)
30. T. Futatsugi et al: Jpn. J. Appl. Phys. 26, L131 (1987)

Index of Contributors

MIX
Papier aus verantwortungsvollen Quellen
Paper from responsible sources
FSC® C105338

If you have any concerns about our products,
you can contact us on
ProductSafety@springernature.com

In case Publisher is established outside the EU,
the EU authorized representative is:
Springer Nature Customer Service Center GmbH
Europaplatz 3, 69115 Heidelberg, Germany

Printed by Libri Plureos GmbH
in Hamburg, Germany